Heat Pipe Applications in Fission Driven Nuclear Power Plants

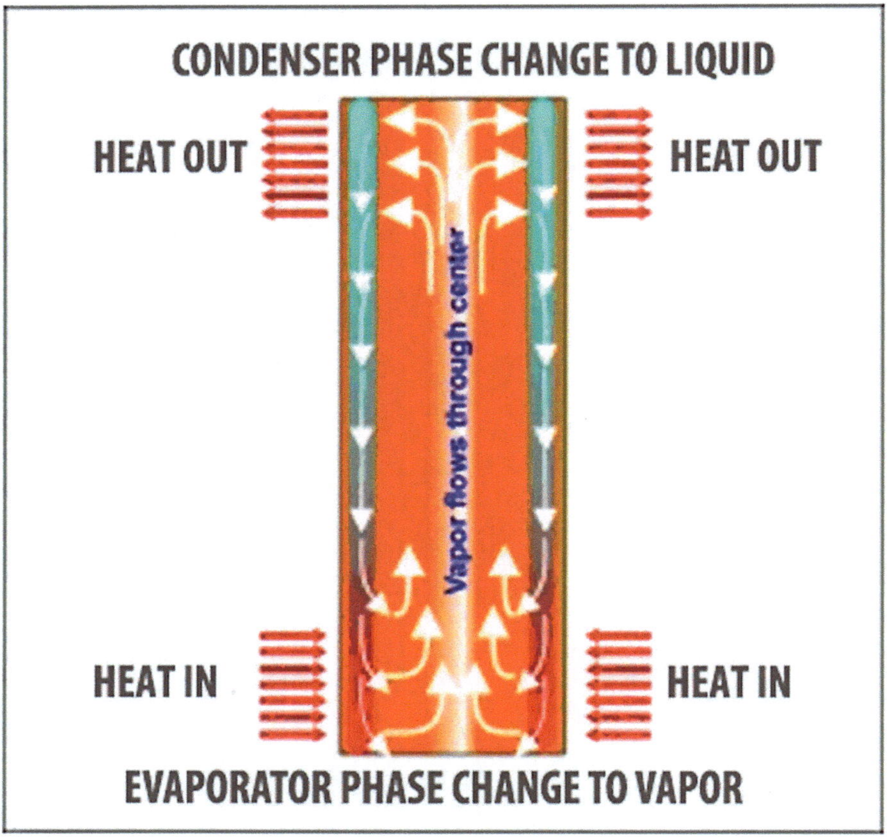
CONDENSER PHASE CHANGE TO LIQUID
HEAT OUT
HEAT OUT
Vapor flows through center
HEAT IN
HEAT IN
EVAPORATOR PHASE CHANGE TO VAPOR

Bahman Zohuri

Heat Pipe Applications in Fission Driven Nuclear Power Plants

Springer

Bahman Zohuri
Department of Electrical and Computer Engineering
University of New Mexico
Galaxy Advanced Engineering, Inc.
Albuquerque, NM, USA

ISBN 978-3-030-05881-4 ISBN 978-3-030-05882-1 (eBook)
https://doi.org/10.1007/978-3-030-05882-1

Library of Congress Control Number: 2018965451

This Springer imprint is published by the registered company Springer Nature Switzerland AG
The registered company address is: Gewerbestrasse 11, 6330 Cham, Switzerland

This book is dedicated to my son Sasha

Preface

As the global population grows, so will the demand for energy to ensure standards of living, health and life expectancy, literacy and opportunity, etc. To cope with this energy demand, nuclear energy, which is believed to be sustainable, clean, and safe, has been extensively advocated. To enhance the future role of nuclear energy systems, a generation of innovative nuclear energy systems, known as Generation IV (Gen IV), has been proposed to replace the current Gen II/Gen III reactors and Gen III+ reactors that will be deployed in the near future.

A new concept involving the use of heat pipes as control devices for nuclear reactors will be investigated in this book. The feature of the concept is that the heat pipe will contain a fissionable material as the working fluid. The primary purpose of the heat pipe will be to change the amount of fuel within a reactor instead of the usual purpose of transferring heat.

In conjunction with heat pipe (HP) in this book, we have also allocated a section on Directed Reactor Auxiliary Cooling System (DRACS), presenting the scalar analysis for it as well in respect to the advanced high-temperature reactors (AHTRs) and small modular reactors (SMRs) of Gen IV such as molten salt reactor and the shape of the pebble-bed reactor (PBR). The pebble-bed reactor (PBR) is a design for a graphite-moderated, gas-cooled nuclear reactor. It is a type of very-high-temperature reactor (VHTR), one of the six classes of nuclear reactors in the Gen IV initiative. The basic design of pebble-bed reactors features spherical fuel elements called pebbles, which you can find more details in Chap. 2 of this book.

Heat pipes offer a passive mechanism to transport heat from one area to another via the evaporation and condensation of a working fluid. The key design goal of the heat pipes in this reactor is to remove the power generated by the core at the highest possible temperature. This study evaluates the range of power levels that can be removed by heat pipes as a function of various design variables including working fluid, operating temperature, wick design, diameter, and length.

Heat pipes are often proposed as cooling system components for small fission reactors. Heat pipes have been used in reactors to cool components within radiation tests; however, no reactor has been built or tested that uses heat pipes solely as the primary cooling system. Heat pipe cooled reactors will likely require the development of a test reactor to determine the main differences in operational behavior from forced cooled reactors.

Albuquerque, NM, USA Bahman Zohuri

Acknowledgment

I am indebted to the many people who aided, encouraged, and supported me beyond my expectations. Some are not around to see the results of their encouragement in the production of this book, yet I hope they know of my deepest appreciations. I especially want to thank my son while he was alive and suffering from depression for encouraging me with this work. He has always kept me going in the right direction. I am also deeply indebted to a lot of friends who were continuously giving me their support without hesitation.

Above all, I offer a very special thanks to my late mother and father and to my children, in particular, my daughters Natalie and Natasha. They have provided constant interest and encouragement, without which this book would not have been written. Their patience with my many absences from home and long hours in front of the computer to prepare the manuscript are especially appreciated.

Contents

About the Author

Bahman Zohuri is currently at the University of New Mexico as Associate Research Professor and Consultant at Sandia National Laboratories as well as Galaxy Advanced Engineering, Inc., a consulting company that he started himself in 1991 when he left both semiconductor and defense industries. After graduating from the University of Illinois in the field of physics and applied mathematics, he then went to the University of New Mexico, where he studied nuclear engineering and mechanical engineering. He joined Westinghouse Electric Corporation, where he performed thermal hydraulic analysis and studied natural circulation in an inherent shutdown heat removal system (ISHRS) in the core of a liquid-metal fast-breeder reactor (LMFBR) as a secondary fully inherent shutdown system acting as secondary loop heat exchange. All these designs were used in nuclear safety and reliability engineering for a self-actuated shutdown system. He designed a mercury heat pipe and electromagnetic pumps for large pool concepts of a LMFBR for heat rejection purposes for this reactor around 1978, when he received a patent for it. He was subsequently transferred to the defense division of Westinghouse, where he oversaw the dynamic analysis and methods of launching and controlling MX missiles from canisters. The results were applied to MX launch seal performance and muzzle blast phenomena analysis (i.e., missile vibration and hydrodynamic shock formation). Dr. Zohuri was also involved in analytical calculations and computations in the study of nonlinear ion waves in rarefying plasma. The results were applied to the propagation of the so-called soliton waves and the resulting charge collector traces in the rarefaction characterization of the corona of laser-irradiated target pellets. As part of his graduate research work at Argonne National Laboratory, he performed computations and programming of multi-exchange integrals in surface physics and solid-state physics. He earned various patents in areas such as diffusion processes and diffusion furnace design while working as a Senior Process Engineer at various semiconductor companies, such as Intel Corp., Varian Medical Systems, and National Semiconductor Corporation. He later joined Lockheed Martin Missile and Aerospace Corporation as Senior Chief Scientist and oversaw research and development (R&D) and the study of the vulnerability, survivability, and both radiation and laser hardening of different components of the Strategic Defense Initiative known as Star Wars. Dr. Zohuri holds PhD in Nuclear Engineering.

Chapter 1
Why Nuclear Power Plant Energy

The major growth in the electricity production industry in the last 30 years has centered on the expansion of natural gas power plants based on gas turbine cycles. The most popular extension of the simple Brayton gas turbine has been the combined cycle power plant with the air-Brayton cycle serving as the topping cycle and the Steam-Rankine cycle serving as the bottoming cycle for new generation of nuclear power plants that are known as GEN-IV. The air-Brayton cycle is an open-air cycle, and the Steam-Rankine cycle is a closed cycle. The air-Brayton cycle for a natural gas-driven power plant must be an open cycle, where the air is drawn in from the environment and exhausted with the products of combustion to the environment. This technique is suggested as an innovative approach to GEN-IV nuclear power plants in the form and type of Small Modular Reactors (SMRs). The hot exhaust from the air-Brayton cycle passes through a heat recovery steam generator (HRSG) prior to exhausting to the environment in a combined cycle. The HRSG serves the same purpose as a boiler for the conventional Steam-Rankine cycle [1, 2].

1.1 Introduction

With the aftermath of the major accidents at Three Mile Island in 1979 and Chernobyl in 1986 and then recent devastated Japan's Fukushima nuclear power plant frailer in March of 2011, we can state that before society can make the decision to shift its source of electric power from its present overwhelming dependence on fossil fuels to a future dependence on nuclear fuels, the following very important questions must be answered [3–5]:

1. How will the cost of electricity be affected, both from short-term and long-term point of view?
2. What impacts will nuclear energy have on the environment?
3. How should its development and site installation be financed and managed?

B. Zohuri, *Heat Pipe Applications in Fission Driven Nuclear Power Plants*,
https://doi.org/10.1007/978-3-030-05882-1_1

4. What is the total cost of ownership (TOC) and return on investment (ROI) for the owner of these power plants to produce electricity by being in the grid?
5. Are nuclear power plants safe, particularly with the advancement of Small Modular Reactors' (SMRs) perspective?
6. What sorts of surveillance are necessary to satisfy both aspects of proliferation and nonproliferation requirements and who should be responsible for them?
7. How are we going to manage our spent fuel and high-level waste from these nuclear power plants? And how our fuel cycle should be dealt with?
8. What is the society and public's attitude toward nuclear power plants and what are their understanding of this fast-moving technology with Generation IV coming on line in a very near future?
9. How are we going to implement these power plants (i.e., Advanced SMRs) to help in the decarbonization of environment by enhancing our green environment and clean air requirements?
10. How will these power plants be used as source of new energy and help us with additional source of renewable energy to match our demand and supply for electricity?

We try our best to address some of these questions in the following sections of this chapter as much as possible, within the scope of this book, yet readers can refer to reference such as *The Future of Nuclear Power: An Interdisciplinary MIT Study* (2003) [6].

Our national investment in the United States for generation energy from nuclear power will be hundreds of billions of dollars before the end of this century, which means everyone in our society has a personal stake of thousands of dollars as taxpayers in the answers to these questions.

Before they can be answered intelligently, we have to make sure and it is necessary for everyone in our society regardless of their education to have some general knowledge and understanding of how the nuclear power plant is working to produce energy in some details.

While we are referring this knowledge and understanding of the working of nuclear energy system in later section of this chapter (i.e., Sect. 1.12), we will elaborate this detail to some degree at the introduction section of this chapter.

Nuclear power plant systems now coming into operation presently drive their energy from *fission* but not from *fusion* reactions yet, in order for them to function and produce enough heat energy to be used to produce steam for steam turbine to drive electrical generator. It is necessary to arrange for fission reactions to occur continuously in a controlled chain reaction mode and at a high rate fashion without requiring much input energy to breakeven. This behavior takes place in today's fission reactors that are in production line and supply electricity into the grid for consumption, which is used in other different industries such as nuclear-powered navy vessels.

Furthermore, futuristic nuclear fusion reactor designs are supposed to work in a very similar fashion to generate heat energy from a fusion reaction rather than fission reactor to drive steam for steam turbine; the very mechanical energy of the steam

Table 1.1 Cross sections for thermal neutrons (in units of 10^{-24} cm^2)

H^1—0.33	Fe^{56}—2.7
H^2 (deuterium)—0.0006	Cd^{113}—20,000
Li^6—940[a]	Xe^{135}—2,700,000
B^{10}—3840[a]	Xe^{136}—0.15
C^{12}—0.0037	U^{235}—590[a]
O^{16}—0.0002	U^{238}—2.7

[a]For Li^6 and B^{10}, these are cross section for (n, α), and for U235 it is for fission. In all other cases, they are for (n, γ)

turbine shaft will ultimately get converted into electricity; however at present technology stage, we are far away from actual breakeven at a continuous pace [7–9].

Basic concept of fission reactor is driven by the fact that a neutron striking a U^{235} nucleus induces a nuclear reaction leading to fission, where in the fission process, an average of 2.5 neutrons are released. If, on average, at least 1.0 of the neutrons, 40%, induces another fission by striking another U^{235} nucleus, the original neutron is replaced, so the processes will continue indefinitely in a chain reaction scenario, so one fission reaction leads to another similar reaction in an endless chain, which we call it *self-sustaining chain reaction* mode. A system in which a chain reaction as such takes place is called a *nuclear reactor* that we know today [10].

The simplest design of nuclear power reactor infrastructure would be a ball of U^{235}. The requirement that at least 40% of the neutrons released in a fission induce another fission means that the ball must have some minimum size, the *critical mass*, for if it is too small, too many neutrons will escape out the surface. For nearly pure U^{235}, this critical mass is about 44 pounds, making it about 5 in. in diameter, the size of a cantaloupe.

However, there is an inherent inefficiency in making a reactor in this way—the neutrons released in fission come out with an energy of about 1 MeV, and at such a high energy, the cross section, which determines the probability of the reaction occurring, is small, about two in the units of Table 1.1.

From that Table 1.1, we see that it would be better to slow the neutrons down to take advantage of the very much larger cross sections for *thermal reactors*—nearly all currently operating reactors are of this type—and the material inserted in a thermal reactor to slow down the neutrons is called a *moderator*, since it moderates their speed within core of the reactor.

In a billiard game table, a rapidly moving cue ball (i.e., white ball) entering a nest of other balls that is set up to start the game is quickly slowed down by collisions.

Its energy of motion is transferred to the balls as they recoil from these collisions, assuming the energy conserved per hour understanding classical mechanics on a frictionless billiards' surface table, so we do not lose any energy due to friction and naturally heat.

The moderator in a reactor works in much the same way.

Neutrons are slowed down by collisions with the nuclei of which the moderator is composed because the struck nuclei recoil away, taking up some of the energy.

However, the question is that what material makes a good moderator?

If the cue ball passes through a net of solid steel balls, it bounces off without being slowed down appreciatively because the steel balls are too heavy to recoil much due to conservation of momentum. A cue ball is more effectively slowed down by passing through a nest of balls not much more massive than itself, such as other billiard balls, since these absorb more energy in recoiling from a collision due to conservation of both energy and momentum.

Carrying this conclusion over to our analogy with neutrons being slowed down by a moderator, it determines the first requirement for a good moderator that must consist of nuclei not much more massive than a neutron, which means nuclei with very few nucleons. A second requirement for a moderator is that it has a small neutron cross section, since neutrons absorbed by the moderator are not available to induced fissions.

Now we should be looking for the best materials which will be satisfying these two above basic requirements that are:

1. A good moderator that consists of nuclei not much more massive than a neutron
2. A good moderator that has a small neutron cross section, so neutrons that we need for continuous chain fission reaction will be there to induce fission reactions

Thus, the best moderator for the first requirement would be ordinary hydrogen (H^1), since its nucleus, a proton, has the same mass as a neutron. From Table 1.1, we see that its rating on the second requirement is only a so-so situation, but it is still a reasonably good moderator as a possibility. The fact that hydrogen is a gas makes the use of water (H^2O) a more practical alternative. The additional oxygen in the water does not have any adverse effect, due to the fact that its cross section is very small.

However, the next best moderator for the first requirement above is H^2 or D for deuterium, and from Table 1.1, we see that it also passes the second requirement with flying colors. "Heavy water" D^2O is therefore the best moderator. But deuterium occurs in nature as only 0.015% of ordinary hydrogen as it can be seen in Fig. 1.1 here with the remaining 99.985% being H^1, and it costs $55 per pound of D^2O to separate it—the best moderator is very expensive [10].

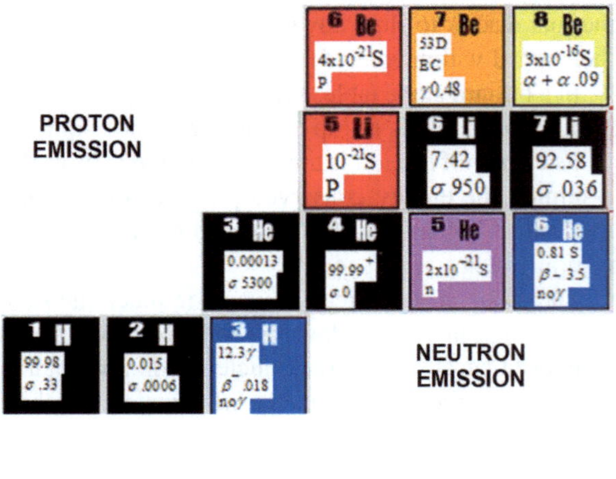

Fig. 1.1 Explanations of sample data

As we consider other elements with few nucleons, the next attractive choice is carbon. This element is 12 times massive than neutron; thus it is heavier than we would like, but its cross section is very small, and it is an inexpensive element in terms of graphite, the material of lead pencil. The three most practical moderators are therefore "light water" H^2O, "heavy water" D^2O, and "graphite" C.

Now, we can start to see what moderators can do and what goal they achieve inside the code of a reactor of our intention here. The most desirable thing to consider would be if they could allow us the use of natural uranium, which consists of 99.3% U^{238} and 0.7% U^{235} as a reactor fuel. In the early stage and days of nuclear energy, separation of uranium isotope technologies was not available to us, and even today this separation is difficult and expensive, so it would be quite advantageous if we could use uranium as it occurs in nature.

Furthermore, it turns out that with ordinary water as a moderator, a natural uranium reactor is *not* possible—the neutron cross section of H^1 and U^{238}, though not large per Table 1.1, is still sufficient to prevent criticality in any size or configuration [10].

However, in order to use water as the moderator, the uranium must be enriched up to a few percent U^{235}, and in fact, that is how most present reactors are made today, i.e., Generations III and III$^+$. In case of Generation IV, some of the Small Modular Reactors (SMRs) fall in this category such as the one designed by NuScale company out of Oregon in cooperation with Oregon State University (OSU); however the drawback of such approach is that the thermal efficiency of this reactor is trapped under vapor dome of saturated steam [1, 2, 11].

The best moderator, heavy water D^2O, can easily make a natural uranium reactor critical such as the modern power reactor in Canada, namely, CANDU is made that way, but large amounts of deuterium were not available in the 1940s and are still very expensive.

Our third moderator, carbon, can also be used to make a natural uranium reactor critical provided the carbon is very pure and the reactor size is very large. Since this avoids the necessity of isotope separation, the first reactor and nearly all of the early reactors were made in this way.

The next concern in running a reactor in a practical way to control the rate at which fissions take place certainly needs to be addressed as well. This is done by inserting *control rods* made of a material with a *large* neutron cross section such as boron as indicated in Table 1.1. When they are inser*ted into the reactor, they absorb neutrons, greatly reducing the number available to cause fissions and thereby stopping the chain reaction. As they are withdrawn, the critical point is reached and passed slightly. At this point of criticality is reached and passed slightly.*

At this point, the fission rate—referred to as the *reactor power*, since the rate of energy release is proportional to the rate at which fission reactions occur—begins to increase. Let us consider how fast this happens. Suppose the fuel, the moderator, and the size of the reactor are such that an average of 1.005 neutrons released in a fission makes another fission, we recall that this number must be at least 1.000 for the reactor to go critical.

However, the time required for a neutron emitted in a fission to be thermalized and cause another fission is about 10^{-4} s. In analogy with biological reproduction, this is called the time for one *generation* of neutrons, and in each succeeding generation, the number of neutrons increased by an additional factor of 1.005.

After 500 generations ($500 \times 10^{-4} = 0.05$), 1/20 of a second, it has increased by $(1.0005)^{500} = 10.6$, so the reactor power and hence the heat generated increased by more than a factor of 10. Since it is not easy to move a control rod in 1/20 of a second, this seems like a very serious situation that could easily lead to mass destruction by melting [10].

The situation is mitigated by the phenomenon of delayed neutron induction. In the fission of U235, about 0.6% of the neutrons are not emitted until a half-second or more after the fission; thus the $1.005 - 0.006 = 0.999$ neutrons that are emitted promptly cannot even sustain a chain reaction, let alone causes its rate to increase. Criticality now depends on the reactor power increases by a factor of even 1.005. This situation greatly eases the control problem, but if the control rods are withdrawn to the point where 1.007 or more neutrons from each fission cause another fission, the reactor is critical on prompt neutrons alone, and the reactor power increases very rapidly. This can be a dangerous situation, and great care must be taken not to reach it.

There are some other problems before we reach to a feasible reactor, and that is cooling and shielding; however there are other books that readers can find to help them with the details of these two problems, which details are beyond the scope of this chapter; however the cooling has been detailed out throughout in a book by Zohuri [12], and for shielding readers can refer to a book by Chilton et al. [13].

This is a very basic approach to a very basic reactor and its concept, so it provides some general idea what is involved with a reactor design and of course is more complex than what it was described here.

1.2 Innovative Approach

In 2007 gas turbine combined cycle plants had a total capacity of 800 GW and represented 20% of the installed capacity worldwide. They have far exceeded the installed capacity of nuclear plants, though in the late 1990s, they had less than 5% of the installed capacity worldwide. There are a number of reasons for this. First natural gas is abundant and cheap. Second combined cycle plants achieve the greatest efficiency of any thermal plant. And third, they require the least amount of waste heat cooling water of any thermal plant.

A typical gas turbine plant consists of a compressor, a combustion chamber, a turbine, and an electrical generator. A combined cycle plant takes the exhaust from the turbine and runs it through a heat recovery steam generator (HRSG) before exhausting to the local environment. The HRSG serves the function of the boiler for a typical closed cycle steam plant. The steam plant consists of a steam turbine, a condenser, a water pump, an evaporator (boiler), and an electrical generator. In a

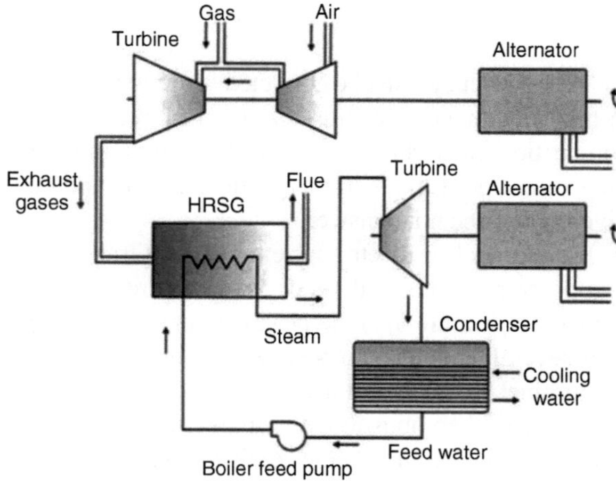

Fig. 1.2 Typical gas turbine combined cycle power plant

combined cycle plant, the gas turbine and steam turbine can be on the same shaft to eliminate the need for one of the electrical generators. However, for the two shafts, two generator systems provide a great deal of more flexibility at a slightly higher cost. In addition to the closed loop for the steam, an open loop circulating water system is required to extract the waste heat from the condenser. The waste heat extracted by this "circulating" water system is significantly less per megawatt for a combined cycle system as the open Brayton cycle exhausts its waste heat directly to the air.

The layout for the components of a typical combined cycle power plant is given below in Fig. 1.2.

General Electric currently markets a system that will produce 61% efficiency at design power and better than 60% efficiency down to 87% of design power [14] for gas turbine combined cycle plants.

An approximate efficiency can be calculated for a combined cycle power plant by the following simple argument [15].

$$\text{Brayton cycle efficiency} = \frac{W_B}{Q_{in}} = \eta_B$$

$$\text{Heat to Rankine cycle} = Q_R = (1 - \eta_B)Q_{in}$$

$$\text{Rankine cycle efficiency} = \frac{W_R}{Q_R} = \eta_R$$

$$\text{Overall efficiency} = \frac{W_B - W_R}{Q_{in}} = \eta_T = \frac{\eta_B Q_{in} + \eta_R Q_R}{Q_{in}} = \frac{\eta_B Q_{in} + \eta_R (1 - \eta_B) Q_{in}}{Q_{in}}$$

$$= \eta_B + \eta_R - \eta_B \eta_R$$

$$\eta_T = \eta_B + \eta_R - \eta_B \eta_R$$

This efficiency must be corrected for pressure losses and assumes that all the heat in the Brayton exhaust is used in the heat recovery steam generator (HRSG). For a combustion gas turbine, this is not usually possible if condensation of the water in the exhaust products is to be avoided. The detailed models developed in this effort give a more accurate answer.

For the nuclear reactor system, the heat transfer is in the opposite direction. All reactor components and fluids in the primary and secondary loops must be at a higher temperature than the peak temperature of the gas exiting the heat exchanger. This severely restricts the peak temperature that can be achieved for the air entering the turbine. However, all is not lost.

In a typical combustion system, there are pressure losses approaching 5% of the total pressure to complete the combustion process [16]. Heat exchangers can be built with significantly lower pressure drops than 5% approaching 1% [16]. Therefore, the most straightforward method to overcome this severe temperature limitation is to borrow a technique from steam power plants and implement multiple reheat cycles. That is the first heat exchanger heats the air to its peak temperature. Then the air is expanded through the first turbine. The air is then reheated to the same peak temperature and expanded through the second turbine. Based on the relative pressure losses that appear possible, up to five turbines might be considered. All five turbines will be driving the same compressor. Multiple compressors on concentric shafts driven by different sets of turbines might be possible, but that has not been considered here.

For a nuclear system to take advantage of combined cycle technology, there are many numbers of changes to the plant components that must be made. The most significant of course is that the combustion chamber must be replaced by a heat exchanger in which the working fluid from the nuclear reactor secondary loop is used to heat the air. The normal Brayton cycle is an internal combustion one where the working fluid is heated by the combustion of the fuel with the air in the combustion chamber. The walls of the combustion chamber can be cooled, and peak temperatures in the working fluid can be significantly above the temperature that the walls of the chamber can tolerate for any length of time.

1.3 Methodology of Combined Cycle

The approach taken in the combined cycle (CC) code developed for this effort is to model the thermodynamics of the components making up the power conversion systems as real components with non-ideal efficiencies. Pressure drops are included for every component except the connected piping. The compressor design is modeled with a small-stage polytropic efficiency to take into account state-of-the-art designs. The gas turbines are likewise modeled with a polytropic efficiency. The steam turbines are modeled with a simple overall thermal efficiency. Pressure drops in each of the heat exchangers are included. The input files specify the pressure drops, and the heat exchangers are designed to meet these specifications if possible [17].

Some scientists are calling the nuclear power plants' source of energy as 100% renewable energy, and of course environmentalists arguably are saying that is a wrong approach, just because in the core of these plants, there exists uranium or plutonium as fuel when we are talking about fission-type nuclear power plants that exist in grid today and are producing electricity to the net. However, on the other side of the spectrum, researchers and scientists at national laboratories and universities around the globe that are working toward fusion program to achieve a breakeven passionately argue that nuclear power plants of fusion type are totally clean so long as the source of energy comes in the form of two hydrogen isotopes such as deuterium (D) and tritium (T) as source of fusion reaction and driving energy from it.

This is a dream that is too far away from the reality of today's need and demand for electricity yet is not out of scope of the near future. Physics of plasma for driving energy via inertial confinement fusion (ICF) [18] or magnetic confinement fusion (MCF) [7] are in agreement with such innovative approaches.

1.4 Why We Still Need Nuclear Power

Nuclear power's track record of providing clean and reliable electricity compares favorably with other energy sources. Low natural gas prices, mostly the result of newly accessible shale gas, have brightened the prospects that efficient gas-burning power plants could cut emissions of carbon dioxide and other pollutants relatively quickly by displacing old, inefficient coal plants, but the historical volatility of natural gas prices has made utility companies wary of putting all their eggs in that basket. Besides, in the long run, burning natural gas would still release too much carbon dioxide. Wind and solar power are becoming increasingly widespread, but their intermittent and variable supply makes them poorly suited for large-scale use in the absence of an affordable way to store electricity. Hydropower, meanwhile, has very limited prospects for expansion in the United States because of environmental concerns and the small number of potential sites.

"The United States must take a number of decisions to maintain and advance the option of nuclear energy. The NRC's initial reaction to the safety lessons of Fukushima must be translated into action; the public needs to be convinced that nuclear power is safe. Washington should stick to its plan of offering limited assistance for building several new nuclear reactors in this decade, sharing the lessons learned across the industry. It should step up its support for new technology, such as SMRs and advanced computer-modeling tools. And when it comes to waste management, the government needs to overhaul the current system and get serious about long-term storage. Local concerns about nuclear waste facilities are not going to magically disappear; they need to be addressed with a more adaptive, collaborative, and transparent waste program."

These are not easy steps, and none of them will happen overnight. But each is needed to reduce uncertainty for the public, the energy companies, and investors. A more productive approach to developing nuclear power—and confronting the mounting risks of climate change—is long overdue. Further delay will only raise the stakes.

1.5 Is Nuclear Energy Renewable Source of Energy

Assuming for time being we are taking fission reaction as the foundation for present (GEN-III) and future (GEN-IV) nuclear power reactors, as nuclear energy source to somewhat degree, we can argue it is a clean source of energy.

Although nuclear energy is considered clean energy, its inclusion in the renewable energy list is a subject of major debate. To understand the debate, we need to understand the definition of renewable energy and nuclear energy first. However, until we manage through future technology of these fission reactors to bring down the price of electricity per kilowatt hours driven by fusion energy to the point of those by gas or fossil fuels, there is no chance to push these reactors beyond GEN-III.

However, efforts toward the reduction of price of electricity driven by nuclear fission power plants, especially using some innovative design of GEN-IV plants with high-temperature baseline in conjunction with some thermodynamics cycles such as Brayton and Rankine, are on the way by so many universities and national laboratories such as Idaho National Laboratory, MIT, UC Berkeley, and the University of New Mexico as well as this author.

Renewable energy is defined as an energy source/fuel type that can regenerate and can replenish itself indefinitely. The five renewable sources used most often are biomass, wind, solar, hydro, and geothermal.

Nuclear energy on the other hand is a result of heat generated through the fission process of atoms. All power plants convert heat into electricity using steam. At nuclear power plants, the heat to make the steam is created when atoms split apart—called fission. The fission releases energy in the form of heat and neutrons. The released neutrons then go on to hit other neutrons and repeat the process, hence generating more heat. In most cases the fuel used for nuclear fission is uranium.

One question we can raise here in order to further understand whether, or not, we need present nuclear technology as a source of energy is that:

What is the difference between clean energy and renewable energy? Put another way, why is nuclear power in the doghouse when it comes to revamping the nation's energy mix?

The issue has come to the forefront the time during the debate over the Waxman-Markey energy and climate bill and its provisions for a national renewable energy mandate.

To simply put it, Republicans have tried—and failed—several times to pass amendments that would christen nuclear power as a "low-emission" power source eligible for all the same government incentives and mandates as wind power and solar power.

Many environmental groups are fundamentally opposed to the notion that nuclear power is a renewable form of energy—on the grounds that it produces harmful waste by-products and relies on extractive industries to procure fuel like uranium.

Even so, the nuclear industry and pronuclear officials from countries including France have been trying to brand the technology as renewable, on the grounds that it produces little or no greenhouse gases. Branding nuclear power as renewable could also enable nuclear operators to benefit from some of the same subsidies and friendly policies offered to clean energies like wind, solar, and biomass.

So far, however, efforts to categorize nuclear power as a renewable source of power are making little headway.

The latest setback came in around August of 2009, when the head of the International Renewable Energy Agency (IRENA)—an intergovernmental group that advises about 140-member countries on making the transition to clean energy—dismissed the notion of including nuclear power among its favored technologies.

"IRENA will not support nuclear energy programs because it's a long, complicated process, it produces waste and is relatively risky," Hélène Pelosse, its interim director general, told in general,

Energy sources like solar power, Ms. Pelosse said, are better alternatives—and less expensive ones, "especially with countries blessed with so much sun for solar plants," she said it in 2009.

1.6 Argument for Nuclear Power as Renewable Energy Source

Most supporters of nuclear energy point out the low-carbon emission aspect of nuclear energy as its major characteristic to be defined as renewable energy. According to nuclear power opponents, if the goal to build a renewable energy infrastructure is to lower carbon emission, then there is no reason for not including nuclear energy in that list [19].

But one of the most interesting arguments for including nuclear energy in the renewable energy portfolio came from Bernard L Cohen, former professor at the University of Pittsburgh. Professor Cohen defined the term "indefinite" (time span required for an energy source to be sustainable enough to be called renewable energy) in numbers by using the expected relationship between the sun (source of solar energy) and the earth. According to Professor Cohen, if the uranium deposit could be proved to last as long as the relationship between the earth and sun is supposed to last (5 billion years), then nuclear energy should be included in the renewable energy portfolio [20].

In his paper Professor Cohen claims that using breeder reactors (nuclear reactor able to generate more fissile material than it consumes), it is possible to fuel the earth with nuclear energy indefinitely. Although the amount of uranium deposit available could only supply nuclear energy for about 1000 years, Professor Cohen believes the actual amount of uranium deposit available is way more than what is considered extractable right now. In his arguments he includes uranium that could be extracted at a higher cost, uranium from the sea water, and also uranium from eroding earth crust by river water. All of those possible uranium resources if used in a breeder reactor would be enough to fuel the earth for another 5 billion years and hence render nuclear energy as renewable energy.

1.7 Argument Against Nuclear Power as Renewable Energy Source

One of the biggest arguments against including nuclear energy in the list of renewables is the fact that uranium deposit on earth is finite, unlike solar and wind. To be counted as renewable, the energy source (fuel) should be sustainable for an indefinite period of time according to the definition of renewable energy.

Another major argument proposed by the opponents of including nuclear energy as renewable energy is the harmful nuclear waste from nuclear power reactors. The nuclear waste is considered as a radioactive pollutant that goes against the notion of a renewable energy source. Yucca Mountain is one of the examples used quite often to prove this point. Most of the opponents in the United States also point at the fact that while most renewable energy source could render the US energy independent, uranium would still keep the country energy dependent as the United States would still have to import uranium.

In conclusion, experts disagree about the future of nuclear power:

- Proponents of Nuclear Power
  - Fund more research and development
  - Pilot—plant testing of potentially cheaper and safer reactors
  - Test breeder fission and nuclear fusion
- Opponents of Nuclear Power
  - Fund rapid development of energy-efficient and renewable energy resources

In general, the above bolt points are what we consider as *advantages* and *disadvantages* of nuclear energy.

1.8 Safety

With the aftermath of the major accidents at Three Mile Island in 1979 and Chernobyl in 1986 and then recent devastated Japan's Fukushima nuclear power plant frailer in March of 2011, pretty much nuclear power fell out of favor, and some countries applied the brakes to their nuclear programs. Concerns about climate change and air pollution, as well as growing demand for electricity, led many governments to reconsider their aversion to nuclear power, which emits little carbon dioxide and had built up an impressive safety and reliability record. Some countries reversed their phaseouts of nuclear power, some extended the lifetimes of existing reactors, and many developed plans for new ones.

Despite all these given concerns and issues in respect to the nuclear energy, still we are facing the fact of why we still need nuclear power as clean source of energy, particularly when we deal with renewable source of energy arguments [5].

Today, roughly 60 nuclear plants are under construction worldwide, which will add about 60,000 megawatts of generating capacity—equivalent to a sixth of the world's current nuclear power capacity; however this movement has been lost after March of 2001 and Japan's Fukushima nuclear power episode.

Nuclear power's track record of providing clean and reliable electricity compares favorably with other energy sources. Low natural gas prices, mostly the result of newly accessible shale gas, have brightened the prospects that efficient gas-burning power plants could cut emissions of carbon dioxide and other pollutants relatively quickly by displacing old, inefficient coal plants, but the historical volatility of natural gas prices has made utility companies wary of putting all their eggs in that basket. Besides, in the long run, burning natural gas would still release too much carbon dioxide. Wind and solar power are becoming increasingly widespread, but their intermittent and variable supply makes them poorly suited for large-scale use in the absence of an affordable way to store electricity.

Hydropower, meanwhile, has very limited prospects for expansion in the United States because of environmental concerns and the small number of potential sites [21].

As part of any nuclear power plant safety that one should consider as part of design and operation of such source of energy is the reactor stability. Understanding time-dependent behaviors of nuclear reactors and the methods of their control is essential to the operation and safety of nuclear power plants. Cohen [20] provides researchers and engineers in nuclear engineering very general yet comprehensive information on the fundamental theory of nuclear reactor kinetics and control and the state-of-the-art practice in actual plants, as well as the idea of how to bridge the two: the dynamics and stability of engineering equipment that affect their economical operation from safety and reliable operation's point of view. In Chap. 13 of this book, we will talk about the existing knowledge that is today's practice for the design of reactor power plants and their stabilities as well as available techniques to designers. However, stable power processes are never guaranteed. An assortment of unstable behaviors wrecks power apparatus, including mechanical vibration, malfunctioning control apparatus, unstable fluid flow, unstable boiling of liquids, or combinations thereof. Failures and weaknesses of safety management systems are the underlying causes of most accidents [22].

The safety and capital cost challenges involved with traditional nuclear power plants may be considerable, but a new class of reactors in the development stage holds promise for addressing them. These reactors, called Small Modular Reactors (SMRs), produce anywhere from 10 to 300 megawatts, rather than the 1000 megawatts produced by a typical reactor. An entire reactor, or at least most of it, can be built in a factory and shipped to a site for assembly, where several reactors can be installed together to compose a larger nuclear power station. SMRs have attractive safety features, too. Their design often incorporates natural cooling features that can continue to function in the absence of external power, and the underground placement of the reactors and the spent-fuel storage pools is more secure.

Since Small Modular Reactors (SMRs) are smaller than conventional nuclear plants, the construction costs for individual projects are more manageable, and thus the financing terms may be more favorable. And because they are factory-assembled, the on-site construction time is shorter. The utility company can build up its nuclear power capacity step by step, adding additional reactors as needed, which means that it can generate revenue from electricity sales sooner. This helps not only the plant owner but also customers, who are increasingly being asked to pay higher rates today to fund tomorrow's plants [2–5, 7, 10, 14–25].

With the US federal budget under tremendous pressure, it is hard to imagine taxpayers funding demonstrations of a new nuclear technology. But if the United States takes a hiatus from creating new clean energy options—be it SMRs, renewable energy, advanced batteries, or carbon capture and sequestration—Americans will look back in 10 years with regret. There will be fewer economically viable options for meeting the US energy and environmental needs, and the country will be less competitive in the global technology market.

Today, nationwide there are about 100 nuclear power plants that are operational in the United States and just about 400 worldwide that altogether are using mostly Light Water Reactor technologies. With the benefit of experience and improved plan toward a more advanced design that is getting ready to go into operation as part of GEN-III+, the performance has improved over time to provide better efficiency as well as unit *capacity factor* of 90% and higher at least in the United States.

Bear in mind that capacity factor is the ratio of actual annual plant electrical production and maximum annual production capability, while worldwide capacity factors are around 75% lower than those recently achieved in the United States; a similar trend of improved capacity factors is observed outside of the United States as well.

Thus, as part of safe operations of the entire nuclear fuel cycle are a parameter of concern and this section we touched upon some of them in above, which should be included the continuation of training personnel for nuclear operations, the threat of terrorist attack as part of non-proliferation and proliferation nuclear plant coming on line or exiting one is concern, as well as nuclear fuel cycle safety, including nuclear fuel reprocessing of these plants [6].

As far as nuclear power installation safety is concerned, we need to take under consideration the infiltration and domestic or international terrorist attack into these facilities and installation. On the one hand, experts have concluded that civil works and security provisions make nuclear plants hard targets. On the other hand, the hazards are on a scale previously considered to be extremely rare in the evaluation of severe reactor accidents. The question is what new security measures, if any, are appropriate? We believe there is no simple, one-size-fits-all answer. It depends on many factors including threat evaluation, plant location, facility design, and government security resources and practices. Thus, nuclear plant safety is a good starting point for the evaluation of security risk [6].

Nuclear plant safety has considered natural external events, such as earthquakes, tornadoes, floods, and hurricanes. Terrorist attack by fire or explosion is analogous to external natural events in its implication for damage and release of radioactivity. The strength of containment buildings and structures presents a major obstacle and hardened target for attack.

The Electric Power Research Institute (EPRI) [26] carried out an evaluation of aircraft crash and nuclear power plant (NPP) structural strength, concluding that US containments would not be breached.

The US NRC is performing its own evaluation, including structural testing at the Sandia National Laboratories (SNL), not yet complete. A broad survey and evaluation of hazards and protective actions are in order to make decisions on adequate protection. Such a survey must begin by identifying possible modes of attack and vulnerabilities associated with designs and locations. It must also identify the cost effectiveness of a range of security options for new designs, old plants near decommissioning, and plants in mid-life. There is also a need for sharing information with governments of countries and supporting institutions that will undertake nuclear power programs in order to provide effective intelligence and security.

1.9 Fuel Cycle

The description of a possible global growth scenario for nuclear power with 1000 or so GWe deployed worldwide must begin with some specification of the nuclear fuel cycles that will be in operation. The nuclear fuel cycle refers to all activities that occur in the production of nuclear energy.

Realization of the global growth scenario entails construction and operation of many fuel cycle facilities around the world. It is important to emphasize that producing nuclear energy requires more than a nuclear reactor steam supply system and the associated turbine-generator equipment required to produce electricity from the heat created by nuclear fission. The process includes ore mining, enrichment, fuel fabrication, waste management and disposal, and finally decontamination and decommissioning of facilities. All steps in the process must be specified, because each involves different technical, economic, safety, and environmental

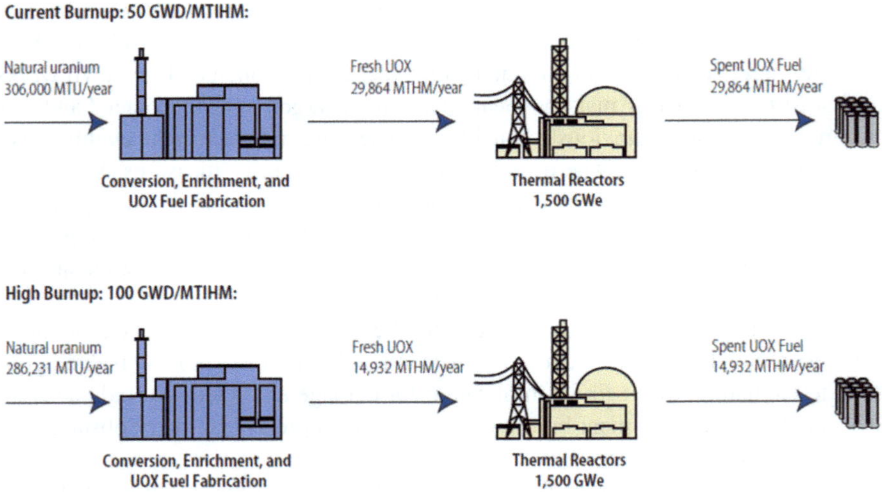

Fig. 1.3 Open fuel cycle: once-through fuel—projected to 2050 [6]

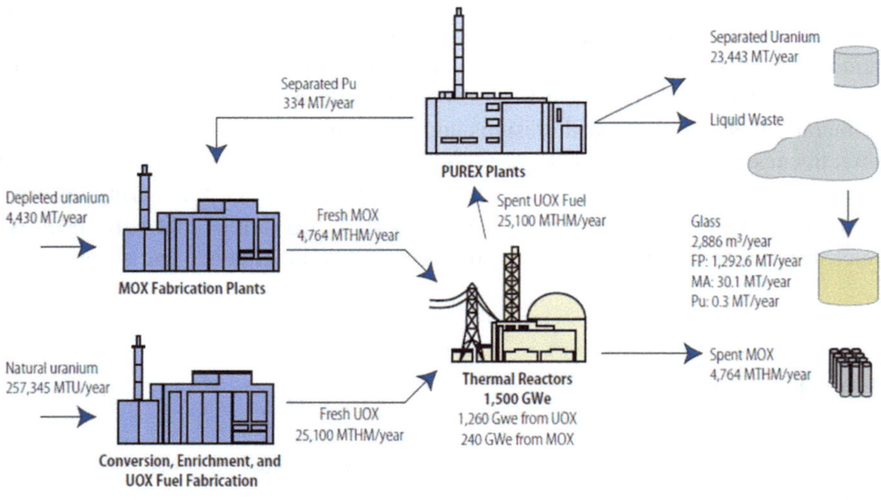

Fig. 1.4 Closed fuel cycle: plutonium recycle (MOX option—one recycle)—projected to 2050 [6]

consequences. A vast number of different fuel cycles appear in the literature [27], and many have been utilized to one degree or another.

We can find two types of classes to the approach to *separation analysis of fuel cycle*. They are, namely, *open* and *closed*, and each one described below while they are demonstrated if form of Fig. 1.3 for open one and Fig. 1.4 is illustration of closed type through fuel cycle.

In open class fuel cycle or once-through fuel cycle, the spent fuel discharged from the reactor is treated as waste. See Fig. 1.3.

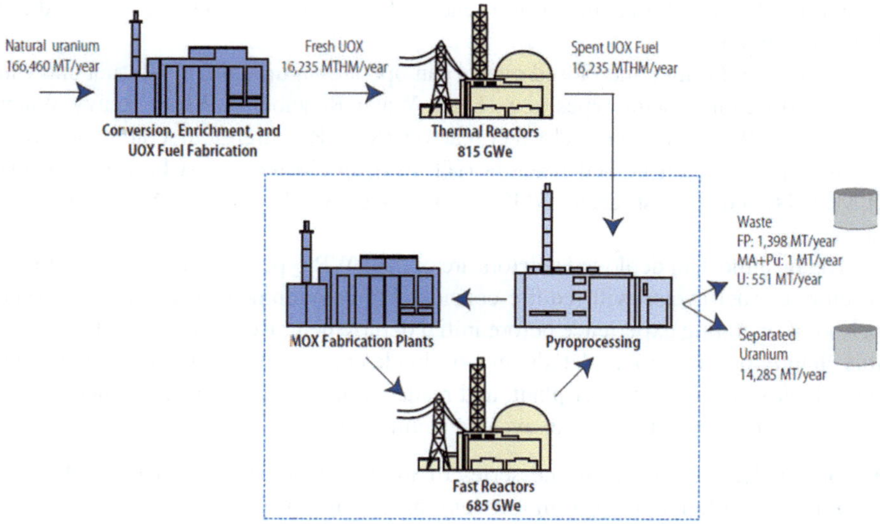

Fig. 1.5 Closed fuel cycle: full actinide recycle—projected to 2050 [6]

In the closed class fuel cycle today, the spent fuel discharged from the reactor is *reprocessed*, and the products are partitioned into uranium (U) and plutonium (Pu) suitable for fabrication into oxide fuel or mixed oxide (MOX) fuel to recycle back into a reactor as illustrated in Fig. 1.4.

The rest of the spent fuel is treated as high-level waste (HLW). In the future, closed fuel cycles could include the use of a dedicated reactor that would be used to transmute selected isotopes that have been separated from spent fuel. See Fig. 1.5.

The dedicated reactor also may be used as a breeder to produce new fissile fuel by neutron absorption at a rate that exceeds the consumption of fissile fuel by the neutron chain reaction. Several countries have explored breeder reactors, notably the United States, France, Russia, Japan, and India.

In a situation depicted in Fig. 1.5, such fuel cycles the waste stream will contain less actinides (i.e., minor actinides are americium (Am), neptunium (Np), and curium (Cm)), which will significantly reduce the long-term radioactivity of the nuclear waste. There are still other options, such as using an accelerator to produce neutrons in subcritical assembly.

However, in general comparing these two classes of fuel cycle open and closed, each one has its own pros and cons, and in summary they can be described here.

We expect the once-through fuel cycle to have an advantage in terms of cost and proliferation resistance (since there is no reprocessing and separation of actinides), compared to the closed cycle. Closed cycles have an advantage over the once-through cycle in terms of resource utilization (since the recycled actinides reduce the requirement for enriched uranium), which in the limit of very high ore prices would be more economical. Some argue that closed cycles also have an advantage for long-term waste disposal, since long-lived actinides can be separated from the fission products and transmuted in a reactor. Our analysis below focused on these key comparisons.

Both once-through and closed cycles can operate on uranium or Th fuel and can involve different reactor types, e.g., Light Water Reactors (LWRs), Heavy Water Reactors (HWRs), supercritical water reactors (SCWRs), high-temperature and very high-temperature gas-cooled reactors (HTGCs), liquid metal fast breeder reactors (LMFBRs) and gas fast reactors (GFRs), or Molten Salt Reactors (MSRs) of various sizes.

Today, almost all deployed reactors are of the LWR type. The introduction of new reactors or fuel cycles will require considerable development resources and some period of operating experience before initial deployment. Reference [6] at the end of this chapter provides more details, in worldwide current characteristic for fuel cycle with deployed nuclear power plant, and reader should refer to this reference.

In summary, the nuclear fuel cycle is summarized here as:

- The nuclear fuel cycle is the series of industrial processes which involve the production of electricity from uranium in nuclear power reactors.
- Uranium is a relatively common element that is found throughout the world. It is mined in a number of countries and must be processed before it can be used as fuel for a nuclear reactor.
- Fuel removed from a reactor, after it has reached the end of its useful life, can be reprocessed so that most is recycled for new fuel.

The various activities associated with the production of electricity from nuclear reactions are referred to collectively as the nuclear fuel cycle. The nuclear fuel cycle starts with the mining of uranium and ends with the disposal of nuclear waste. With the reprocessing of used fuel as an option for nuclear energy, the stages form a true cycle.

To prepare uranium for use in a nuclear reactor, it undergoes the steps of mining and milling, conversion, enrichment, and fuel fabrication. These steps make up the "front end" of the nuclear fuel cycle.

After uranium has spent about 3 years in a reactor to produce electricity, the used fuel may undergo a further series of steps including temporary storage, reprocessing, and recycling before wastes are disposed. Collectively these steps are known as the "back end" of the fuel cycle. See Fig. 1.6.

Figure 1.6 demonstration is based on the following assumptions: enrichment to 4.5% U-235 with 0.22% tails assay, hence 182,000 SWU (separative work units) of enrichment needed (one SWU requires about 50 kWh of electricity at an enrichment plant); refueling so that 24.3 tU/year is replaced annually (or 36.5 tU every 18 months) for core load 73 tU; and operation—45,000 MWday/t (45 GWd/t) burnup, 33% thermal efficiency.

In fact, a nuclear power reactor cannot be expected to run at 100% load factor—90% is more typical for good performance, so an output of around 7.9 TWh/year is more realistic, but this simply means scaling back the inputs accordingly, e.g., to 190 tU/year or 147 tU/year for the modern unit.

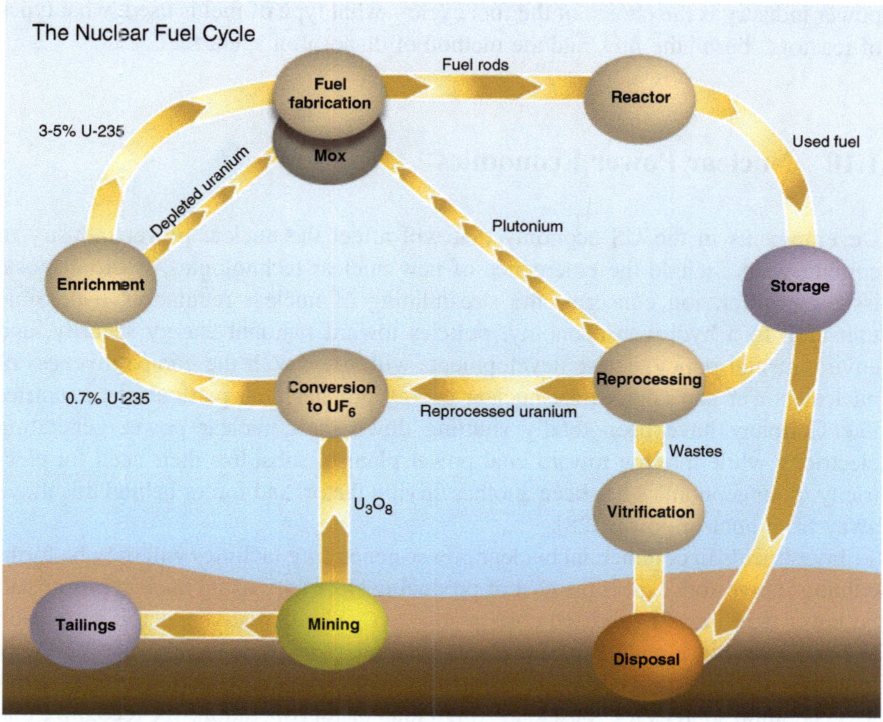

Fig. 1.6 The nuclear fuel cycle (Courtesy of the World Nuclear Association)

With the higher (5%) enrichment and burnup in the second (AP1000 or EPR) set of figures, enrichment input rises to 198,000 SWU. In the used fuel, transuranic and fission product numbers will be slightly lower due to high thermal efficiency.

Canadian figures for tU/GWe/year suggest slightly lower uranium requirements and utilization for PHWRs than for Light Water Reactors. An International Atomic Energy Agency technical report [1] gives 157 tU at typical 7.5 GWd/t burnup and 31% thermal efficiency or 142 tU at 90% capacity factor, hence 80% of the input compared with a typical LWR above. This is 17.9 tU/TWh.

Considering just how much of the original uranium is actually used (0.7% fissile U-235 is in natural U (Unat)), on the above "typical" figures, 0.49% of Unat goes into fuel as the fissile part, 0.394% is actually fissioned, and in addition about half of U-238 that turned into Pu-239 is fissioned, giving about a 0.6% utilization of the original Unat.

With the evolutionary power reactor (EPR) figures, 0.538% of Unat goes into fuel as the fissile part, 0.452% of that is actually fissioned, and in addition about half of U-238 that turned into Pu-239 is fissioned, giving about a 0.67% utilization of the original Unat.

In conclusion, nuclear power has unresolved challenges in long-term management of radioactive wastes. A critical factor for the future of an expanded nuclear power industry is the choice of the fuel cycle—what type of fuel is used, what types of reactors "burn" the fuel, and the method of disposal of spent fuel.

1.10 Nuclear Power Economics

Developments in the US economy that will affect the nuclear power industry in coming years include the emergence of new nuclear technologies, waste disposal issues, proliferation concerns, the streamlining of nuclear regulation, a possible transition to a hydrogen economy, policies toward national energy security, and environmental policy. These developments will affect both the competitiveness of nuclear power and appropriate nuclear energy policies, and particularly, countries like Germany have been totally shutting down their nuclear power generating electricity while moving toward coal power plant to subsidize their need for electricity. Poor economics has been another driving factor, and forces behind this move away from nuclear power [28].

Investments in commercial nuclear power-generating facilities will only be forthcoming if investors expect the cost of producing electricity using nuclear power will be lower than the risk-adjusted costs associated with alternative electric generation technologies. Since nuclear power plants have relatively high capital costs and very low marginal operating costs, nuclear energy will compete with alternative electricity generation sources for "baseload" (high load factor) operation. We recognize that over the next 50 years, some significant but uncertain fraction of incremental electricity supplies will come from renewable energy sources (e.g., wind) either because these sources are less costly than alternatives or because government policies (e.g., production tax credits, high mandated purchase prices, and renewable energy portfolio standards) or consumer choice favors renewable energy investments. Despite the efforts to promote renewable energy options, however, it is likely that a large fraction of the incremental and replacement investments in electricity-generating capacity needed to balance the supply and demand over the next 50 years will, in the absence of a nuclear power generation option, rely on fossil fuels—primarily natural gas or coal. This is particularly likely in developing countries experiencing rapid growth in income and electricity consumption. Accordingly, we focus on the costs of nuclear power compared to these fossil fuel-generating alternatives in baseload applications [6].

Any analysis of the costs of nuclear power must take into account a number of important considerations. First, all of the nuclear power plants operating today were developed by state-owned or regulated investor-owned vertically integrated utility monopolies. Although, in the United States and the United Kingdom, some nuclear plants were subsequently sold or transferred to merchant companies generating electricity, many developed countries and an increasing number of developing countries are in the process of moving away from an electric industry structure built upon vertically integrated regulated monopolies to an industry structure that relies primarily on competitive generation power plant investors.

We assume that in the future, nuclear power will have to compete with alternative generating technologies in competitive wholesale markets—as merchant plants, where these plants sell their output under short-, medium-, and longer-term supply contracts negotiated competitively with distribution companies and wholesale and retail market. The power plant developers take on permitting, development, construction cost, and operating performance risks but may transfer some or all risks associated with market price volatility to buyers for a price through the terms of their contracts.

These changes in the structure of the electric power sector have important implications for investment in generating capacity. Under traditional industry and regulatory arrangements, many of the risks associated with construction costs, operating performance, fuel price changes, and other factors were borne by consumers rather than suppliers. It is often assumed that regulated monopolies were subject to "cost-plus" regulation which insulated utilities from all of these risks. This is an extreme and inaccurate characterization of the regulatory process, at least in the United States. Several US utilities were faced with significant cost disallowances associated with nuclear power plants they completed or abandoned, a result inconsistent with pure cost-plus regulation. Nevertheless, it is clear that a large fraction of these cost and market risks were shifted to consumers from investors when the industry was governed by regulated monopolies.

The insulation of investors from many of these risks necessarily had significant effects on the cost of capital they used to evaluate alternative generation options and on whether and how they took extreme contingencies into account. Specifically, the process reduced the cost of capital and led investors to give less weight to regulatory (e.g., construction and operating licenses) and construction cost uncertainties, operating performance uncertainties, and uncertainties associated with future oil, gas, and coal prices than if they had to bear these cost and performance risks.

The commercial/civilian nuclear industry has been in operation for over 50 years so far. During such a long period of time, it would be a usual and natural normal process for technological improvement cycle and experience to result in learning and subsequently to enhancement in economic efficiency, when it comes to deal with new generation of these power plants known as Generation IV. However, the nuclear industry did not follow this pattern except in past few years that we start seeing some innovative approaches suggested by some experts in this field in particular, in Advanced Small Modular Reactors (ASMRs), including these authors [2, 3].

With these considerations in mind, we now proceed to examine the relative costs of new nuclear power plants, pulverized coal plants, and Combined Cycle Gas Turbine (CCGT) plants in baseload operations in the United States. The combined open-air Brayton cycle has been suggested by these authors and colleagues as an innovative solution to nuclear power plants of SMRs as well [2, 3].

Note that also we are not considering competition between new nuclear plants and existing coal and gas plants, whose construction costs are now sunk costs. We recognize there may be economical opportunities to increase the capacity of some existing nuclear plants and to extend their commercial lives. We do not consider these opportunities here.

The analysis is not designed to produce precise estimates either but rather a "reasonable" range of estimates under a number of different assumptions reflecting uncertainties about future construction and operating costs. Similar analysis for Europe and especially Japan and Korea would be somewhat more favorable to nuclear since gas and coal costs are typically higher than in the United States.

Elements associated as variables that are driving the function of nuclear power economics are listed below, and they are:

1.10.1 Rising Construction Costs

Country after country, the nuclear construction programs are getting to be over the budget considerably. In the United States, an assessment of 75 of the country's reactors is predicting costs to be as high as $45 billion; however the real costs were almost over tripled, and they got to be around $145 billion, and in some countries such as India, the cost of completion for ten reactors were averaged at least 300% over the budget. Figure 1.7 is an illustration of history of timeline cost overruns.

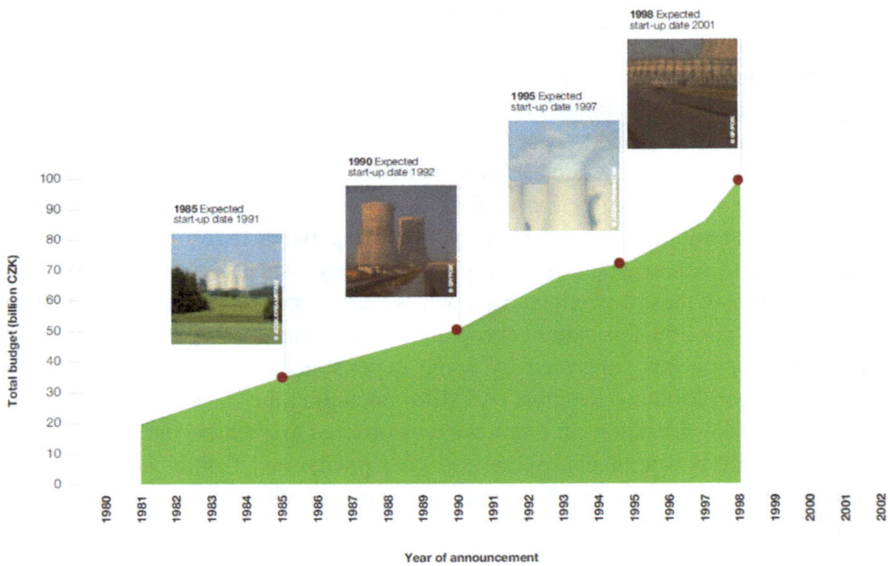

Fig. 1.7 History of timeline cost overrun

1.10.2 Rising Construction Time

The average construction time for nuclear power plants has also increased from 66 months for end-to-end completions in at least the mid-1970s to almost 116 months, roughly speaking equivalent of 10 years' time period for similar in-completion projects between 1995 and 2000.

1.10.3 Falling Construction Demand

Presently, 50 nuclear reactors worldwide are under construction, and out of that, 2 of them are in the United States. Figure 1.8, statistically, shows the number of nuclear reactors under construction worldwide as of February 2018, by country. There were 18 nuclear reactors under construction in China, compared to the two reactors in the United States that are currently being built.

Construction started on 5 of the reactors over 20 years ago, and consequently the likelihood of the reactors being built to their current timetable is open to question.

There are further 14 reactors on which construction has started but is currently suspended, 10 of which are in Central and Eastern Europe. This low level of nuclear construction provides little relevant experience on which to build confidence in cost forecasts [28].

1.10.4 Untested Technology

The nuclear industry is promoting a new generation of reactors (Generations III and III+) and hoping that a wave of orders will be placed for them in the next few years. Also at least, at the design stage, some nuclear manufactures are looking at six different and major types of Generation IV, which are briefly listed in Table 1.2 here as well.

Generation III+ reactors: No Generation III+ plant has yet been completed, and only one is under construction. The most widely promoted of these latest designs are the new generation of pressurized water reactors (PWRs) and in particular AREVA's European Pressurized Water Reactor (EPR) and the Westinghouse AP1000.

Among of them are considered as Advanced Small Modular Reactors (ASMRs), and they are the focus of this book as well, namely:

(a) Molten Salt Reactor (MSR) system
(b) Sodium-cooled fast reactor (SFR) system
(c) Very high-temperature reactor (VHTR) system

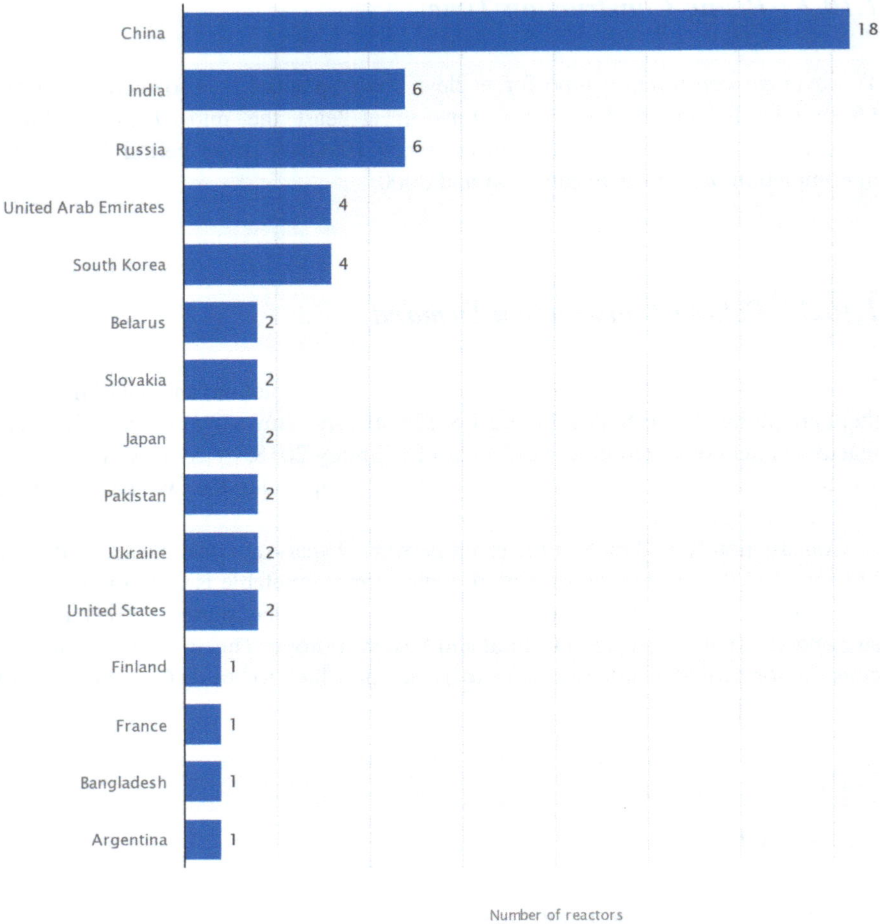

Fig. 1.8 Number of nuclear reactors under construction worldwide as of February 2018, by country

The nuclear present technology is up on Generation III and going forward with the buildup of Generation III+ reactors, and each is described as follows:

Generation III reactors: The only Generation III reactors currently in operation are the advanced boiling water reactors (ABWR) developed in Japan. By the end of 2006, four ABWR were in service and two under construction in Taiwan. Total construction costs for the first two units were well above the forecast range.

Generation IV reactors: Even more speculative are the "paper" designs for Generation IV plutonium-fueled reactors. While several designs are being produced, technical difficulties make it unlikely that they will be deployed for at least two decades, if at all, while the economics of fuel reprocessing also remains unproven.

Table 1.2 Six major designs of the Generation IV international program

GFR—Gas-cooled fast reactor system: The GFR system is a helium-cooled reactor with fast-neutron spectrum and closed fuel cycle. It uses helium as coolant, due to extreme temperatures (850 °C outlet; compared to 300 °C for PWRs and 500 °C for FBRs). Consequently, "High temperatures and extreme radiation conditions are difficult challenges for fuels and materials." It will use plutonium and burn actinides
LFR—Lead-cooled fast reactor system: LFR systems are reactors cooled by liquid metal (lead or lead/bismuth) with a fast-neutron spectrum and closed fuel cycle system. A full actinide recycle fuel cycle with central or regional facilities is envisaged. A wide range of unit sizes is planned, from "batteries" of 50–150 MW(e), and modular units of 300–400 MW(e), to large single plants of 1200 MW(e). The LFR battery option is a small factory-built turnkey plant with very long core life (10–30 years). It is designed for small grids and for developing countries that may not wish to deploy a fuel cycle infrastructure. Among the LFR concepts, this battery option is regarded as the best, concerning the fulfillment of Generation IV goals. However, it also has the largest research needs and longest development time
MSR—Molten Salt Reactor system: The MSR system is based on a thermal neutron spectrum and a closed fuel cycle. The uranium fuel is dissolved in the sodium fluoride salt coolant that circulates through graphite core channels. The heat, directly generated in the molten salt, is transferred to a secondary coolant system and then through a tertiary heat exchanger to the power conversion system. It is primarily envisioned for electricity production and waste burn-down. The reference plant has a power level of 1000 MW(e). Coolant temperature is 700 °C at very low pressure. Of all six reactor systems, MSR requires the highest costs for development ($1 billion)
SCWR—Supercritical-water-cooled reactor system: The SCWRs are high-temperature, high-pressure water-cooled reactors that operate above the thermodynamic critical point of water (i.e., at pressures and temperatures at which there is no difference between liquid and vapor phase). The reference plant has a 1700 MW(e) power level, an operating pressure of 25 MPa, and a reactor outlet temperature of 550 °C. Fuel is uranium oxide
SFR—Sodium-cooled fast reactor system: The SFR system consists of a fast-neutron reactor and a closed fuel cycle system. There are two major options: One is a medium-sized (150–500 MW(e)) reactor with metal alloy fuel, supported by a fuel cycle based on pyrometallurgical reprocessing in collocated facilities. The second is a medium to large (500–1500 MW(e)) reactor with MOX fuel, supported by a fuel cycle based upon advanced aqueous reprocessing at a centralized location serving a number of reactors. According to GIF, the SFR has the broadest development base of all the Generation IV concepts
VHTR—Very high-temperature reactor system: The VHTR system uses a thermal neutron spectrum and a once-through uranium fuel cycle. The reference reactor concept has a 600 MW(th) graphite-moderated helium-cooled core based on either the prismatic block fuel of the GT-MHR or the pebble bed of the PBMR

1.10.5 Unfavorable Market Place

The economics of nuclear power have always been questionable. The fact that consumers or governments have traditionally borne the risk of investment in nuclear power plants meant that utilities were insulated from these risks and were able to borrow money at rates reflecting the reduced risk to investors and lenders.

However, following the introduction of competitive electricity markets in many countries, the risk that the plant would cost more than the forecast price was transferred to the power plant developers, which are constrained by the views of financial organizations such as banks, shareholders, and credit rating agencies. Such organizations view investment in any type of power plant as risky, raising the cost of capital to levels at which nuclear is less likely to compete.

The logic of this transfer to competitive electricity markets was that plant developers possessed better information and had direct control over management and so had the means as well as the incentive to control costs. Builders of nonnuclear power plants were willing to take these risks, as were vendors of energy efficiency services. Consequently, when consumers no longer bore the economic risk of new plant construction, nuclear power, which combines uncompetitively high prices with poor reliability and serious risks of cost overruns, had no chance in countries that moved to competitive power procurement.

However, this issue is no longer a concerned risk, going forward with Small Modular Reactors of Generation IV, due to modulization construction and installation of each individual module as the first element of power plant facility, and when demand for electricity is ascending, so as the number of module installation accordingly.

1.10.6 Unreliable Forecasts

In recent years there have been numerous studies of the economics of nuclear power. The values of the key parameters used to generate the forecast cost of nuclear power vary significantly from one study to another. For example, the assumed cost of construction ranges from \$856 to 3600/kW, while the assumed construction time varies from 60 to 120 months. The resultant price of electricity consequently also varies significantly, producing a range of between \$21 and 76/MWh.

1.10.7 Subsidies Needed

It is now 29 years since the last order for a new nuclear power plant in the United States and 34 years since the last order for a plant that was actually completed. Utilities suffered heavy losses in the 1980s as economic regulators became increasingly unwilling to pass huge cost overruns from nuclear projects onto consumers, forcing utilities to bear the extra costs. The introduction of power markets has meant that plant owners are now fully exposed not just to the risk of cost overruns but also to plant unreliability. The nuclear provisions of the US Energy Policy Act of 2005 (EPACT 2005) are an effort to reverse these changes and protect investors from that large economic risk.

The most important nuclear provisions of EPACT 2005 offer three types of support:

- A limited number of new nuclear power plants can receive an \$18/MWh (€13.7) production tax credit for up to \$125 m per 1000 MW (or about 80% of what the plant could earn if it ran 100% of the time).
- A provision for federal loan guarantees covering up to 80% of project costs.
- Up to \$500 m in risk insurance for the first 2 units and \$250 m for units 3–6. This insurance is to be paid if delays that are not the fault of the licensee slow the licensing of the plant.

These subsidies are said to be worth between \$2 and 20/MWh. Without these subsidies, it is unlikely that any US company would be considering investing in a new nuclear plant. Government financial or contractual guarantees would effectively take nuclear power out of the market so that it is paid for, as in the past, by electricity consumers and taxpayers. If nuclear power is to be subsidized in this way, there needs to be clear and compelling evidence that this is a cost-effective and worthwhile way to use taxpayers' and electricity consumers' money.

As part of prospects and status of technology for nuclear power cost is *the cost of energy production and use from all technologies are falling systematically with innovation and scale economies in manufacturing and use, apart from nuclear power since the 1970s.*

Various reasons have been put forward for the relatively low learning rate of nuclear power, including the relatively small post-1970s' reactor ordering rate, the interface between the complexity of nuclear power plants and the regulatory and political processes, and the variety of designs deployed. While some of these factors may be overcome in the future, for example, the UK Government's Performance and Innovation Unit also highlighted a number of areas in which future nuclear power plants may not exhibit comparable learning rates to other technologies, including:

- Nuclear power is a relatively mature technology, and therefore dramatic "technological stretch" is less likely than in other technologies.
- The relatively long lead times for the construction and commissioning mean that improvements derived by feeding back information from operating and design experiences on the first units are necessarily slow.
- The scope for economies of scale is less in the nuclear case than for renewables, due to the latter's smaller initial scale and wider potential application.

Figure 1.9 is a depiction of such learning rates of selected energy technologies as indicated.

For further details, readers should refer to Reference [28] at the end of this chapter.

1.10.8 A Nuclear Revival or Decline

The current list of plants under construction (see Fig. 1.10) is a short one. Of the 22 units, 16 are being supplied by vendors from China, Russia, and India. It seems

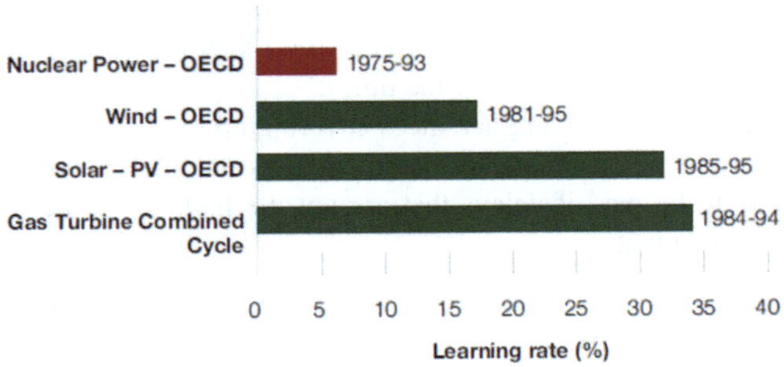

Fig. 1.9 Learning rates of selected energy technologies (Courtesy of McDonald, A., and Schrattenholzer, L.)

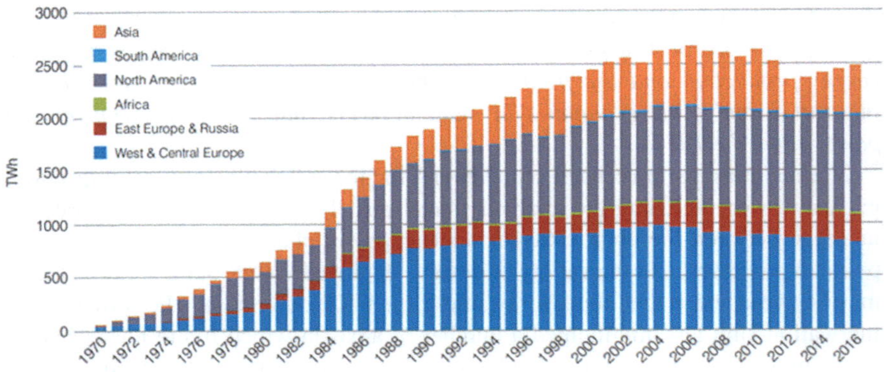

Fig. 1.10 Installation of new nuclear capacity onto grid (Source: World Nuclear Association, IAEA Power Reactor Information Service (PRIS))

unlikely that any of these vendors would be considered in Western Europe or North America, the markets that would need new orders if a global nuclear revival were to take place. Most of India's plants are largely based on a Canadian design from the 1960s, long since superseded in Canada. China's plants are also closely modeled on old Western designs—prior to the AP1000 order—albeit not so out of date as the Indian plants. China will probably continue to supply mainly its home market, with one or two exports to Pakistan.

As it can be seen from Fig. 1.10, around 11% of the world's electricity is generated by about 450 nuclear power reactors. About 60 more reactors are under construction, equivalent to 16% of existing capacity, while an additional 150–160 are planned, equivalent to nearly half of existing capacity.

In 2016 nuclear plants supplied 2477 TWh of electricity, up from 2441 TWh in 2015. This is the fourth consecutive year that global nuclear generation has risen, with output 130 TWh higher than in 2012.

Figure 1.11 is indication of world electricity production by source in 2017.

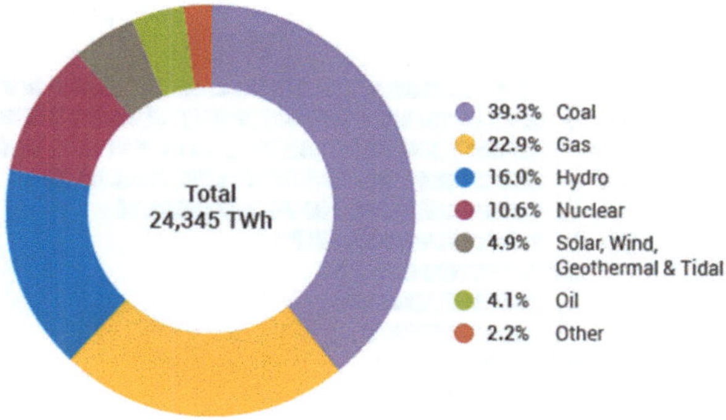

Fig. 1.11 2017 world electricity production by source (Source: IEA Electricity Information 2017)

Sixteen countries depend on nuclear power for at least one-quarter of their electricity. France gets around three-quarters of its electricity from nuclear energy; Hungary, Slovakia, and Ukraine get more than half from nuclear, while Belgium, the Czech Republic, Finland, Sweden, Switzerland, and Slovenia get one-third or more. South Korea and Bulgaria normally get more than 30% of their electricity from nuclear power, while in the United States, the United Kingdom, Spain, Romania, and Russia, about one-fifth of electricity is from nuclear power. Japan is used to relying on nuclear power for more than one-quarter of its electricity and is expected to return to somewhere near that level (Fig. 1.12).

1.11 Spent Fuel and High-Level Waste Management

In this section we focus on spent fuel and reprocessed high-level waste, since these waste types contain most of the radioactivity generated in the nuclear power fuel cycle and pose the greatest technical and political challenges for final disposal. We also include in the discussion the so-called transuranic (TRU) waste—non-high-level waste contaminated with significant quantities of long-lived transuranic radionuclides—which because of its longevity will likely be disposed of in the same facilities as high-level waste. Other types of nuclear waste, including low-level waste and uranium mill tailings, are generated in larger volumes in the nuclear fuel cycle but pose fewer technical challenges for disposal, although localized opposition to disposal facilities for these materials has sometimes been intense.

Transuranic waste consists of waste that is contaminated with man-made radioactive elements which are heavier than uranium (meaning the elements have higher atomic numbers than uranium on the periodic table of elements). Because they come after uranium on the periodic table, they are referred to as "transuranic." The concentration of these transuranic elements in the waste determines whether it is transuranic (TRU) waste or low-level waste. See Fig. 1.13.

Furthermore, material contaminated with transuranic elements—artificially made, radioactive elements, such as neptunium, plutonium, americium, and

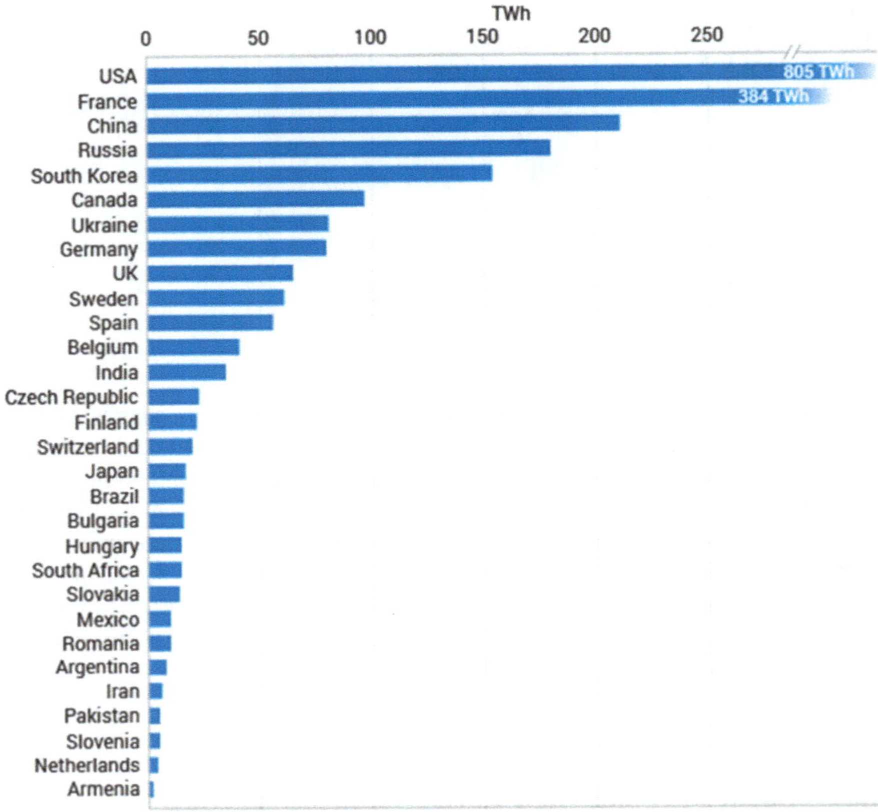

Fig. 1.12 2016 nuclear power generation by countries (Source: IAEA PRIS Database)

Fig. 1.13 Transuranic waste package

others—that have atomic numbers higher than uranium in the periodic table of elements. Transuranic waste is primarily produced from recycling spent fuel or using plutonium to fabricate nuclear weapons.

Defined by the Waste Isolation Pilot Plant (WIPP) Land Withdrawal Act as "waste containing more than 100 nanocuries of alpha-emitting transuranic isotopes per gram of waste with half-lives greater than 20 years, except for:

(a) High-level radioactive waste.
(b) Waste that the Secretary of Energy has determined, with concurrence of the Administrator of the Environmental Protection Agency, does not need the degree of isolation required by the disposal regulations.
(c) Waste that the Nuclear Regulatory Commission has approved for disposal on a case-by-case basis in accordance with part 61 of title 10 Code of Federal Regulations (CFR)."

More than 70,000 containers of this waste (sometimes referred to as suspect TRU waste) were stored under a layer of dirt in the 1970s and 1980s, in the 200 area low-level burial grounds of the Hanford Site. The intention was to retrieve the waste (which is why sometimes it is also referred to as retrievably stored waste) at a later date when a national repository was established to accept transuranic waste.

The management and disposal of radioactive waste from the nuclear fuel cycle is one of the most difficult problems currently facing the nuclear power industry. Today, more than 40 years after the first commercial nuclear power plant entered service, no country has yet succeeded in disposing of high-level nuclear waste—the longest-lived, most highly radioactive, and most technologically challenging of the waste streams generated by the nuclear industry.

With the 1999 opening of the Waste Isolation Pilot Plant (WIPP) outside of Carlsbad, New Mexico, TRU waste from Hanford is now being retrieved, packaged to meet WIPP's acceptance criteria, and shipped to WIPP for permanent disposal. See Fig. 1.14.

The waste was stored in boxes and 55-gallon and 85-gallon drums. The waste consists of tools, clothing, laboratory equipment, and other materials needed during the plutonium production mission at Hanford. As the waste was expected to be dug up when it was originally stored in the 1970s and 1980s, a typical TRU waste trench

Fig. 1.14 Storage facility (Courtesy of DOE Hanford Site)

consists of several levels of drums stacked on asphalt pads, separated with plywood, draped with tarps, and then covered with dirt.

During retrieval, crews unearth the drums and boxes, check for the stability of the container holding the waste, and determine the level of radioactivity within the container. If the container is damaged or corroded, it must be carefully placed into a second container called an overpack, so that the materials do not spill onto the ground during removal from trenches and to allow for future characterization for disposal. The stability of the containers holding the waste varies. In some trenches, most of the containers are in good shape and can be safely removed without the need for an overpack. In other trenches, most containers are damaged or corroded requiring an overpack.

Most transuranic elements emit alpha particles, which are the least penetrating form of radiation and can be blocked (shielded) using a sheet of paper or the outer layer of skin. However, because alpha particles present an internal hazard, damaged or corroded TRU waste drums require special handling to prevent workers from inhaling the particles. Because of this, crews often work in protective clothing and breathing equipment before they can put such containers into overpacks for safe removal from the trench.

Because the definition of transuranic waste (based on the concentration of transuranic elements) has changed over time, only about half of the retrievably stored waste is considered a TRU waste. The other half is low-level waste. The TRU waste is shipped to WIPP for disposal. The low-level waste is treated as necessary and disposed at Hanford.

The Waste Isolation Pilot Plant (WIPP) is a deep geologic repository for the permanent disposal of radioactive waste. WIPP safely disposes of the nation's defense-related transuranic waste (TRU waste). Located in the Chihuahuan Desert, outside Carlsbad, NM, WIPP began disposal operations in March 1999. WIPP is managed by the Carlsbad Field Office. See Fig. 1.15.

It is noteworthy to say that international cooperation in the field of high-level waste management and disposal is presently underdeveloped. Stronger international coordination of standards and regulations for waste transportation, storage, and disposal will be necessary in order to strengthen public confidence in the safety of these activities. There is also considerable potential for international sharing of waste storage and disposal facilities. This might not only reduce proliferation risks from the fuel cycle but could also yield significant economic and safety benefits, although formidable political obstacles will have to be overcome first.

1.12 Proliferation and Nonproliferation

Given geopolitical circumstances that globally, exits today, between the countries with nuclear power plant technologies and capabilities of developing and nations that with the possibility wishing to acquire or enhance their nuclear weapons capability will use commercial nuclear power as a source of technological of

Fig. 1.15 Aerial photo of Chihuahuan Desert, outside Carlsbad, NM

know, how or nuclear weapons usable material, notably plutonium (i.e., Pu^{239}) at enriched weapon grade.

Although this has not proved to be the preferred pathway to nuclear weapons' capability, the possession of a complete nuclear fuel cycle, including enrichment, fuel fabrication, reactor operation, and reprocessing, certainly moves any nation closer to obtaining such a capability. The key step for achieving nuclear weapons' capability is the acquisition of sufficient weapons-usable fissionable material, either high-enriched uranium or plutonium [6].

Unfortunately, reprocessing of spent fuel for the fuel cycle operation in Europe, Russia, and Japan has led to the accumulation of about 200 tons of separated plutonium. The associated risks have been viewed with increased alarm since the 9/11 events that demonstrated the reach of international terrorism. Radiation exposure from spent fuel that is not reprocessed is a strong, but not certain, barrier to theft and misuse.

Nuclear weapons' proliferation has been prominent in discussions about nuclear power since its earliest days. The birth of nuclear technology that began with the production of the first weapons-usable fissionable material—plutonium production in nuclear reactors and high-enriched uranium (HEU) by isotope enrichment—assured that this would be so. *Today, the objective is to minimize the proliferation risks of nuclear fuel cycle operation.* We must prevent the acquisition of weapons-usable material, either by diversion in the case of plutonium or by misuse of fuel cycle facilities including related facilities, such as research reactors or hot cells and control, to the extent possible, the know-how about how to produce and process either HEU (enrichment technology) or plutonium [6].

This proliferation concern has led, over the last half century, to an elaborate set of international institutions and agreements, none of which have proved entirely satisfactory. The Nuclear Nonproliferation Treaty (NPT) is the foundation of the control regime, since it embodies the renunciation of nuclear weapons by all signatories except for the declared nuclear weapons states—the P-5 (the United States, Russia, the United Kingdom, France, China)—and a commitment to collaborate on developing peaceful uses of nuclear energy. However, non-signatories India and Pakistan tested nuclear weapons in 1998, and signatories, such as South Africa and North Korea, have admitted to making nuclear weapons [6].

The International Atomic Energy Agency (IAEA) has the responsibility for verifying NPT compliance with respect to fuel cycle facilities through its negotiated safeguard agreements with NPT signatories. The IAEA's safeguard efforts, however, are seriously constrained by the scope of their authorities (as evidenced in Iraq, Iran, and North Korea during the last decade), by their allocation of resources, and by the growing divergence between responsibilities and funding.

For further information, please refer to MIT report [6] or IAEA web site.

1.13 Society Attitude and Public Understanding of Nuclear Power

Nuclear energy is a controversial and sensitive subject among the public because it strongly relates public health and safety. Because this mass media is handling this subject in a diverse manner, thereby it creates strong impact among the people. Public's acceptance and support to the nuclear energy are also preoccupied. Monitoring the effectiveness of campaigns to change public attitudes to nuclear power required a tracking study based on a high-quality, representative national sample with high-sensitivity indicators.

We need to have a means and indicator to study and assess the level of public knowledge of nuclear energy, that will explore, public understanding of the arguments for and against the construction of new nuclear power plant nationwide and worldwide as well, and the fact that how much the news media helps the ordinary people have knowledge and desire to understand the new generation of the nuclear power energy regardless of negative impacts recent accident such as the one on as recent as 2012 in Fukushima Daiichi in Japan, which was based on over 20 years ago technology of plant. See Fig. 1.16.

One strategy that could be implemented to help such indicator of study is a methodology to allow not only the monitoring of attitudes toward nuclear power but also the formulation of an effective communications strategy. The success of the study relied on the quality of the sample and its representativeness of the entire population, both nationwide and worldwide among the countries with nuclear power plant technology, and the idea of going forward with such clean and renewable source of energy based on new advanced technology in particular with the dawn of Generation IV of these plants and Small Modular Reactors (SMRs).

Fig. 1.16 Aerial picture of Fukushima Daiichi in Japan

The sample data collection of such methodology should be structured on the formulation of an effective communications strategy and not just only the monitoring toward the nuclear power as well as the participation of the government of each nation with such technology to back up such methodology and strategy in both short and long term. This is an effort that certain agencies of each government are involved in licensing in cooperation and collaboration with industries designing the new generation of nuclear energy power plants.

One of the key indicators for public awareness about nuclear technology and benefits from it is that as stakeholder, we should create an awareness for all the benefits that *nuclear technology brings to mankind.* We should put our standing point with the fact that "*The public are often unaware of the extent to which aspects of their everyday life involve products and processes originated from the application of nuclear technology via the nuclear industry.*"

With the fifth anniversary of Fukushima having just passed last month and the 30th anniversary of Chernobyl this month, we have a steady reminder of the issues that never seem to go away for the nuclear industry. It is our nature. However, the key question of why is it that the safest source of large-scale electricity generation we have ever come up with is considered so dangerous by enough people that in several countries there is an effort to stop using nuclear energy.

But for those who may not get there quickly enough, here is a summary of the benefits that nuclear technology brings to society each and every day. As stated at the beginning of this chapter, "*Nuclear technology is vital for more than just providing reliable, low-carbon energy. It also has life-saving medical application; improves manufacturing, mining, transport and agriculture; and help us discover more about the planet we live on and how we can sustainably live with it.*"

So, for example, did you know that:

- Nuclear technology saves lives through the use of radioisotopes for screening, diagnosis, and therapy of various medical conditions. According to the WNA, over 10,000 hospitals worldwide use radioisotopes. Radioisotopes are used in therapy to control and damage cancerous growths. Iodine-131 is used to treat thyroid cancer and phosphorus-32 to treat leukemia. Nuclear techniques are used for neonatal screening for sickle cell disease, hypothyroidism, and cystic fibrosis, as well as childhood cancers.
- Radiation is used to preserve seeds and food products and breed disease-resistant plants. In plant breeding, some 1800 new crop varieties have been developed through mutation induced by ionizing radiation.
- Irradiation technology is increasingly being used to preserve food—spices, grains, fruit, vegetables, and meat. It avoids the use of potentially harmful chemical fumigants and insecticides.
- The use of the IAEA's sterile insect technique irradiates the eggs of these insects to sterilize them before hatching. The IAEA estimates that, by suppressing insect pest populations with SIT, pesticide use worldwide has been reduced by 600,000 L annually.
- In industrial radiography, nuclear substances are used for the nondestructive examination and testing of new materials. Radiation from the substances passes through the material and allows defects in welds or constituency to be recorded on film or a digital imager.

This list does not do justice to the report itself which I strongly suggest you read. It is time to stop being defensive and make sure that we no longer have to write reports that start with "The public are often unaware of the extent to which aspects of their everyday life involve products and processes originated from the application of nuclear technology via the nuclear industry." It is time to celebrate our successes and not just talk about where we need to improve. We are proud to be part of the nuclear industry, and we are confident that we are making a difference that helps to make the world a better place [29].

1.14 Conclusion

It seems like at the heart of debate lies the confusion over the exact definition of renewable energy and the requirements that need to be met in order to be one. The recent statement of Hélène Pelosse, the interim director general of the International Renewable Energy Agency (IRENA), saying IRENA will not support nuclear energy programs because it is a long, complicated process and it produces waste and is relatively risky, proves that their decision has nothing to do with having a sustainable supply of fuel [21]. However, if that is the case, then nuclear proponents would have to figure out a way to deal with the nuclear waste management issue and other political implications of nuclear power before they can ask IRENA to reconsider including nuclear energy in the renewable energy list [25].

One more strong argument against fission nuclear power plants as a source of renewable energy comes from Dr. James Singmaster in August 3, 2009, and has been republished here as follows:

> *The basic problem of the climate crisis is the ever-expanding overload of heat energy in the closed biosphere of earth. Temperatures going up indicate the increasing heat energy overload. Everyone reading this should check out Dr. E. Chaisson's article titled 'Long-Term Global Warming from Energy Usage' in EOS, Trans. Amer. Geophys. Union, V. 89, No. 28, Pgs. 253-4(2008) to learn that nuclear energy, be it fission or fusion, being developed should be dropped with money put into it being put to developing renewable energy supplies using the sun, wind and hydrogen.*
>
> *The hydrogen needs to be generated from splitting water using sunlight with the best one or two of seven catalysts reported in the last two years. Or with excess solar or wind collection generating electricity, that could be used to generate hydrogen by electrolysis of water.*
>
> *There is no way that nuclear power can avoid releasing trapped energy to increase the energy overload, so it should be forgotten.*
>
> *To remove some of the energy as well as some of the carbon overload in the biosphere, we need to turn to pyrolysis of massive ever-expanding organic waste streams to remake charcoal that will be removing some of both overloads. It will require using renewable energy and the pyrolysis process expels about 50% of the carbon as small organic chemicals that can be collected, refined and used foe fuel that is a renewable one. For more about using pyrolysis, search my name on GreenInc blog or google it for other blog comments on pyrolysis.*
>
> *Dr. J. Singmaster*

References

1. Zohuri, B. (2015). *Combined cycle driven efficiency for next generation nuclear power plants: An innovative design approach* (1st ed.). New York: Springer.
2. Zohuri, B., & McDaniel, P. (2017). *Combined cycle driven efficiency for next generation nuclear power plants: An innovative design approach* (2nd ed.). New York: Springer.
3. Zohuri, B. (2018). *Small modular reactors as renewable energy sources*. New York: Springer.
4. Zohuri, B. (2018). *Hydrogen energy: Challenges and solutions for a cleaner future*. New York: Springer.

5. Zohuri, B. (2017). *Hybrid energy systems: Driving reliable renewable sources of energy storage*. New York: Springer.
6. *The future of nuclear power, an interdisciplinary MIT study*. (2003).
7. Zohuri, B. (2016). *Plasma physics and controlled thermonuclear reactions driven fusion energy*. New York: Springer.
8. Zohuri, B. (2016). *Inertial confinement fusion driven thermonuclear energy*. New York: Springer.
9. Zohuri, B. (2017). *Magnetic confinement fusion driven thermonuclear energy*. New York: Springer.
10. Cohen, B. L. (1974). *Nuclear science and society*. Garden City, NY: Anchor Books.
11. Zohuri, B., & McDaniel, P. (2015). *Thermodynamics in nuclear power plant systems*. New York: Springer.
12. Zohuri, B. (2017). *Thermal-hydraulic analysis of nuclear reactors* (2nd ed.). New York: Springer.
13. Chilton, A. B., Kenneth Shults, J., & Faw, R. E. (1983). *Principle of radiation shielding* (1st ed.). Upper Saddle River: Prentice Hall.
14. *GE Energy Flex Efficiency 50 Combined Cycle Power Plant, e-brochure*. (2012).
15. Horlock, J. H. (1997). *Cogeneration-combined heat and power (CHP)*. Malabar, FL: Krieger Publishing.
16. Mattingly, J. D. (1996). *Elements of gas turbine propulsion*. New York: McGraw-Hill.
17. Zohuri, B., McDaniel, P., & De Oliveira, C. (2015). Advanced nuclear open-air-Brayton cycles for highly efficient power conversion. *Nuclear Technology Journal.*
18. Zohuri, B. (2017). *Inertial confinement fusion driven thermonuclear energy*. New York: Springer.
19. Johnson, K. (2009, May 21). Is nuclear power renewable energy. *Wall Street Journal.*
20. Cohen, B. L. (1983). Breeder reactors: A renewable energy source. *American Journal of Physics, 51*, 75.
21. Moniz, E. Retrieved from http://energy.mit.edu/news/why-we-still-need-nuclear-power/.
22. Zohuri, B. (2016). *Neutronic analysis for nuclear reactor systems* (1st ed.). New York: Springer.
23. Fraas, A. P. (1989). *Heat exchanger design* (2nd ed.). New York: Wiley.
24. Kanter, J. (2009, August 3). Is nuclear power renewable. *New York Times.*
25. Chowdhury, D. (2012, March 22). *Is nuclear energy renewable energy*. Stanford Physics Department.
26. *Deterring terrorism—Aircraft crash impact analyses demonstrate nuclear power plant's structure strength. EPRI study*. Retrieved December 2002, from www.nei.org.
27. OECD Nuclear Energy Agency. (2001). *Trends in the nuclear fuel cycle*. ISBN 92-64-19664-1; Nuclear Science Committee. (1998, October). *Summary of the workshop on advanced reactors with innovative fuel*. NEA/NSC/DOC (99) 2.
28. Bradford, T. S., Froggatt, A., & Milborrow, D. (2007). *The economics of nuclear power*. Research report 2007. Reenpeace.org.
29. Retrieved from http://www.theenergycollective.com/mzconsulting/2376894/lets-create-awareness-for-all-the-benefits-that-nuclear-technology-brings-to-mankind.

Chapter 2
Small Modular Reactors and Innovative Efficient Enhancement Design

The smaller-sized nuclear reactors were becoming instrumental during the pioneering days of commercial nuclear power to facilitate the development and demonstration of early reactor technologies and to establish operational experience for the fledgling nuclear power industry starting with first powered US Navy nuclear submarine. As part of innovative approach in design of Small Modular Reactors (SMRs) in new era of nuclear power energy to meet the demand for electricity due to growth of population, researchers are in quest of a more efficient electrical out power by these types of reactors known as Advanced Small Modular Reactors (AdvSMRs) suggested as the air-Brayton cycle is an open-air cycle and the Steam-Rankine cycle is a closed cycle. The air-Brayton cycle for a natural gas-driven power plant must be an open cycle, where the air is drawn in from the environment and exhausted with the products of combustion to the environment.

2.1 Introduction

As we have learned so far, nuclear technology is one of the main base-load electricity-generating sources available in the world today, producing around 11.2% of the global power production. According to the International Atomic Energy Agency (IAEA), the use of nuclear energy for electricity generation is expected to grow around the world, particularly in Asia and the Pacific Rim, as demand for electricity increases in that region as foreseen. It is also expected that the types of nuclear power reactors to be used in the future in several countries will not be as large as today, more than 1000 MW of capacity, opening the possibility to the introduction of Small Modular Reactors (SMRs) with a maximum capacity of 300 Me, which could be built in factories and transported to the different sites, by truck, train, or ship.

According to the IAEA designation of plant sizes, 139 of the 442 commercial power reactors that are currently in operation worldwide are SMRs. However, most

B. Zohuri, *Heat Pipe Applications in Fission Driven Nuclear Power Plants*,
https://doi.org/10.1007/978-3-030-05882-1_2

of these are merely scaled-down versions of large plant designs and are not the focus of this review. Instead, the current excitement in smaller-sized reactors is for reactor designs that are deliberately small, i.e., designs that do not scale to large sizes but rather capitalize on their smallness to achieve specific performance characteristics. An integral primary system reactor is an example of a Deliberately Small Reactor (DSR) and will be described later in this paper. These DSRs may be appropriate for base-load electricity generation, especially in regions with smaller electricity grids, but are especially well suited for more specialized non-electrical energy applications.

Smaller-sized nuclear reactors were instrumental during the pioneering days of commercial nuclear power to facilitate the development and demonstration of early reactor technologies and to establish operational experience for the fledgling nuclear power industry. As the United States embarks on its "second nuclear era," the question becomes: Will smaller-sized plants have a significant role in meeting the nation's needs for electricity and other energy demands?

The major growth in the electricity production industry in the last 30 years has centered on the expansion of natural gas power plants based on gas turbine cycles. The most popular extension of the simple Brayton gas turbine has been the combined cycle power plant with the air-Brayton cycle serving as the topping cycle and the Steam-Rankine cycle serving as the bottoming cycle for new generation of nuclear power plants that are known as GEN-IV. The air-Brayton cycle is an open-air cycle and the Steam-Rankine cycle is a closed cycle. The air-Brayton cycle for a natural gas-driven power plant must be an open cycle, where the air is drawn in from the environment and exhausted with the products of combustion to the environment. This technique is suggested as an innovative approach to GEN-IV nuclear power plants in form and type of Small Modular Reactors (SMRs). The hot exhaust from the air-Brayton cycle passes through a heat recovery steam generator (HSRG) prior to exhausting to the environment in a combined cycle. The HRSG serves the same purpose as a boiler for the conventional Steam-Rankine cycle [1–3].

Nuclear energy can play a very significant long-term role for meeting the world's increasing energy demands while simultaneously addressing challenges associated with global climate and environmental impact. Many nations of the world, particularly the Asia/Pacific Rim countries, are actively engaged in a major expansion of their nuclear energy complex. The degree to which nuclear energy can address long-term energy needs, either globally or regionally, will be dictated by the pace and adequacy of technical and policy solutions for waste, safety, security, and nonproliferation issues, as well as the capital cost of construction. Small Modular Reactors (SMRs) could successfully address several of these issues. SMRs offer simpler, standardized, and safer modular design by being factory-built, requiring smaller initial capital investment and having shorter construction times. The SMRs could be small enough to be transportable, could be used in isolated locations without advanced infrastructure and without power grid, or could be clustered in a single site to provide a multi-module, large-capacity power plant.

Uniqueness of the subject and benefits of GEN-IV power plants and argument to defend why we need the nuclear power plant for source of renewable energy, efficiency, cost-effectiveness, as well as safety-related issues are discussed in this book with every detail oriented both from an experimental and technical point of view. This book also goes over different types of GEN-IV nuclear power plants by giving general configurations and specifications of each modular reactor that are considered as part of the new generation of these nuclear plants that fall into the Small Modular Reactors category.

As part of this chapter, we envision a world where almost every country has one affordable type of GEN-IV power plant, such as Molten Salt Reactors burning thorium, uranium, and spent fuel actinides, producing electricity, hydrogen, and desalinized water without any serious accidents.

Small Modular Reactors (SMRs) are highlighted as a viable alternative to the nuclear power plants (NPPs) that have been used as desalination plant energy sources for the following reasons:

SMRs have lower investment costs.
Almost all SMR concepts appear to show increased availability ($90%).
Because of inherent safety features, most SMRs have good potential for location near population centers, hence lowering the transport costs.

This chapter also discusses some angles of nuclear energy generated by Small Modular Reactors (SMRs) specifically and why we need nuclear power plants from SMRs' point of view that are the new generation and design for nuclear plants. As we know, the major growth in the electricity production industry in the last 30 years has centered on the expansion of natural gas power plants based on gas turbine cycles. The most popular extension of the simple Brayton gas turbine has been the combined cycle power plant with the air-Brayton cycle serving as the topping cycle and the Steam-Rankine cycle serving as the bottoming cycle for new generation of nuclear power plants that are known as GEN-IV. The air-Brayton cycle is an open-air cycle and the Steam-Rankine cycle is a closed cycle. The air-Brayton cycle for a natural gas-driven power plant must be an open cycle, where the air is drawn in from the environment and exhausted with the products of combustion to the environment. This technique is suggested as an innovative approach to GEN-IV nuclear power plants in form and type of Small Modular Reactors (SMRs).

In addition, this chapter shows the uniqueness of these Small Modular Reactors (SMRs) and presents innovative approaches to the implementation of these reactors as part of GEN-IV technologies, and it summarizes some of the basic features of SMRs for early deployment and several Advanced Small Modular Reactor (AdvSMR) concepts and points out the benefits and challenges in regulatory, economical, safety, and security issues

2.2 Generation IV Drive New Nuclear Reactor Concepts

Concerns over energy resource availability, climate change, air quality, and energy security suggest an important role for nuclear power in future energy supplies. While the current Generation II and III nuclear power plant designs provide a secure and

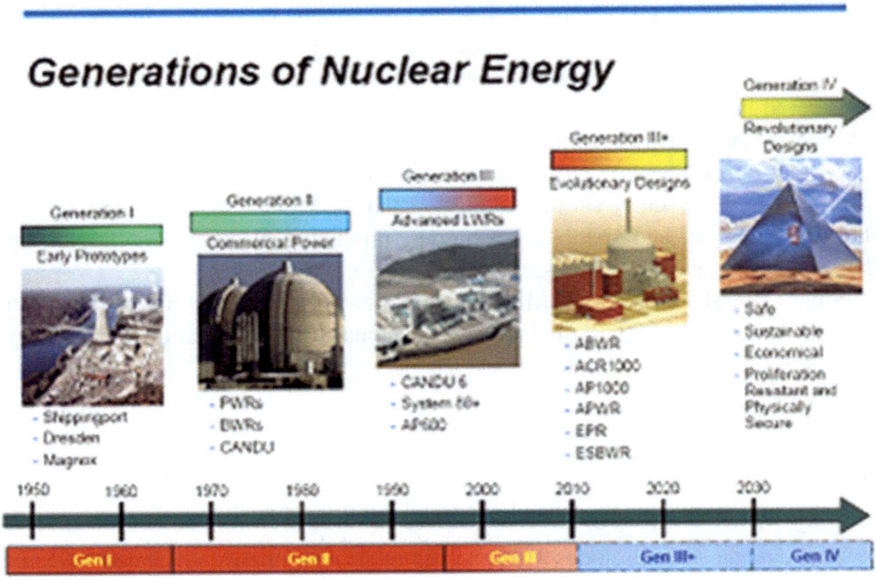

Fig. 2.1 The evolution of nuclear power

low-cost electricity supply in many markets, further advances in nuclear energy system design can broaden the opportunities for the use of nuclear energy. To explore these opportunities, the US Department of Energy's Office of Nuclear Energy has engaged governments, industry, and the research community worldwide in a wide-ranging discussion on the development of next-generation nuclear energy systems known as "Generation IV" (see Fig. 2.1).

The goal of the Gen-IV nuclear energy systems is to address the fundamental research and development (R&D) issues necessary to establish the viability of next-generation nuclear energy system concepts to meet tomorrow's needs for clean and reliable electricity and nontraditional applications of nuclear energy. Successfully addressing the fundamental research and development (R&D) issues will allow Gen-IV concepts that excel in safety, sustainability, cost-effectiveness, and proliferation risk reduction to be considered for future commercial development and deployment by the private sector.

GEN-IV reactor concepts are being developed to use advanced fuels, fashioned from recycled reactor fuel and capable of high burnups. The corresponding fuel cycle strategies allow for efficient utilization of domestic uranium resources while minimizing waste. Reduction of proliferation risk and improvements in physical protection are being designed into Gen IV concepts to help thwart those who would target nuclear power plants for terrorist acts or use them improperly to develop materials for nuclear weapons. GEN-IV concepts will feature advances in safety and reliability to improve public confidence in nuclear energy while providing enhanced investment protection for plant owners. Competitive life-cycle costs and acceptable financial risk are being factored into GEN-IV concepts with high-efficiency electricity generation systems, modular construction, and shortened development schedules before plant startup [4].

GEN-IV is also an active participant in the International Project on Innovative Nuclear Reactors and Fuel Cycles (INPRO). INPRO was established in 2001 in response to a resolution by the IAEA General Conference to help to ensure that nuclear energy is available to contribute, in a sustainable manner, to meeting the energy needs of the twenty-first century and to bring together technology holders and users so that they can consider jointly the international and national actions required for achieving desired innovations in nuclear reactors and fuel cycles. INPRO provides a forum for discussion for experts and policy-makers from industrialized and developing countries on all aspects of nuclear energy planning as well as on the development and deployment of innovative nuclear energy systems in the twenty-first century.

The Generation IV International Forum (GIF) was chartered in May 2001, to lead the collaborative efforts of the world's leading nuclear technology nations to develop the next generation of nuclear energy systems. The initial efforts of GIF resulted in the identification of the six most promising reactor concepts to be investigated by this international research community and are documented in the Generation IV technology roadmap. Thirteen members have signed the GIF Charter: Argentina, Brazil, Canada, People's Republic of China, Euratom, France, Japan, Republic of Korea, the Russian Federation, Republic of South Africa, Switzerland, the United Kingdom, and the United States. This unique international effort reached a major milestone on February 28, 2005, as five of the Forum's member countries (Canada, France, Japan, the United Kingdom, and the United States) signed the world's first multilateral agreement aimed at the international development of advanced nuclear energy systems—the Framework Agreement for International Collaboration on Research and Development of Generation IV Nuclear Energy Systems. Subsequent signatories to the Framework Agreement included People's Republic of China, Euratom, Republic of Korea, Republic of South Africa, and Switzerland. The United Kingdom is a signatory of the Framework but is currently a non-active member. Argentina and Brazil have not ratified the Framework Agreement and are therefore considered non-active. The Russian Federation is working on the necessary approvals for its accession to the Framework [4] (see Fig. 2.2).

As detailed in its Charter and subsequent GIF policy statements, GIF is led by the Policy Group (PG), which is responsible for the overall coordination of GIF's research and development (R&D) collaboration and policy formation and for interactions with other organizations. France with currently chairs the Policy Group vice chairs from the United States and Japan. An Experts Group and the Senior Industry Advisory Panel advises the Policy Group on (R&D) strategy, priorities, and methodology and on evaluating research plans for each Generation IV system. The Framework Agreement establishes two levels of implementing arrangements in order to conduct the joint (R&D). The first level consists of a System Arrangement for each Generation IV reactor concept directed by a System Steering Committee (SSC). Under each SSC, Project Arrangements are established with Project Management Boards to manage and implement the joint (R&D).

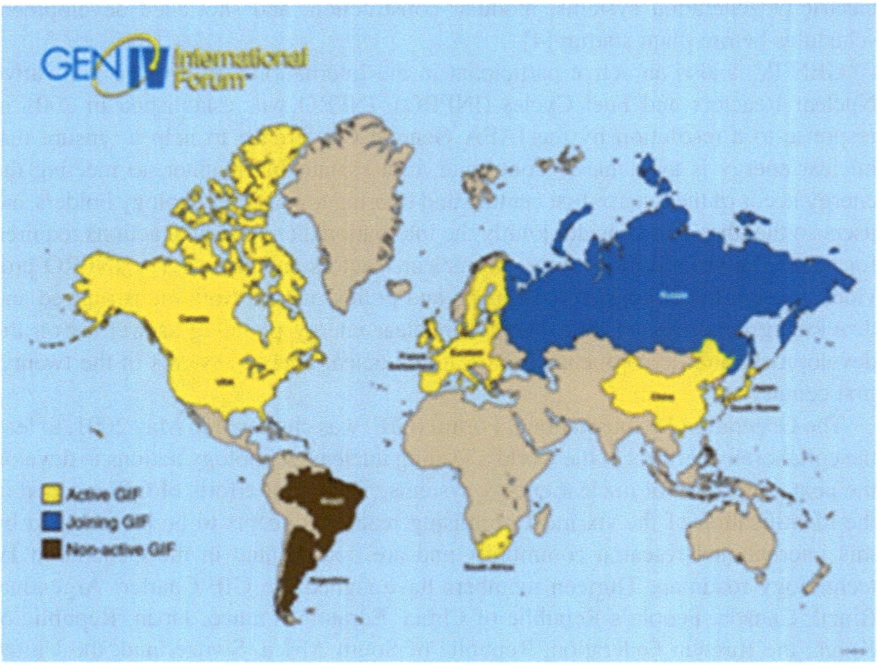

Fig. 2.2 Map of member countries

2.3 Technological State-of-the-Art and Anticipated Developments

It has been demonstrated that Generation II plants can be safely and economically operated for up to 60 years through the development of improved harmonized Plant-Life Management technologies and Plant License Extension practices (PLIM/PLEX) and that developments in fuel technologies can still lead to improvements in reactor performance [5]. The first Generation III reactors, which are an evolution of thermal reactors with even further improved safety characteristics and economy, are now being built. In the coming decades, nuclear electricity generation should increase or at least maintain its current level by a combination of lifetime extension and power upgrades of Generation II reactors and new build of Generation III reactors. Two 1.6 GWe Generation III reactors are presently under construction in Finland and France, targeted for connection to the grid in 2012.

The Finnish reactor was a First-of-a-Kind (FOAK), and the construction has suffered delays with the overnight cost increasing from 2000 to 3100 €/kW(e), whereas the overnight cost for the second reactor in France is now 2400 €/kW(e). In series production, the industry expects the cost to be 2000 ± 500 €/GWe, which is in line with recent international studies. An additional capacity of 100 GWe of Generation III reactors over the next 25 years is a reasonable estimate, which would require an investment in the range of 200–280 billion Euros. The capital costs represent typically 60–70% of the levelized cost for nuclear electricity, operation and maintenance 20–25%, and fuel 10–15%. The front-loaded cost profile means that the levelized cost is very sensitive to construction time and the financial schemes for the investment. Estimates in 2007 for the United Kingdom resulted in range of 31–44 £/MWh (37–53 €/MWh).

Though uranium is relatively abundant in the earth's crust and oceans, estimates of natural reserves are always related to the cost of mineral extraction. As the price of uranium increases on world markets, the number of economically exploitable deposits also increases. The most recent estimates [6] identified 5.5 million tons of uranium (MtU) that could be exploited below 130 $/kg. The total amount of undiscovered resources (reasonably assured and speculative) available at an extraction cost below 130 $/kgU is estimated at 10.5 MtU. Unconventional resources, from which uranium is extracted as a by-product only, e.g., in phosphate production, lie between 7 and 22 MtU, and reserves in sea water are estimated to be 4000 MtU. Japanese studies suggest that uranium from sea water can be extracted at 300 €/kg [8]. At a conservative estimate, 25,000 tons of the uranium are required to produce the fuel to generate 1000 TWhe in an open fuel cycle. The global electricity supplied by nuclear is 2600 TWhe, which means that the conventional resources below 130 $/kgU at the current rate of consumption would last for at least 85 years with the already identified resources (5.5 MtU) and 246 years, if the undiscovered are also included (5.5 + 10.5 MtU). In addition to uranium, it is also possible to use thorium, which is three times more abundant in the earth's crust, though would require different reactors and fuel cycles. Nonetheless, natural resources are plentiful and do not pose an immediate limiting factor for the development of nuclear energy.

However, in a scenario with a large expansion of nuclear energy, resources will become an issue much earlier, especially since new plants have at least a 60-year lifetime and utilities will need assurances when ordering new build that uranium supply can be maintained for the full period of operation. Eventually, known conventional reserves will all be earmarked for current plants or those under construction, and this could happen by the middle of this century. This underlines the need to develop the technology for a new generation, the so-called Generation IV, of reactors and fuel cycles that are more sustainable. In particular, fast-neutron breeder reactors could produce up to 100 times more energy from the same quantity of uranium than current designs and may significantly reduce the amount of ultimate radioactive waste for disposal.

Fast reactors convert non-fissile material (U-238) in the fuel into fissile material (Pu-239) during reactor operation so that the net amount of fissile material increases (breeding). After reprocessing of the spent fuel, the extracted fissile materials are then recycled as new fuels. Reduction of the radiotoxicity and heat load of the waste is achieved by separating some long-lived radionuclides, the minor actinides, which could then be "burned" in fast reactors or alternatively in accelerator-driven systems (ADS), through transmutation. The fast reactor concept has been demonstrated in research programs and national prototypes in the past, but further R&D is needed to make it commercially viable and to develop the designs in compliance with true

Generation IV criteria. Major issues involve new materials that can withstand higher temperatures, higher burnups and neutron doses, corrosive coolants, reactor designs that eliminate severe accidents, and development of fuel cycles for waste minimization and elimination of proliferation risks. Fast reactors are expected to be commercially available from 2040.

So far nuclear power has primarily been used to produce electricity, but it can also be used for process heat applications [7, 8]. Currently, LWRs are already being used to a limited extent for some lower temperature applications (200 °C), such as district heating and desalination of seawater. Existing designs of high-temperature reactors (HTRs) that can reach 800 °C can be deployed in the coming decades, and very-high-temperature reactors (VHTRs) that can reach gas coolant temperatures beyond 1000 °C are being studied as a Generation IV concept for later deployment. Process heat applications include petroleum refinery applications (400 °C), recovery of oil from tar sands (600–700 °C), synthetic fuel from CO_2 and hydrogen (600–1000 °C), hydrogen production (600–1000 °C), and coal gasification (900–1200 °C). Small reactors that can be inherently safe and used to support specific high-energy applications and often in remote areas are another very interesting application that is receiving more attention, in particular in the IAEA INPRO Initiative.

The management of radioactive waste, as either spent fuel or ultimate waste, depending on the national strategy, is a key issue for public acceptance of nuclear energy. There is scientific consensus that geological disposal is the only safe long-term solution for the management of ultimate waste. After a long period of intensive research and development coupled with in-depth political and social engagement, the world's first deep geological repositories for nuclear waste will be in operation in Sweden and Finland by 2020, with France following a few years later, demonstrating that practical solutions exist for the safe long-term management of hazardous waste from the operation of nuclear power plants. Though there will also be ultimate waste from Generation IV fast reactor fuel cycles after reprocessing, the volumes and heat loads will be greatly reduced thereby facilitating disposal operations and optimizing use of space in available geological repositories.

2.4 Next Generation Nuclear Plant (NGNP)

The Next Generation Nuclear Plant (NGNP) demonstration project forms the basis for an entirely new generation of advanced nuclear plants capable of meeting the nation's emerging need for greenhouse gas-free process heat and electricity. The NGNP is based on the very-high-temperature gas-cooled reactor (VHTR) technology, which was determined to be the most promising for the United States in the medium term. The determination is documented as part of the Generation IV implementation strategy in a report submitted to Congress in 2013 following an extensive international technical evaluation effort. The VHTR technology incorporates substantive safety and operational enhancements over existing nuclear technologies. As required by the Energy Policy Act (EPAct) of 2005, the NGNP will be a prototype nuclear power plant, built at the Idaho National Laboratory (INL). Future commercial versions of the NGNP will meet or exceed the reliability, safety, proliferation resistance, and economy of existing commercial nuclear plants [9].

It is envisioned that these advanced nuclear plants would be able to supply cost-competitive process heat that can be used to power a variety of energy-intensive industries, such as the generation of electricity, hydrogen, enhanced oil recovery, refineries, coal-to-liquids and coal-to-gas plants, chemical plants, and fertilizer plants [9].

The US Nuclear Regulatory Commission (NRC) is responsible for licensing and regulating the construction and operation of the NGNP. The EPAct authorizes the US Department of Energy (DOE) to build the NGNP at the Idaho National Laboratory and charges INL with responsibility for leading the project development. The project's completion depends on the collaborative efforts of DOE and its national laboratories, commercial industry participants, US universities, and international government agencies as well as successful licensing by the NRC. At present and pending further evaluation as the NGNP proceeds through Phase 1 in cost-shared collaboration with industry as required by the EPAct, DOE has not made a final determination on whether the license applicant will be DOE or one or more entities that reflect a partnership between DOE and private sector firms [9].

Under the provisions of Section 644 of the EPAct, the Secretary of Energy and the Chairman of the Nuclear Regulatory Commission are to jointly submit to Congress a licensing strategy for the New Generation Nuclear Power (NGNP) within 3 years of the enactment of the Act on August 8, 2005. This report addresses the requirement by outlining a NGNP licensing strategy jointly developed by the NRC and DOE. The scope of the document includes all four elements of the NGNP licensing strategy described in Section 644 (b) of the EPAct:

1. A description of the ways in which current NRC Light Water Reactor (LWR) licensing requirements need to be adapted for the types of reactors considered for the project.
2. A description of the analytical tools that the NRC will need to develop in order to independently verify the NGNP design and its safety performance.
3. A description of other research or development activities that the NRC will need to conduct for the review of an NGNP license application.
4. A budget estimate associated with the licensing strategy.

DOE has determined that the NGNP nuclear reactor will be a very-high-temperature gas-cooled reactor (VHTR) for the production of electricity, process heat, and hydrogen. The VHTR can provide high-temperature process heat (up to 950 °C) that can be used as a substitute for the burning of fossil fuels for a wide range of commercial applications. Since the VHTR is a new and unproven reactor design, the NRC will need to adapt its licensing requirements and process, which have historically evolved around Light Water Reactor (LWR) designs, for licensing the NGNP nuclear reactor. Thus, Section 644 of the EPAct recognized the need for an alternative licensing strategy. This report provides the recommended NGNP licensing strategy, jointly developed by the NRC and DOE. As the technology matures, the government/industry partnership evolves, and input is provided by the general public, revisions to the strategy may be necessary and appropriate [9].

The report addresses the four elements of the licensing strategy set forth in Section 644(b) of the EPAct. These elements are summarized above [9].

2.5 Generation IV Systems

The world's population is expected to expand from 6.7 billion people today to over 9 billion people by the year 2050, all striving for a better quality of life. As the earth's population grows, so does the demand for energy and the benefits that it brings: improved standards of living, better health and longer life expectancy, improved literacy and opportunity, and many others. Simply expanding the use of energy along the same mix of today's production options, however, does not satisfactorily address concerns over climate change and depletion of fossil resources. For the earth to support its population while ensuring the sustainability of humanity's development, we must increase the use of energy supplies that are clean, safe, and cost-effective and which could serve for both basic electricity production and other primary energy needs. Prominent among these supplies is nuclear energy.

There is currently 370 GWe of nuclear power capacity in operation around the world, producing 3000 TWh each year—15% of the world's electricity—the largest share provided by any non-greenhouse gas-emitting source. This reduces significantly the environmental impact of today's electricity generation and affords a greater diversity of electricity generation that enhances energy security.

For more than a decade, Generation IV International Forum (GIF) has led international collaborative efforts to develop next-generation nuclear energy systems that can help meet the world's future energy needs. Generation IV designs will use fuel more efficiently, reduce waste production, be economically competitive, and meet stringent standards of safety and proliferation resistance.

As, we said the Generation IV International Forum (GIF) was initiated in May 2001 and formally chartered in mid-2001. It is an international collective representing government of 13 countries where nuclear energy is significant now and also seen as vital for the future. Most are committed to joint development of the next generation of nuclear technology. Led by the United States, Argentina, Brazil, Canada, China, France, Japan, Russia, South Korea, South Africa, Switzerland, and the United Kingdom are charter members of the GIF, along with the EU (Euratom). Most of these are party to the Framework Agreement (FA), which formally commits them to participate in the development of one or more Generation IV systems selected by GIF for further R&D. Argentina and Brazil did not sign the FA, and the United Kingdom withdrew from it; accordingly, within the GIF, these three are designated as "inactive members." Russia formalized its accession to the FA in August 2009 as its tenth member, with Rosatom as implementing agent. In 2011 the 13 members decided to modify and extend the GIF charter indefinitely.

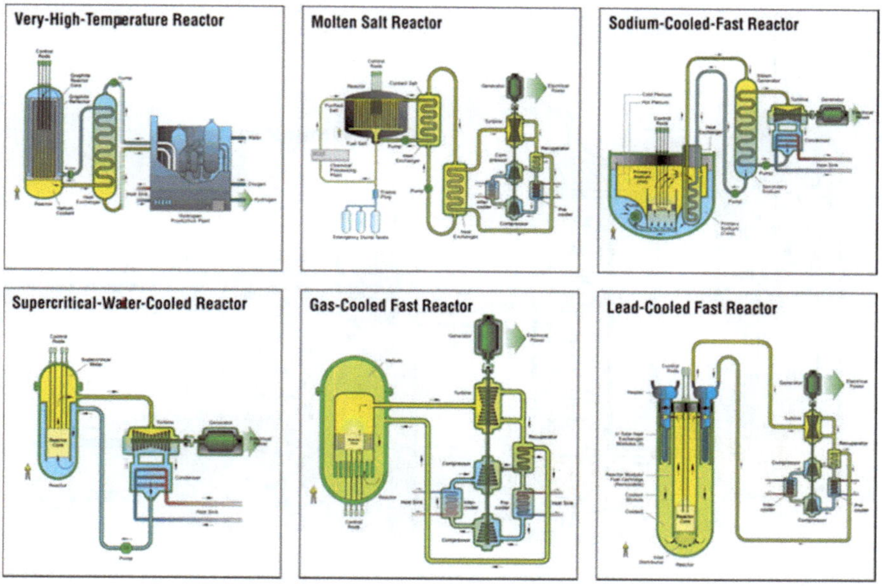

Fig. 2.3 Six reactor technologies of Generation IV (Courtesy of the Generation IV International Forum)

With these goals in mind, some 100 experts evaluated 130 reactor concepts before GIF selected six reactor technologies for further research and development.

These include:

1. Very-high-temperature reactor (VHTR)
2. The Molten Salt Reactor (MSR)
3. The sodium-cooled fast reactor (SFR)
4. The Supercritical Water-cooled Reactor (SCWR)
5. The gas-cooled fast reactor (GFR)
6. The lead-cooled fast reactor (LFR)

Figure 2.3 is an illustration of the six types of reactors that are considered as part of Generation IV power plant. More details of each of these reactors are provided in later sections.

2.5.1 Very-High-Temperature Reactor (VHTR)

Among the six candidates of the GEN-IV nuclear systems in the technical roadmap of GEN-IV International Forum (GIF), the very-high-temperature reactor (VHTR) is primarily dedicated to the cogeneration of electricity and hydrogen, the latter being extracted from water by using thermochemical, electrochemical, or hybrid processes. Its high outlet temperature makes it attractive also for the chemical, oil,

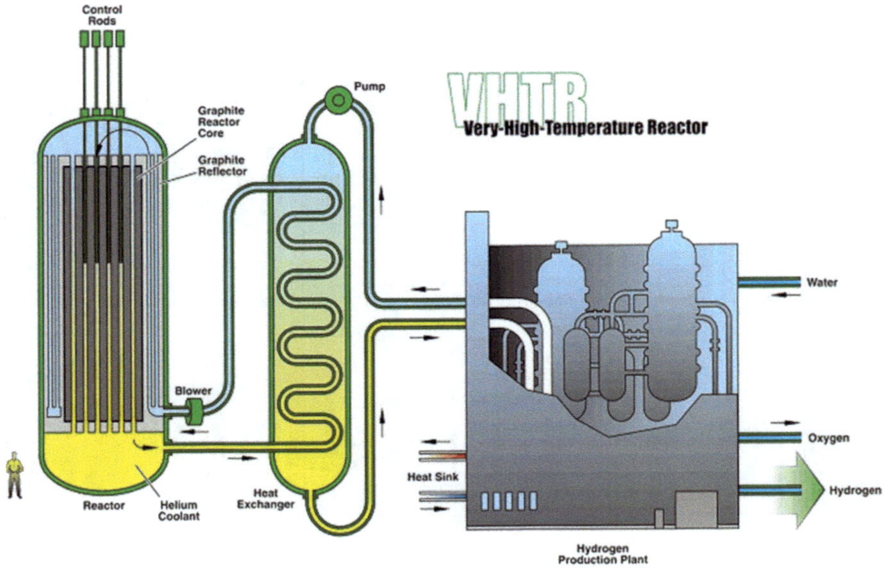

Fig. 2.4 Very-high-temperature reactor (Courtesy of the Generation IV International Forum)

and iron industries (Fig. 2.4). Original target of outlet temperature of 1000 °C from VHTR can support the efficient production of hydrogen by thermochemical processes. The technical basis for VHTR is the tristructural-isotropic (TRISO)-coated particle fuel, the graphite as the core structure, helium coolant, as well as the dedicated core layout and lower power density to removal decay heat in a natural way. The VHTR has potential for inherent safety, high thermal efficiency, process heat application capability, low operation and maintenance costs, and modular construction.

The VHTR is a next step in the evolutionary development of high-temperature gas-cooled reactors. It is a graphite-moderated, helium-cooled reactor with thermal neutron spectrum. It can supply nuclear heat and electricity over a range of core outlet temperatures between 700 and 950 °C or more than 1000 °C in future. The reactor core type of the VHTR can be a prismatic block core such as the Japanese HTTR or a pebble-bed core such as the Chinese HTR-10. For electricity generation, a helium gas turbine system can be directly set in the primary coolant loop, which is called a direct cycle, or at the lower end of the outlet temperature range, a steam generator can be used with a conventional Rankine cycle. For nuclear heat applications such as process heat for refineries, petrochemistry, metallurgy, and hydrogen production, the heat application process is generally coupled with the reactor through an Intermediate Heat Exchanger (IHX), the so-called indirect cycle. The VHTR can produce hydrogen from only heat and water by using thermochemical processes (such as the sulfur-iodine (S–I) process or the hybrid sulfur process), high-temperature steam electrolysis (HTSE), or from heat, water, and natural gas by applying the steam reformer technology.

While the original approach for VHTR at the start of the Generation IV program focused on very high outlet temperatures and hydrogen production, current market assessments have indicated that electricity production and industrial processes based on high-temperature steam that require modest outlet temperatures (700–850 °C) have the greatest potential for application in the next decade. This also reduces technical risk associated with higher outlet temperatures. As a result, over the past decade, the focus has moved from higher outlet temperature designs such as GT-MHR and PBMR to lower outlet temperature designs such as HTR-PM in China and the NGNP in the United States.

The VHTR has two typical reactor configurations, namely:

1. The pebble-bed type
2. The prismatic block types

Although the shape of the fuel element for two configurations are different, the technical basis for both configurations is same, such as the TRISO-coated particle fuel in the graphite matrix, full ceramic (graphite) core structure, helium coolant, and low power density.

This will allow achieving high outlet temperature and the retention of fission production inside the coated particle under normal operation condition and accident condition. The VHTR can support alternative fuel cycles such as U–Pu, Pu, MOX, and U–Th.

2.5.2 *Molten Salt Reactor (MSR)*

The MSR is distinguished by its core in which the fuel is dissolved in molten fluoride salt. The technology was first studied more than 50 years ago. Modern interest is on fast reactor concepts as a long-term alternative to solid-fueled fast-neutron reactors. The on-site fuel-reprocessing unit using pyro-chemistry allows breeding plutonium or uranium-233 from thorium. R&D progresses toward resolving feasibility issues and assessing safety and performance of the design concepts. Key feasibility issues focus on a dedicated safety approach and the development of salt redox potential measurement and control tools in order to limit corrosion rate of structural materials. Further work on the batchwise online salt processing is required. Much work is needed on molten salt technology and related equipment (see Fig. 2.5).

Molten Salt Reactor (MSR) technology was partly developed, including two demonstration reactors, in the 1950s and 1960s in the United States (Oak Ridge National Laboratory). The demonstrations MSRs were thermal neutron-spectrum graphite-moderated concepts. Since 2005, R&D has focused on the development of fast-spectrum MSR concepts (MSFR) combining the generic assets of fast-neutron reactors (extended resource utilization, waste minimization) with those relating to molten salt fluorides as fluid fuel and coolant (low pressure and high-boiling temperature, optical transparency).

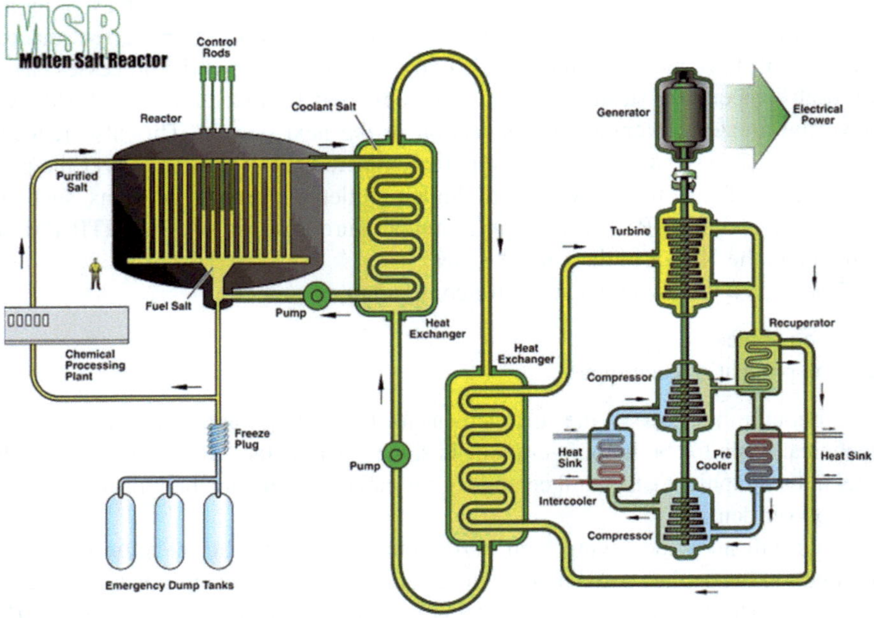

Fig. 2.5 Molten Salt Reactor (Courtesy of the Generation IV International Forum)

In contrast to most other Molten Salt Reactors previously studied, the MSFR does not include any solid moderator (usually graphite) in the core. This design choice is motivated by the study of parameters such as feedback coefficient, breeding ratio, graphite life-span, and ^{233}U initial inventory. MSFR exhibit large negative temperature and void reactivity coefficients, a unique safety characteristic not found in solid-fuel fast reactors.

Compared with solid-fuel reactors, MSFR systems have lower fissile inventories, no radiation damage constraint on attainable fuel burnup, no requirement to fabricate and handle solid fuel, and a homogeneous isotopic composition of fuel in the reactor. These and other characteristics give MSFRs potentially unique capabilities for actinide burning and extending fuel resources.

MSR developments in Russia on the Molten Salt Actinide Recycler and Transmuter (MOSART) aim to be used as efficient burners of transuranic (TRU) waste from spent UOX and MOX Light Water Reactor (LWR) fuel without any uranium and thorium support and also with it. Other advanced reactor concepts are being studied, which use the liquid salt technology, as a primary coolant for fluoride salt-cooled high-temperature reactors (FHRs), and coated particle fuels similar to high-temperature gas-cooled reactors.

More generally, there has been a significant renewal of interest in the use of liquid salt as a coolant for nuclear and non-nuclear applications. These salts could facilitate heat transfer for nuclear hydrogen production concepts, concentrated solar electricity generation, oil refineries, and shale oil processing facilities among other applications.

2.5.3 *Sodium-Cooled Fast Reactor (SFR)*

The sodium-cooled fast reactor (SFR) uses liquid sodium as the reactor coolant, allowing high-power density with low coolant volume fraction and operation at low pressure. While the oxygen-free environment prevents corrosion, sodium reacts chemically with air and water and requires a sealed coolant system (see Fig. 2.6).

Plant-sized options under considerations are ranging from small, 50–300 MW(e), modular reactors to larger plants up to 1500 MW(e). The outlet temperature is 500–550 °C for the options, which allows the use of the materials developed and proven in prior fast reactor programs.

The SFR closed fuel cycle enables regeneration of fissile fuel and facilitates management of minor actinides. However, this requires that recycle fuels be developed and qualified for use. Important safety features of the Generation IV system include a long thermal response time, a reasonable margin to coolant boiling, a primary system that operates near atmospheric pressure, and an intermediate sodium system between the radioactive sodium in the primary system and the power conversion system. Water/steam, supercritical carbon dioxide, or nitrogen can be considered as working fluids for the power conversion system to achieve high performance in terms of thermal efficiency, safety, and reliability. With innovations to reduce capital cost, the SFR is aimed to be economically competitive in future electricity markets. In addition, the fast-neutron spectrum greatly extends the

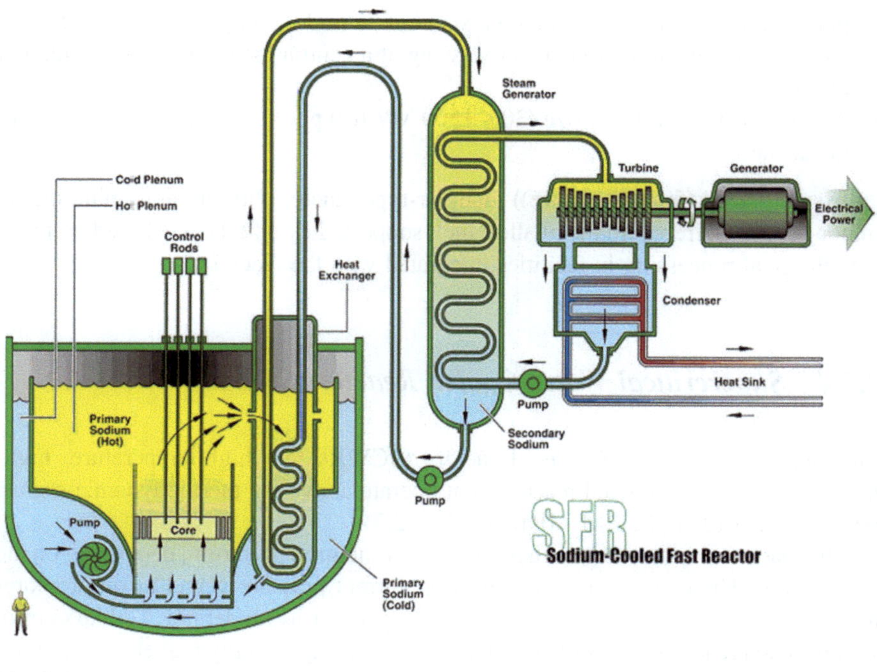

Fig. 2.6 Sodium-cooled fast reactor (Courtesy of the Generation IV International Forum)

uranium resources compared to thermal reactors. The SFR is considered to be the nearest-term deployable system for actinide management.

Much of the basic technology for the SFR has been established in former fast reactor programs and is being confirmed by the Phenix end-of-life tests in France, the restart of Monju in Japan, and the lifetime extension of BN-600 in Russia. New programs involving SFR technology include the Chinese experimental fast reactor (CEFR) which was connected to the grid in July 2011 and India's prototype fast-breeder reactor (PFBR) which is currently planned to go critical in 2013.

The SFR is an attractive energy source for nations that desire to make the best use of limited nuclear fuel resources and manage nuclear waste by closing the fuel cycle.

Fast reactors hold a unique role in the actinide management mission because they operate with high-energy neutrons that are more effective at fissioning actinides. The main characteristics of the SFR for actinide management mission are:

- Consumption of transuranic in a closed fuel cycle, thus reducing the radiotoxicity and heat load, which facilitates waste disposal and geologic isolation
- Enhanced utilization of uranium resources through efficient management of fissile materials and multi-recycle

High level of safety achieved through inherent and passive means also allows accommodation of transients and bounding events with significant safety margins.

The reactor unit can be arranged in a pool layout or a compact loop layout. Three options are considered:

- A large-sized (600–1500 MW(e)) loop-type reactor with mixed uranium-plutonium oxide fuel and potentially minor actinides, supported by a fuel cycle based upon advanced aqueous processing at a central location serving a number of reactors
- An intermediate-to-large size (300–1500 MW(e)) pool-type reactor with oxide or metal fuel

A small-sized (50–150 MW(e)) modular-type reactor with uranium–plutonium–minor-actinide–zirconium metal alloy fuel, supported by a fuel cycle based on pyro-metallurgical processing in facilities integrated with the reactor.

2.5.4 Supercritical Water-Cooled Reactor (SCWR)

The Supercritical Water-Cooled Reactors (SCWRs) are high-temperature, high-pressure, light-water-cooled reactors that operate above the thermodynamic critical point of water (374 °C, 22.1 MPa) (see Fig. 2.7).

The reactor core may have a thermal or a fast-neutron spectrum, depending on the core design. The concept may be based on current pressure-vessel or on pressure-tube reactors and thus use light water or heavy water as moderator. Unlike current water-cooled reactors, the coolant will experience a significantly higher enthalpy rise in the core, which reduces the core mass flow for a given thermal power and

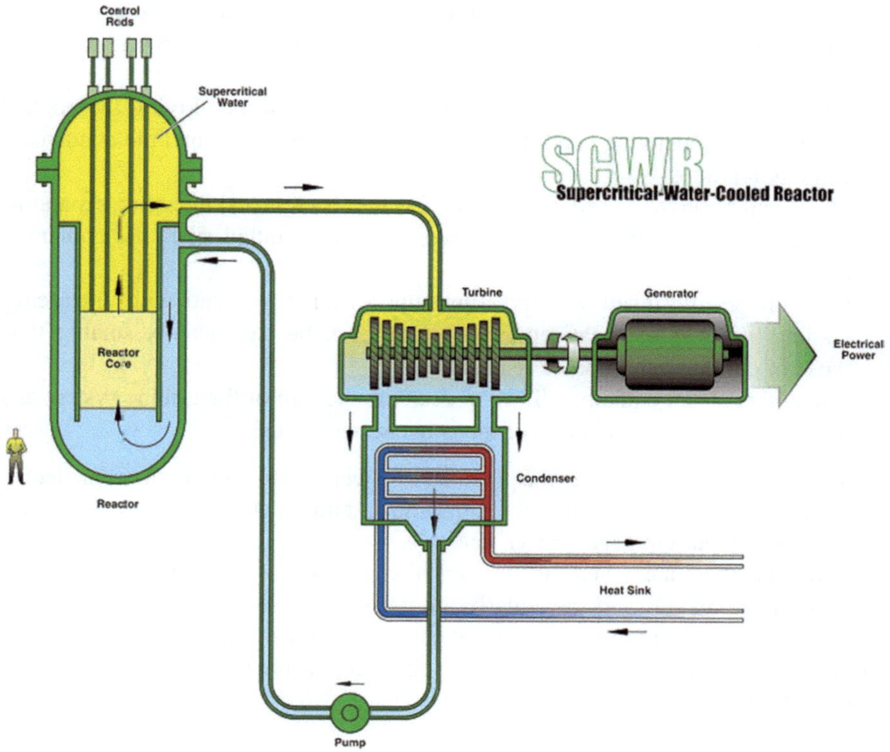

Fig. 2.7 Supercritical Water-Cooled Reactor (Courtesy of the Generation IV International Forum)

increases the core outlet enthalpy to superheated conditions. For both pressure-vessel and pressure-tube designs, a once-through steam cycle has been envisaged, omitting any coolant recirculation inside the reactor. As in a boiling water reactor, the superheated steam will be supplied directly to the high-pressure steam turbine, and the feed water from the steam cycle will be supplied back to the core. Thus, the SCWR concepts combine the design and operation experiences gained from hundreds of water-cooled reactors with those experiences from hundreds of fossil-fired power plants operated with Supercritical Water (SCW). In contrast to some of the other Generation IV nuclear systems, the SCWR can be developed incrementally step-by-step from current water-cooled reactors.

2.5.4.1 Advantage and Challenges

Such SCWR designs have unique features that offer many advantages compared to state-of the-art water-cooled reactors:

- SCWRs offer increases in thermal efficiency relative to current-generation water-cooled reactors. The efficiency of a SCWR can approach 44% or more, compared to 34–36% for current reactors.
- Reactor coolant pumps are not required. The only pumps driving the coolant under normal operating conditions are the feed water pumps and the condensate extraction pumps.
- The steam generators used in pressurized water reactors and the steam separators and dryers used in boiling water reactors can be omitted since the coolant is superheated in the core.
- Containment, designed with pressure suppression pools and with emergency cooling and residual heat removal systems, can be significantly smaller than those of current water-cooled reactors can.
- The higher steam enthalpy allows to decrease the size of the turbine system and thus to lower the capital costs of the conventional island.

These general features offer the potential of lower capital costs for a given electric power of the plant and of better fuel utilization and thus a clear economic advantage compared with current Light Water Reactors.

However, there are several technological challenges associated with the development of the SCWR and particularly the need to validate transient heat transfer models (for describing the depressurization from supercritical to subcritical conditions), qualification of materials (namely, advanced steels for cladding), and demonstration of the passive safety systems.

2.5.4.2 GIF Progress Up to 2012

Preconceptual core design studies for a core outlet temperature of more than 500 °C have been performed in Japan, assuming either a thermal neutron spectrum or a fast-neutron spectrum. Both options are based on a coolant heat-up in two steps with intermediate mixing underneath the core. Additional moderator for a thermal neutron spectrum is provided by feed water inside water rods. The fast-spectrum option uses zirconium-hydride (ZrH_2) layers to minimize hardening of the neutron spectrum in case of core voiding. A preconceptual design of safety systems for both options has been studied with transient analyses.

A preconceptual plant design with 1700 MW net electric power based on a pressure-vessel-type reactor has been studied by Yamada et al. and has been assessed with respect to efficiency, safety, and cost. The study confirms the target net efficiency of 44% and estimates a cost reduction potential of 30% compared with current pressurized water reactors. Safety features are expected to be similar to advanced boiling water reactors.

A preconceptual design of a pressure-vessel-type reactor with a 500 °C core outlet temperature and 1000 MW electric power has been developed in Europe, as summarized by Schulenberg and Starflinger. The core design is based on coolant heat-up in three steps. Additional moderator for the thermal neutron spectrum is provided in water rods and in gaps between assembly boxes. The design of the nuclear island and of the balance of the plant confirms results obtained in Japan, namely, an efficiency improvement up to 43.5% and a cost reduction potential of 20–30% compared with latest boiling water reactors. Safety features as defined by the stringent European Utility Requirements are expected to be met.

Canada is developing a pressure-tube-type SCWR concept with a 625 °C core outlet temperature at the pressure of 25 MPa. The concept is designed to generate 1200 MW electric power (a 300 MW concept is also being considered). It has a modular fuel channel configuration with separate coolant and moderator. A high-efficiency fuel channel is incorporated to house the fuel assembly. The heavy-water moderator directly contacts the pressure tube and is contained inside a low-pressure calandria vessel. In addition to providing moderation during normal operation, it is designed to remove decay heat from the high-efficiency fuel channel during long-term cooling using a passive moderator cooling system. A mixture of thorium oxide and plutonium is introduced as the reference fuel, which aligns with the GIF position paper on thorium fuel. The safety system design of the Canadian SCWR is similar to that of the ESBWR. However, the introduction of the passive moderator cooling system coupled with the high-efficiency channel could reduce significantly the core damage frequency during postulated severe accidents such as large-break loss-of-coolant or station blackout events.

Preconceptual designs of three variants of pressure-vessel supercritical reactors with thermal, mixed, and fast-neutron spectrum have been developed in Russia, which joined the SCWR System Arrangement in 2011.

Outside of the GIF framework, two conceptual SCWR designs with thermal and mixed neutron spectrum cores have been established by some research institutes in China. This is done, under framework of the Chinese national R&D projects from 2007 to 2012, covering some basic research projects on materials and thermohydraulics, the core/fuel design, the main system design (including the conventional part), safety systems design, reactor structure design, and fuel assembly structure design. The related feasibility studies have also been completed and show that the design concept has promising prospects in terms of the overall performance, integration of design, component structure feasibility, and manufacturability.

Prediction of heat transfer in SCW can be based on data from fossil-fired power plants as discussed by Pioro et al. Computational tools for more complex geometries like fuel assemblies are available but still need to be validated with bundle experiments. System codes for transient safety analyses have been upgraded to include SCW, including depressurization transients to subcritical conditions. Flow stability in the core has been studied numerically. As in boiling water reactors, flow stability can be ensured using suitable inlet orifices in fuel assemblies.

A number of candidate cladding materials have been tested in capsules, autoclaves, and recirculating loops up to 700 °C at a pressure of 25 MPa. Stainless steels with more than 20% chromium (Cr) are expected to have the required corrosion resistance up to a peak cladding temperature of 650 °C. More work is needed to develop alloys suitable for use at the design peak cladding temperatures of 850 °C for the Canadian SCWR concept. Further work is also needed to better identify the coolant conditions that lead to stress corrosion cracking. It has been shown that the creep resistance of existing alloys can be improved by adding small amounts of elements, such as zirconium (Zr), as reported by Kaneda et al. In the longer term, the steel experimental oxide dispersion strengthened (ODS) alloys offer an even higher potential, whereas nickel-based alloys that are being considered for use in ultra-supercritical fossil-fired plants are less favorable for use in SCWRs due to their high neutron absorption and associated swelling and embrittlement.

Key water chemistry issues have been identified by Guzonas et al.; predicting and controlling water radiolysis and corrosion product transport (including fission products) remain the major R&D areas. In this regard, the operating experience using nuclear steam reheat at the Beloyarsk nuclear power plant in Russia is extremely valuable.

2.5.5 Gas-Cooled Fast Reactor (GFR)

The gas-cooled reactor (GFR) system is a high-temperature helium-cooled fast-spectrum reactor with a closed fuel cycle. It combines the advantages of fast-spectrum systems for long-term sustainability of uranium resources and waste minimization (through fuel multiple reprocessing and fission of long-lived actinides), with those of high-temperature systems (e.g., high thermal cycle efficiency and industrial use of the generated heat, for hydrogen production) (see Fig. 2.8).

The GFR uses the same fuel recycling processes as the SFR and the same reactor technology as the VHTR. Therefore, its development approach is to rely, in so far as feasible, on technologies developed for the VHTR for structures, materials, components, and power conversion system. Nevertheless, it calls for specific R&D beyond the current and foreseen work on the VHTR system, mainly on core design and safety approach.

The reference design for GFR is based around a 2400 MW(th) reactor core contained within a steel pressure vessel. The core consists of an assembly of hexagonal fuel elements, each consisting of ceramic-clad, mixed-carbide-fueled pins contained within a ceramic hex-tube. The favored material at the moment for the pin clad and hex-tubes is silicon carbide fiber-reinforced silicon carbide. The figure below shows the reactor core located within its fabricated steel pressure vessel surrounded by main heat exchangers and decay heat removal loops. The whole of the primary circuit is contained within a secondary pressure boundary, the guard containment.

As it is illustrated in Fig. 2.9, the coolant is helium, and the core outlet temperature will be of the order of 850 °C. A heat exchanger transfers the heat from the primary helium coolant to a secondary gas cycle containing a helium-nitrogen mixture, which in turn drives a closed-cycle gas turbine. The waste heat from the gas turbine exhaust is used to raise steam in a steam generator, which is then used to drive a steam turbine. Such a combined cycle is common practice in natural gas-fired

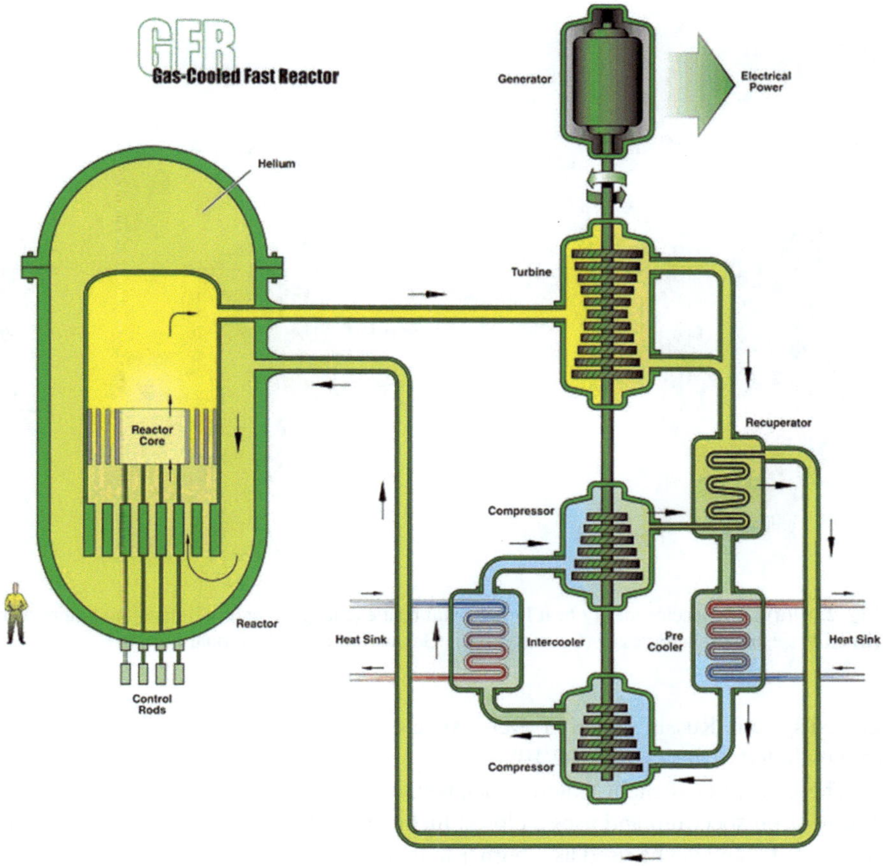

Fig. 2.8 Gas-cooled fast reactor (Courtesy of the Generation IV International Forum)

power plant so represents an established technology, with the only difference in the GFR case being the use of a closed cycle gas turbine.

2.5.6 Lead-Cooled Fast Reactor (LFR)

The lead-cooled fast reactor (LFR) features a fast-neutron spectrum, high-temperature operation, and cooling by molten lead or lead-bismuth eutectic (LBE), low-pressure, chemically inert liquids with very good thermodynamic properties. It would have multiple applications including production of electricity, hydrogen, and process heat. System concepts represented in plans of the Generation IV International Forum (GIF) System Research Plan (SRP) are based on Europe's ELFR lead-

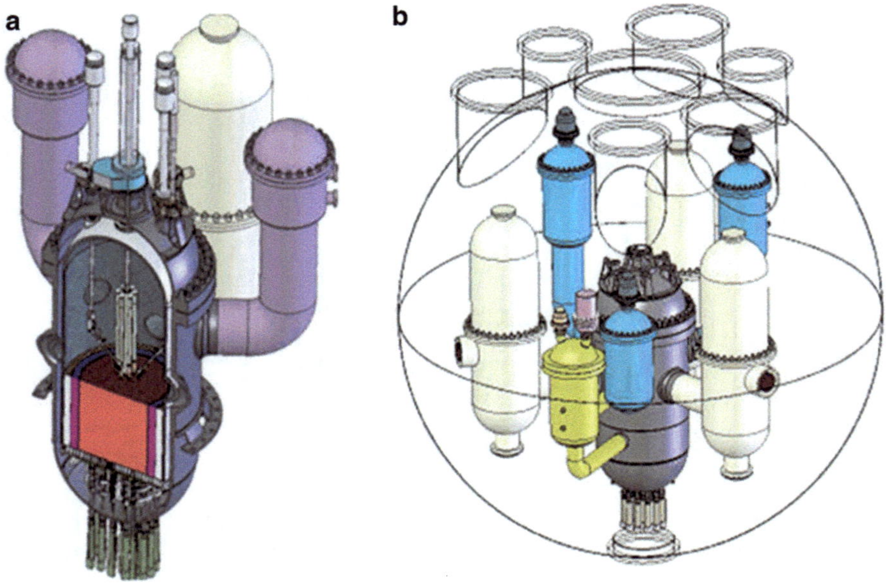

Fig. 2.9 (**a**) GFR-reactor, decay heat loops, main heat exchangers, and fuel handling equipment. (**b**) GFR spherical guard vessel (Courtesy of the Generation IV International Forum)

cooled system, Russia's BREST-OD-300, and the SSTAR system concept designed in the United States (see Fig. 2.10).

The LFR has excellent material management capabilities since it operates in the fast-neutron spectrum and uses a closed fuel cycle for efficient conversion of fertile uranium. It can also be used as a burner to consume actinides from spent LWR fuel and as a burner/breeder with thorium matrices. An important feature of the LFR is the enhanced safety that results from the choice of molten lead as a chemically inert and low-pressure coolant. In terms of sustainability, lead is abundant and hence available, even in case of deployment of a large number of reactors. More importantly, as with other fast systems, fuel sustainability is greatly enhanced by the conversion capabilities of the LFR fuel cycle. Because they incorporate a liquid coolant with a very high margin to boiling and benign interaction with air or water, LFR concepts offer substantial potential in terms of safety, design simplification, proliferation resistance, and the resulting economic performance. An important factor is the potential for benign end state to severe accidents.

The LFR has development needs in the areas of fuels, materials performance, and corrosion control. During the next 5 years, progress is expected on materials, system design, and operating parameters. Significant test and demonstration activities are underway and planned during this period.

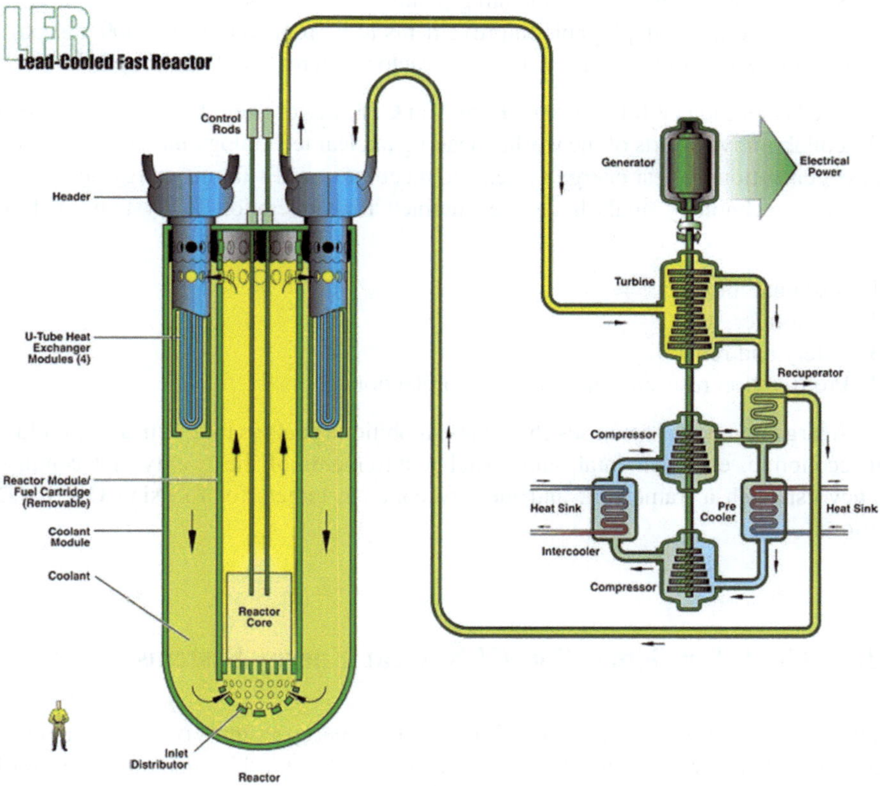

Fig. 2.10 Lead-cooled fast reactor (Courtesy of the Generation IV International Forum)

2.6 Next Generation of Nuclear Power Reactors for Power Production

Experts are projecting worldwide electricity consumption will increase substantially in the coming decades, especially in the developing world, accompanying economic growth and social progress that has direct impact on rising electricity prices, and have focused fresh attention on nuclear power plants. New, safer, and more economical nuclear reactors could not only satisfy many of our future energy needs but could combat global warming as well. Today's existing nuclear power plants on line in the United States provide fifth of the nation's total electrical output.

Taking into account the expected increase in energy demand worldwide and the growing awareness about global warming, climate change issues, and sustainable development, nuclear energy will be needed to meet future global energy demand.

Nuclear power plant technology has evolved as distinct design generations as we mentioned in previous section and briefly summarized here again as follows:

- First generation: prototypes and first realizations (~1950–1970)

- Second generation: current operating plants (~1970–2030)
- Third generation: deployable improvements to current reactors (~2000 and on)
- Fourth generation: advanced and new reactor systems (2030 and beyond)

The **G**eneration IV **I**nternational **F**orum, or **GIF**, was chartered in July 2001 to lead the collaborative efforts of the world's leading nuclear technology nations to develop next-generation nuclear energy systems to meet the world's future energy needs.

Eight technology goals have been defined for Generation IV systems in four broad areas:

1. Sustainability
2. Economics
3. Safety and reliability
4. Proliferation resistance and physical protection

A large number of countries share these ambitious goals as they aim at responding to economic, environmental, and social requirements of the twenty-first century. They establish a framework and identify concrete targets for focusing GIF R&D efforts.

2.7 Goals for Generation IV Nuclear Energy Systems

The next generation ("Generation IV") of nuclear energy systems is intended to meet the below goals (while being at least as effective as the "third" generation in terms of economic competitiveness, safety, and reliability) in order to provide a sustainable development of nuclear energy.

In principle, the Generation IV systems should be marketable or deployable from 2030 onward. The systems should also offer a true potential for new applications compatible with an expanded use of nuclear energy, in particular in the fields of hydrogen or synthetic hydrocarbon production, seawater desalination, and process heat production.

It has been recognized that these objectives, widely and officially shared by a large number of countries, should be at the basis of an internationally shared R&D program, which allows keeping open and consolidating the technical options and avoiding any early or premature down selection.

In fact, because the next-generation nuclear energy systems will address needed areas of improvement and offer great potential, many countries share a common interest in advanced R&D that will support their development. The international research community should explore such development benefits from the identification of promising research areas and collaborative efforts. The collaboration on R&D by many nations on the development of advanced next-generation nuclear energy systems will in principle aid the progress toward the realization of such systems, by leveraging resources, providing synergistic opportunities, avoiding unnecessary duplication, and enhancing collaboration (Table 2.1).

Table 2.1 Areas of improvement for Generation IV

Sustainability—1	Generation IV nuclear energy systems will provide sustainable energy generation that meets clean air objectives and provides long-term availability of systems and effective fuel utilization for worldwide energy production
Sustainability—2	Generation IV nuclear energy systems will minimize and manage their nuclear waste and notably reduce the long-term stewardship burden, thereby improving protection for the public health and the environment
Economics—1	Generation IV nuclear energy systems will have a clear life cycle cost advantage over other energy sources
Economics—2	Generation IV nuclear energy systems will have a level of financial risk comparable to other energy projects
Safety and reliability—1	Generation IV nuclear energy systems operations will excel in safety and reliability
Safety and reliability—2	Generation IV nuclear systems will have a very low likelihood and degree of reactor core damage
Safety and reliability—3	Generation IV nuclear energy systems will eliminate the need for offsite emergency response
Proliferation resistance and physical protection	Generation IV nuclear energy systems will increase the assurance that they are very unattractive and the least desirable route for diversion or theft of weapons usable materials and provide increased physical protection against acts of terrorism

As it is illustrated by Fig. 2.1, we are able to see the evolution of nuclear power plant life cycles.

In 2009, the Experts Group published an outlook on Generation IV R&D, to provide a view of what GIF members hope to achieve collectively in the period 2010–2014. All Generation IV systems have features aiming at performance improvement, new applications of nuclear energy, and/or more sustainable approaches to the management of nuclear materials. High-temperature systems offer the possibility of efficient process heat applications and eventually hydrogen production. Enhanced sustainability is achieved primarily through adoption of a closed fuel cycle with reprocessing and recycling of plutonium, uranium, and minor actinides using fast reactors; this approach provides significant reduction in waste generation and uranium resource requirements.

2.8 Why We Need to Consider the Future Role of Nuclear Power Now

The following reasons are some arguments that show why we need to consider the future role in design of new nuclear power plant:

1. Nuclear power has been part of the global energy need mix for the past five decades. Currently it provides about 18% of the electricity we use in our homes and workplaces. For example, in the United Kingdom, about one-third of our emissions of carbon dioxide come from electricity generation [10]. The vast majority of those emissions come from coal and gas power plants.
2. Energy companies will need to invest in around 30–35 GW of new electricity-generating capacity—as coal and nuclear plants retire—over the next two decades, with around two-thirds needed by 2020. This is equivalent to about one-third of our existing capacity. The world needs a clear and stable regulatory framework to reduce uncertainty for business to help ensure sufficient and timely investment in technologies that contribute to our energy goals.
3. Of the capacity that is likely to close over the two decades, two-thirds is from carbon-intensive fossil fuel generation and about 10 GW is nuclear and therefore low carbon. So, companies' decisions on the type of power stations they invest in to replace this capacity will have significant implications for the level of carbon emissions. As an illustration, if our existing nuclear power stations were all replaced with fossil fuel-fired power stations, our emissions would be between eight and sixteen million tons of carbon (MtC) a year higher as a result (depending on the mix of gas and coal-fired power stations). This would be equivalent to about 30–60% of the total carbon savings we project to achieve under our central scenario from all the measures we are bringing forward in the reference that is known as Energy White Paper [11]. Our gas demand would also be higher, at a time when we are becoming more dependent on imported sources of fossil fuels.
4. Electricity demand in the United States is expected to grow significantly in the future. Over the past decade, Americans used 17% more electricity, but domestic capacity rose only 2.3% (National Energy Policy, May 2001). Unless the United States significantly increases its generating capacity, the country will face an energy shortage that is projected to adversely affect our economy, our standard of living, and our national security. Coupled with this challenge is the need to improve our environment.
5. New nuclear power stations have long lead times. This time is necessary to secure the relevant regulatory and development consents, which must be obtained before construction can begin, and there is also a long construction period compared to other generating technologies. Our conservative assumption is that for the first new nuclear plant, the pre-construction period would last around 8 years (to secure the necessary consents), and the construction period would last around 5 years. For subsequent plants, this is assumed to be 5 and 5 years, respectively. New nuclear power stations are therefore unlikely to make a significant contribution to the need for new capacity before 2020.
6. Even with our expectations that the share of renewable will grow, it is likely that fossil fuel generation will meet some of this need. However, beyond that date, there are still significant amounts of new capacity needed; for example, in 2023, one-third or 3 GW of our nuclear capacity will still be operational, based on published lifetimes. Given the likely increase in fossil fuel generation before this

date, it is important that much of this capacity is replaced with low-carbon technologies. New nuclear power stations could make an important contribution, as outlined in this consultation document, to meeting our needs for low-carbon electricity generation and energy security in this period and beyond to 2050. Because of the lead times, without clarity, now we will foreclose the opportunity for nuclear power.

7. The existing approach on new nuclear build was set out in 200311: "Nuclear power is currently an important source of carbon-free electricity. However, its current economics make it an unattractive option for new, carbon-free generating capacity and there are also important issues of nuclear waste to be resolved. These issues include our legacy waste and continued waste arising from other sources. This white paper does not contain specific proposals for building new nuclear power stations. However, we do not rule out the possibility that at some point in the future new nuclear build might be necessary if we are to meet our carbon targets. Before any decision to proceed with the building of new nuclear power stations, there will need to be the fullest public consultation and the publication of a further white paper setting out our proposals."
8. Since 2003 there have been a number of developments, which have led the government to consider afresh the potential contribution of new nuclear power stations. Firstly, there has been significant progress in tackling the legacy waste issue:

   - We have technical solutions for waste disposal that scientific consensus and experience from abroad suggest could accommodate all types of wastes from existing and new nuclear power stations.
   - There is now an implementing body (the Nuclear Decommissioning Authority), with expertise in this area, and government is reconstituting the Committee on Radioactive Waste Management (CoRWM) in order to provide continued independent scrutiny and advice.
   - A framework for implementing long-term waste disposal in a geological repository will be consulted on in the coming months.

9. The government has also made progress in considering the issue of waste management in relation to potential new nuclear power stations:

   - This consultation provides the opportunity to discuss the ethical, intergenerational, and public acceptability issues associated with a decision to allow the private sector to invest in new nuclear power stations and generate new nuclear waste.
   - The government is developing specific proposals to protect the taxpayer. Under these proposals, private sector developers would meet the full decommissioning costs and full share of waste management costs. The proposals would be implemented in the event that we conclude that energy companies should be allowed to invest in new nuclear power stations. They would need to be in place before proposals for new power stations could go ahead.

10. Secondly, the high-level economic analysis of nuclear power, prepared for the energy review, concluded that under likely scenarios for gas and carbon prices and taking prudent estimates of nuclear costs, nuclear power would offer general economic benefit to the United Kingdom in terms of reduced carbon emissions and security of supply benefits. Therefore, the government believes that it has a potential contribution to make alongside other low-carbon generating technologies.
11. Thirdly, some energy companies have expressed a strong interest in investing in new nuclear power stations. They assess that new nuclear power stations could be an economically attractive low-carbon investment, which could help diversify their generation portfolios. Their renewed interest reflects assessments that with carbon being priced to reflect its impacts and gas prices likely to be higher than previously expected, the economics of new nuclear power stations are becoming more favorable.
12. Nuclear power stations have long lead times. If they are to be an option to replace the capacity closing over the next two decades, and in particular after 2020, a decision on whether allowing energy companies the option of investing in new nuclear power stations would be in the public interest needs to be taken now. Energy companies would need to begin their initial preparations in the near future in order to have a reasonable prospect of building new generation in this period. Not taking the public interest decision now would foreclose the option of new nuclear being one of our options for tackling climate change and achieving energy security.

2.9 The Generation IV Roadmap Project

As the Generation IV goals were being finalized, preparations were made to develop the Generation IV technology roadmap. The organization of the roadmap is shown in the Fig. 2.11 below. The Roadmap Integration Team (RIT) is the executive group. Groups of international experts were organized to undertake identification and evaluation of candidate systems and to define R&D to support them [12].

In a first step, an Evaluation Methodology Group was formed to develop a process to systematically evaluate the potential of proposed Generation IV nuclear energy systems to meet the Generation IV goals. A discussion of the Evaluation Methodology Group's evaluation methodology is included in this report. At the same time, a solicitation was issued worldwide, requesting that concept proponents submit information on nuclear energy systems that they believe could meet some or all of the Generation IV goals. Nearly 100 concepts and ideas were received from researchers in a dozen countries [12].

Technical Working Groups (TWGs) were formed—covering nuclear energy systems employing water-cooled, gas-cooled, liquid-metal-cooled, and non-classical reactor concepts—to review the proposed systems and evaluate their potential using the tools developed by the Evaluation Methodology Group. Because

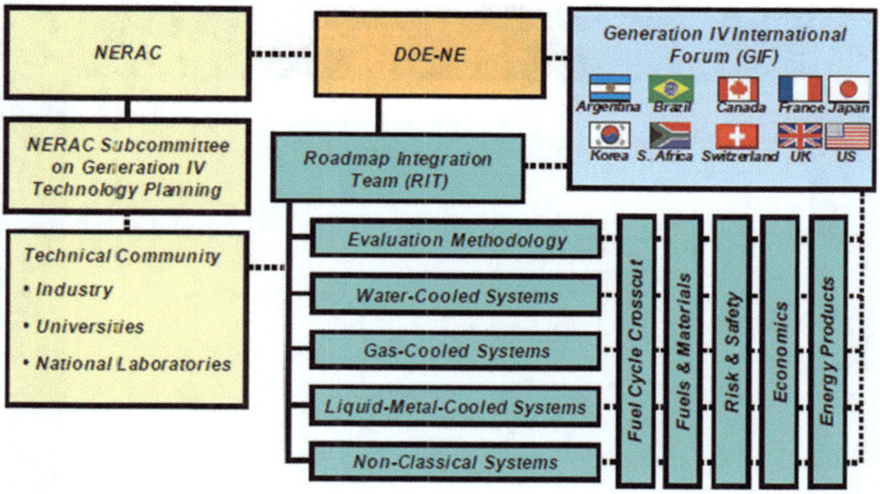

Fig. 2.11 The Roadmap Organization

of the large number of system concepts submitted, the TWGs collected their concepts into sets of concepts with similar attributes. The TWGs conducted an initial screening, *termed screening for potential*, to eliminate those concepts or concept sets that did not have reasonable potential for advancing the goals or were too distant or technically infeasible [12].

A Fuel Cycle Crosscut Group (FCCG) was also formed at a very early stage to explore the impact of the choice of fuel cycle on major elements of sustainability—especially waste management and fuel utilization. Their members were equally drawn from the working groups, allowing them to compare their insights and findings directly. Later, other Crosscut Groups were formed covering economics, risk and safety, fuels and materials, and energy products. The Crosscut Groups reviewed the TWG reports for consistency in the technical evaluations and subject treatment and continued to make recommendations regarding the scope and priority for crosscutting R&D in their subject areas. Finally, the TWGs and Crosscut Groups worked together to report on the R&D needs and priorities of the most promising concepts [12].

The international experts that contributed to this roadmap represented all ten GIF countries, the Organization for Economic Cooperation and Development Nuclear Energy Agency, the European Commission, and the International Atomic Energy Agency [12].

At least in United States under Department of Energy (DOE), with Idaho SMRs project, with a roadmap, federal agencies may wish to follow, when making power purchase decisions that may involve a Small Modular Reactor (SMR), the key steps in the decision process may be summarized in Fig. 2.12. SMRs are designed to provide valuable resilience services as a secure, reliable, and flexible source of primary and backup power. SMRs, coupled with transmission hardening, could

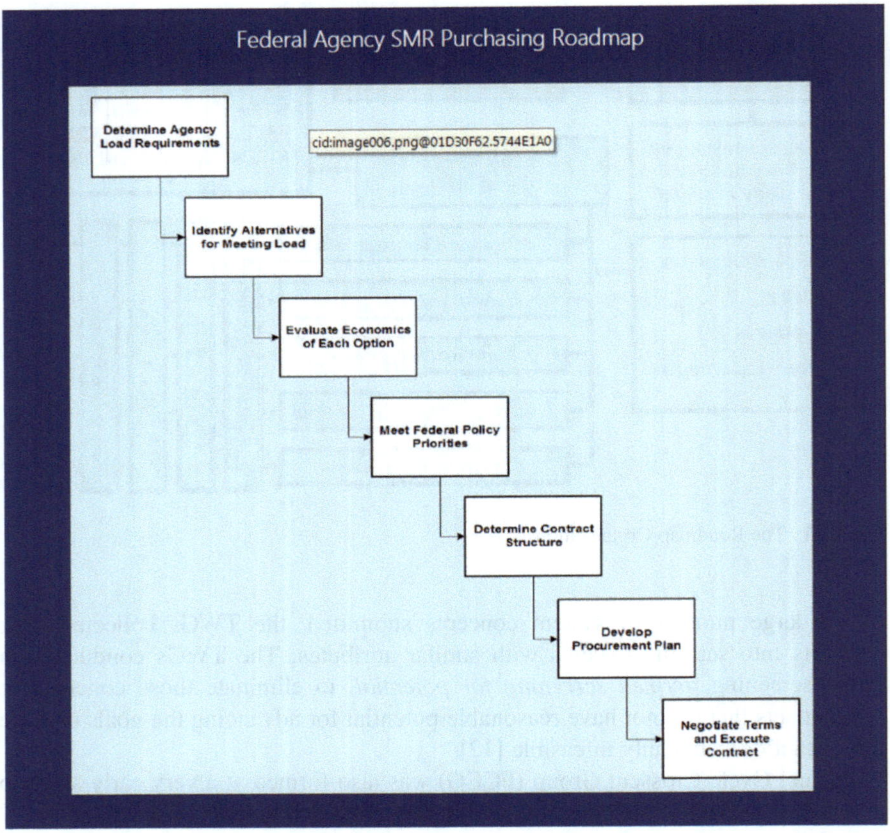

Fig. 2.12 Federal Agency SMR Purchasing Roadmap

provide highly reliable, non-intermittent, clean, and carbon-free power. SMRs can also easily store 2 years' worth of fuel on-site. Certain SMR designs allow for output to be varied over days, hours, or minutes, thereby enabling the SMR to adjust to be in line with changing load demands.

However, implementing SMR projects around the country is difficult due to the FOAK technology, construction challenges, and licensing requirements. SMRs introduce significant expenses and risks that may be challenging for a project to bear without any financial support from the intended end user. The recommendations offered in this chapter, along with the potential solutions described elsewhere in this report, may be considered to assist with overcoming these challenges and advance the deployment of SMRs in the United States while improving grid resiliency at the Oak Ridge Reservation.

More specifically, this chapter identifies how the federal government can assist with making the financing and development of SMRs easier—both in its role as a customer and as a governing body—which permits federal agencies to enter into agreements for a term of up to 30 years to purchase power produced by Small Modular Reactors.

Congress may wish to consider enacting legislation which would permit federal agencies to share in the risks associated with the construction of SMRs. Leveraging the federal government's strong credit standing as a purchaser of the power and its continual need for base-load power is important in the development of SMRs. Federal agency purchasers can help to set the market and offer more certainty to other initial buyers. While there are a range of legal authorities, federal agencies may use to purchase power, most often GSA's 40 U.S.C. § 501 is used, limiting PPA terms to 10 years. This 10-year limitation impacts a party's ability to take advantage of government purchasing for financing.

Federal agencies should be able to purchase power produced by an SMR for a term of up to 30 years. Currently, only DOD (pursuant to 10 U.S.C. § 2922a) has the authority to purchase power for a term of up to 30 years in limited circumstances. By creating an authority that permits federal agencies to purchase power for up to 30 years, SMR developers will be able to use traditional financing to repay a project financed project or a long-term bond over an up to 30-year term, making the financing more affordable. Depending upon the size of the federal agency's offtake, as compared to the size of the power source being funded, this discrepancy may make it difficult for financing.

2.10 Licensing Strategy Components

A DOE and NRC working group was formed to develop the licensing strategy. This group conducted an in-depth analysis of LWR licensing process and technical requirements options, which was performed by the experienced senior staff of the two agencies. The methodology used in formulating the NGNP licensing strategy alternatives also included development of a Phenomena Identification and Ranking Table (PIRT) for a prototypical NGNP by subject matter experts in the nuclear field. The PIRT results assisted in the identification of key R&D needs. Based on the detailed analysis of these alternatives and balancing schedule considerations with licensing risk and other pertinent factors, the Secretary of Energy and the Commission concluded that the following NGNP licensing strategy provides the best opportunity for meeting the 2021 date for initial operation of a prototype NGNP, which details of such analysis can be found in NGNP report to Congress [22].

NGNP reactor technology will differ from that of commercial LWRs currently used for electric power generation. LWRs have a well-established framework of regulatory requirements, a technical basis for these requirements, and supporting regulatory guidance on acceptable approaches an applicant can take to show that NRC requirements are met. The NRC uses a Standard Review Plan to review licensing applications for these reactor designs. Additionally, the NRC has a well-established set of validated analytical codes and methods and a well-established infrastructure for conducting safety research needed to support its independent safety review of an LWR plant design and the technical adequacy of a licensing application.

New nuclear power plants can be licensed under either of two existing regulatory approaches. The first approach is the traditional "two-step" process described in Title 10, Part 50, "Domestic Licensing of Production and Utilization Facilities," of the *Code of Federal Regulations* (10 CFR Part 50), which requires both a Construction Permit (CP) and a separate Operating License (OL). The second approach is the new "one-step" licensing process described in 10 CFR Part 52, "Licenses, Certifications, and Approvals for Nuclear Power Plants," which incorporates a combined Construction and Operating License (COL). Both of these processes allow a deterministic or risk-informed performance-based approach to technical requirements.

Many of the regulatory requirements and supporting review guidance for LWRs are technology-neutral; that is, they are applicable to non-LWR designs as well as LWR designs. However, certain LWR requirements may not apply to the unique aspects of a VHTR design. Accordingly, in developing the NGNP licensing strategy, the NRC and DOE considered the various options available to the NRC staff for adapting current NRC LWR licensing requirements for the NGNP VHTR. These options related to legal, process, technical, research, and regulatory infrastructure matters and included an examination of historical licensing activities. These considerations led to selection of a licensing strategy that would comply best with the considerations identified in the EPAct.

The licensing strategy outlined in this report is composed of two distinct aspects. The first aspect is a recommended approach for how the NRC will adapt the current LWR technical requirements to apply to a VHTR. The second aspect is a recommended licensing process alternative that identifies which of the procedural alternatives in the NRC regulations would be best for licensing the NGNP. To arrive at these recommendations, NRC and DOE evaluated a number of options and alternatives.

2.11 Market and Industry Status and Potentials

Europe plays a leading role in the development of nuclear energy and has 35% of the globally installed capacity. The reactors in Europe have been in operation for 27 years on average. Current plans in most EU member countries are to extend their lifetime on a case by case basis beyond 40 years, and even beyond 60 years in some cases, in combination with power upgrades. The first two Generation III reactors, European Pressurized Water Reactor (EPR) are currently being constructed.

The global growth of the nuclear energy can be measured by the increasing number of reactors [9, 10] (3 more in 2005 and 2006, 7 in 2007, and 10 in 2008) but with a strong concentration in Asia. Nevertheless, a number of these reactors are of European design. There are presently four reactors under construction in Europe: the EPRs in Finland and France and two smaller reactors of Generation II type

(VVER 440) in Slovakia and with plans to build new reactors in France, Romania, Bulgaria, and Lithuania. Perhaps more importantly the United Kingdom has taken concrete steps toward new build with bidding beginning in 2009 from leading utilities, and Italy has declared that it intends to start a nuclear program with a target to produce 25% of the electricity by 2030. The estimated maximum potential installed capacities of nuclear fission power for the EU-27 (150 GWe by 2020 and 200 GWe by 2030) appear more realistic than the baseline (115 GWe in 2020 and 100 GWe in 2030).

Programs to build fast reactor and high-temperature reactor demonstrators are being implemented in Russia and several Asian countries. Although these are not Generation IV designs, transfer of knowledge and experience from operation will contribute significantly to future Generation IV development. In Europe, a concerted effort is proposed in the form of a European Industrial Initiative in sustainable nuclear fission as part of the Community's SET-Plan. The EII has singled out the Sodium Fast Reactor (SFR) as its primary system with the basic design selected by 2012 and construction of a prototype of 250–600 MW(e) that is connected to the grid and operational by 2020.

In parallel, a gas- or lead-cooled fast reactor (GFR/LFR) will also be investigated. The selection of the alternative fast reactor technology is scheduled for 2012 on the basis of a current program of preconceptual design research. The reactor will be a 50–100 MW(th) demonstrator reactor that should also be in operation by 2020. The SFR prototype and LFR/GFR demonstrator will be complemented by a fuel fabrication workshop that should serve both systems and by a range of new or refurbished supporting experimental facilities for qualification of safety systems, components, materials, and codes. A commercial deployment for a SFR reactor is expected from 2040 and for the alternative design a decade later.

High-temperature reactors dedicated to cogeneration of process heat for the production of synthetic fuels or industrial energy products could be available to meet market needs by 2025, which would trigger requirements to construct "first-of-a-kind" demonstrators in the next few years. Indeed, such programs are currently being set up in some countries (United States, Japan, South Africa, and China). The key aspect is the demonstration of the coupling with the conventional industrial plant. Supercritical water reactors and Molten Salt Reactors, as well as accelerator-driven subcritical systems dedicated to transmutation of nuclear waste, are currently being assessed in terms of feasibility and performance, though possible industrial applications have yet to be clearly identified.

2.12 Barriers

The high capital cost of nuclear energy in combination with uncertain long-term conditions constitutes a financial risk for utilities and investors. The lack of widespread support in the EU Member States may undermine the strength of EU industry for the development of new technologies. Harmonized regulations, codes, and standards at the EU level would strengthen the competitiveness of Europe's nuclear sector and promote deployment of Generation III technology in the near term. The industry, infrastructures, and services that support nuclear power have shrunk significantly during the last decades. This situation in Europe is not unique, but it may pose a bottleneck for the deployment of reactors in the relatively near future. One example is large forgings needed for pressure-vessel heads. World capacity is limited, and even at the present new build construction rate, there is a waiting list for delivery of these components.

Public acceptance remains an important issue, but even though opinion is not very favorable in a number of Member States, there are signs that the mood is changing. Nevertheless, concerted efforts are still required, based on objective and open dialogue among all stakeholders. International cooperation currently exists at the level of research, and this is being facilitated in the area of Generation IV systems by the Generation IV International Forum (GIF). However, EU industry is facing stiff competition, especially in Asia where strong corporate support for R&D is putting industry in a better position to gain leadership in the near future. Another significant potential barrier for nuclear fission is the shortage of qualified engineers and scientists as a result of the lack of interest in nuclear careers during the 1990s and the reduced availability of specialist courses at universities. Preservation of nuclear knowledge remains a major issue, especially since most of the current generations of nuclear experts are nearing retirement.

2.13 Needs

The high initial capital investments and sensitive nature of the technology involved means that renewed deployment of currently available nuclear technology can only take place in a stable (or, at least, predictable) regulatory, economic, and political environment. In June 2009, the EU established a common binding framework on nuclear safety with the adoption of the Council Directive establishing a community framework for the safety of nuclear installations [13, 14]. This is the first binding EU legislation in this field.

In order to retain its leading position and to overcome bottlenecks in the supply chain, Europe also needs to reinvigorate the industrial supply chains supporting the nuclear sector. Apart from this overriding requirement for a clear European strategy on nuclear energy, a new research and innovation system is needed that can assure additional funding, especially for the development of Generation IV technology. In this context the Sustainable Nuclear Energy Technology Platform [11] plays a key role. The timescales involved, and the fact that key political and strategic decisions are yet to be taken regarding this technology, mean that a significant part of this additional funding must be public. The launch of the European Sustainable Nuclear Industrial Initiative under the community's SET-Plan, bringing together key industrial and R&D organizations, would be a very significant step toward the construction and operation of the necessary demonstrators and prototypes.

High-temperature reactors based on existing technology can also be deployed in the near future with the aim of demonstrating the cogeneration of process heat and the coupling with industrial processes. This would need to be built and funded through a European or international consortium, which should also include the process heat end-user industries. In the meantime, an enhanced research effort is needed to ensure Europe's leadership in sustainable nuclear energy technologies that include continuous innovation in LWRs, qualification and development of materials, closed fuel cycle with U–Pu multi-recycling and (very) high-temperature reactors, and related fuel technology.

Breakthroughs are especially sought in the fields of materials to enhance safety, nuclear fuels, and fuel cycle processes. Additionally, there is a need for harmonization of European standards and a strategic planning of national and European research infrastructures for use by the European research community. The implementation of geological disposal of high-level waste is also being pursued as part of national waste management programs, though some countries are not as advanced as others. The new Implementing Geological Disposal Technology Platform, launched in November 2009, is coordinating the remaining necessary applied research in Europe leading up to the start of operation of the first geological repositories for high-level and long-lived waste around 2020 and will facilitate progress in and technology transfer with other national programs.

More effort is needed to inform and interact with the public and other stakeholders, and the education and training of a new generation of nuclear scientists and engineers and transfer of knowledge from the generation that designed and built reactors in the seventies and eighties needs urgent attention. The European Nuclear Energy Forum (ENEF) provides a unique platform for a broad open discussion on the role nuclear power plays today and could play in the low-carbon economy of the future. ENEF analyzes and discusses the opportunities (competitiveness, financing, grid, etc.) and risks (safety, waste) and need for education and training associated with the use of nuclear power and proposes effective ways to foster communication with and participation of the public.

2.14 Synergies with Other Sectors

Nuclear energy provides a very stable base-load electricity supply and can therefore work in synergy with renewable energies that are more intermittent. Nuclear energy should also contribute significantly to a low-carbon transport sector as high-temperature application can provide synthetic fuel and hydrogen, while generated electricity could provide a large share of the energy for electrical cars. Interactions are anticipated with activities in "Hydrogen Energy and Fuel Cells" through the potential of nuclear hydrogen production and with "grids" from the characteristics of nuclear electricity generation. With respect to basic materials research, there should be synergies with other applications, such as "biofuels" and "clean coal," where materials are subjected to extreme environments. In addition, the opportunities for important common research with the fusion program, especially in the area of materials, need to be fully exploited. The European Energy Research Alliance under the SET-Plan is also expected to provide opportunities for synergies and collaborative work in the area of nuclear materials. In general, crosscutting research would benefit from more clearly defined channels of interaction, responsibilities, and increased flexibility regarding funding and programming.

2.15 Combined Cycles for Efficiency of New-Generation Nuclear Power Plants

In Chap. 1, Sect. 1.1 of this book, we briefly explained an innovative design approach to combined cycle-driven efficiency for next generation of nuclear power plants as part of Generation IV, which includes all six types of Small Modular Reactors (SMRs) as well as references by this author [1–3].

Out of the six chosen rector designs of Generation IV systems that are described in Sect. 2.4 and more details of each of these reactors are given in its subsections, while they are summarized in Table 2.2, three to four of them perfectly fit into combined cycles of open-air Brayton cycle due to the nature of their high-temperature design requirements, namely, very-high-temperature gas reactor (VHTR), sodium-cooled fast reactor(SFR), gas-cooled fast reactor (GFR), and Molten Salt Reactor (MSR), where their temperature design ranges anywhere from 550 to 900 °C, and that is type of temperature range which is required by innovative open-air Brayton combined cycle, where the steady-state thermal analysis computer code is developed by this author and his collaborator [1–3].

For this combined cycle to work in more efficient way, we pay our attention and are more interested in a very-high-temperature reactor (VHTR) within this generation (i.e., GEN-IV) for open-air Brayton cycle implementation as part of new innovative design. Historically speaking, the development of high-temperature gas-cooled reactor (HTGR), also known as very high-temperature gas-cooled reactor (VHTR) for its Generation IV designs, has continued for over half a century. Several reactors have been built or being constructed. These are identified in Table 2.3. Still others are being developed at various stages, including more units of high-temperature reactor-pebble bed module (HTR-PM) power reactor in China, multipurpose GTHTR300C in Japan, NuH2 for nuclear hydrogen and process heat in Korea, Next Generation Nuclear Plant (NGNP) cogenerating reactor in the United States, and an experimental power reactor in Indonesia.

Dragon, the first reactor built, pioneered the use of tri-isotropic (TRISO)-coated particle fuel, still the standard fuel form today. The Arbeitsgemeinschaft Versuchsreaktor (AVR) tested additional fuel designs and accumulated extensive performance data. The prototypical Fort St. Vrain (FSV) validated the prismatic core physics design with high burnup (90 GWd/t) on thorium fuel and demonstrated steam turbine power generation at 39% thermal efficiency and easy load following.

Table 2.2 Summary of the main characteristics of the six Generation IV systems

System	Neutron spectrum	Coolant	Temp. (°C)	Fuel cycle	Size (MW (e))
VHTR (Very-high-tempera-ture gas reactor)	Thermal	Helium	900–1000	Open	250–300
SFR (Sodium-cooled fast reactor)	Fast	Sodium	550	Closed	30–150 300–1500 1000–2000
SCWR (Supercritical Water-Cooled Reactor)	Thermal/fast	Water	510–625	Open/closed	300–700 1000–2000
GFR (Gas-cooled fast reactor)	Fast	Helium	850	Closed	1200
LFR (Lead-cooled fast reactor)	Fast	Lead	480–800	Closed	20–180 300–1200 600–1000
MSR (Molten Salt Reactor)	Epithermal	Fluoride salt	700–800	Closed	1000

As part of Advanced Small Modular Reactor (AdvSMR) safety and licensing progress is concerned, an overview of this roadmap is demonstrated in Fig. 2.13.

Yet the component failures, such as with the primary coolant circulator, forced excess outage and undermined its economics. The thorium high-temperature reactor-300 (THTR-300) of a pebble-bed core design encountered technical problems after only a brief period of operation, and their scrutiny led to protracted shutdown. The FSV and THTR-300 were prematurely decommissioned largely as business decision [15].

Asia then became home to the latest builds. The High Temperature Engineering Test Reactor (HTTR) in Japan and the high-temperature test reactor (HTR-10) in China were constructed and started up around the turn of the millennium. Both remain operational today. The 30 MW(th) HTTR demonstrated operation of 950 °C reactor outlet coolant and export of 863 °C process heat. Such high-temperature capability would raise reactor thermal efficiency and support advanced applications as reported by the plant design of GTHTR300 by Japan Atomic Energy Agency (JAEA) [16, 17]. The Generation IV system employs a 600 MW(th) reactor with outlet coolant temperature of 950 °C to power a gas turbine for electricity generation and a thermochemical process for hydrogen production, yielding thermal efficiency of 50% or higher [15].

Based on the experience of HTR-10 and extensive engineering development of the reactor components, China is constructing the world's first prototype modular reactor plant HTR-PM in the northeastern Shandong province [18]. Although not a VHTR by coolant temperature, the twin-unit (2 × 250 MW(th)) power plant with reactor coolant temperature of 750 °C shares some of the design approaches of VHTR, including passive safety features and high-temperature heat application

Table 2.3 High-temperature gas-cooled reactor built worldwide (Courtesy of X.L. Yan, Japan Atomic Agency, Oarai-Machi, Ibaraki-ken, Japan)

	Test HTGRs				Prototype HTGRs			

	Dragon	AVR	HTTR	HTR-10	Peach bottom	FSV	THTR-300	HTR-PM
Country	UK (OECD)	Germany	Japan	China	USA	USA	Germany	China
Period of operation	1963–1976	1967–1988	1998–Present	2000-Present	1967–1974	1976–1989	1986–1989	2017 planned
Reactor core type	Tube	Pebble bed	Prismatic	Pebble bed	Tube	Prismatic	Pebble bed	Pebble bed
Thermal power MW(th)	21.5	46	30	10	115	842	750	2 × 250
Coolant outlet temperature, °C	750	950	950	700	725	775	750	750
Coolant pressure, MPa	2	1.1	4.0	3.0	2.25	4.8	3.9	7.0
Electrical output, MW(e)	–	13	–	2.5	40	330	300	211
Process heat output, MW(th)	–	–	10	–	–	–	–	–
Process heat temperature, °C	–	–	863	–	–	–	–	–
Core power density, W/cm^3	14	2.6	2.5	2	8.3	6.3	6.0	3.2
Fuel design	UO_2 TRISO	(Th/U, U)O_2, C_2 BISO	UO_2 TRISO	UO_2 TRISO	ThC_2 BISO	(Th/U, Th)C_2 TRISO	(Th/U)O_2 BISO	UO_2 TRISO

BISO Bi-isotropic coating of fuel particle

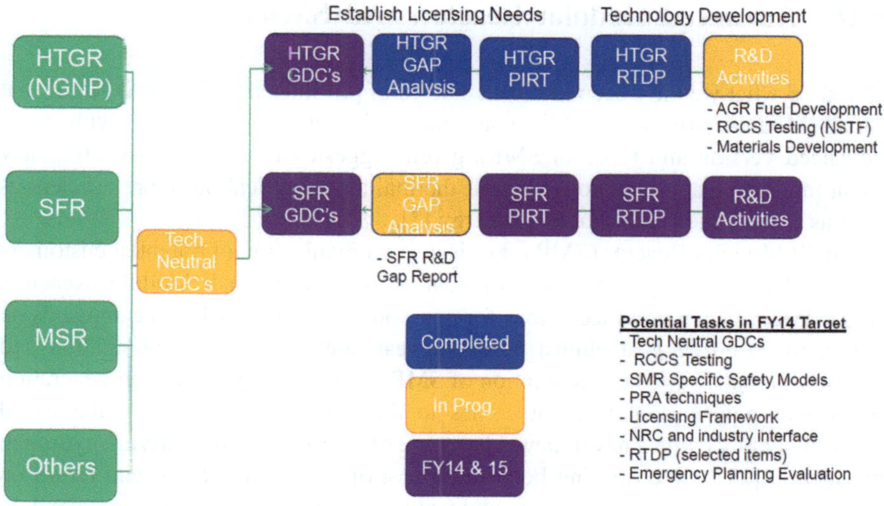

Fig. 2.13 Overview of Advanced Small Modular Reactor safety and licensing progress (Courtesy of Idaho National Laboratory). *BISO* Bi-isotropic coating of fuel particle

potential. The construction began in December 2012 and the operation is expected in 2017.

Note that any Small Modular Reactor of VHTR type is an excellent candidate for innovative design approach for open-air Brayton combined cycle [1–3]. These types of reactors are considered as Advanced Small Modular Reactor (AdvSMR), which is part of family Generation IV nuclear power reactors.

In 2001, the Generation IV International Forum (GIF) endorsed six nuclear system concepts, which will deliver affordable energy products while satisfactorily addressing the issues of nuclear safety, waste, and proliferation [19]. Recognizing the VHTR to be nearest term deployable and exceptionally suitable, not only for electricity generation, but also for hydrogen production and other industrial applications, the US Department of Energy (DOE) has placed the Generation IV priority on the VHTR. The Energy Policy Act of 2005 formally established the NGNP as a DOE project to demonstrate commercial high-efficiency generation of electricity and hydrogen (The US Energy Policy Act of 2005, 2005). At present, the advanced gas reactor (AGR) fuel development and qualification program at the US Idaho National Laboratory is qualifying uranium oxide/uranium carbide (UCO) TRISO fuel [19] and the NGNP.

Industry Alliance, a consortium of HTGR designers, utility plant owner/operators, suppliers, and end users, is promoting the reactor commercialization and industrial applications (ngnpalliance). In 2012, the Alliance selected AREVA's prismatic SC-HTGR of 625 MW(th) that provides steam and electricity cogeneration as its primary choice of reactor design for prototype implementation in mid-2020s [20].

For further information, reader should refer to reference written by Yan [15].

2.16 Advanced Modular Reactors (AdvSMRs)

Based on what we have seen in recent studies of pro-nuclear industry, it seems that Small Modular Reactors (SMRs) are nuclear's future, particularly with more advanced version and types of SMRs given suggested combined drive efficiency as an innovative approach to raise their thermal and electrical out more efficient as well as a new renewable source of energy [3].

Small Modular Reactor (SMR) start is a new organization of potential customers and vendors investing in the development of Advanced Small Modular Reactors (AdvSMRs), which provided a hope for new source of energy when the demands on electricity is growing at almost 17% per year due to growth in population and industry globally. The new generation of SMRs is also a very promising generation of nuclear power plant when it comes to the idea of decarbonizing the world environment by being a new renewable source of energy as well as driving hydrogen production [21] while offering better total cost of ownership (TCO) and return on investment (ROI) for the customers and vendors of this newly formed organization.

This newly established organization was formed with one main goal in mind, and that was to get SMRs over the initial market humps that plague all new technologies by using proven public/private partnerships that have succeeded with other technologies like renewable energies. Because of their small size of 300 MW or less, SMRs are economic, factory-built, and shippable, flexible enough to desalinate, refine oil, load follow wind, produce hydrogen and provide something we have all been waiting for—a reactor that cannot meltdown and most important aspect of these AdvSMRs by utilizing the open-air Brayton combined cycles that have no need for water as source of cooling down, while they are not constraint under *steam dome* as the Light Water Reactors (LWRs) are [1–3].

Note that some even newly chosen design of SMRs such as the design approach taken by NuScale is following LWR technology; therefore, they are limited to the steam dome constraint.

SMR Start's goal is to ensure that SMRs are a cost-competitive option in the future, with the first units operating by the mid-2020s.

There is increasing interest in Small Modular Reactors (SMRs) and their applications. SMRs are newer-generation reactors designed to generate electric power up to 300 MW, whose components and systems can be shop-fabricated and then transported as modules to the sites for installation as demand arises. Most of the SMR designs adopt advanced or even inherent safety features and are deployable either as a single or multi-module plant. SMRs are under development for all principal reactor lines: water-cooled reactors; high-temperature gas-cooled reactors; liquid-metal, sodium-, and gas-cooled reactors with fast-neutron spectrum; and Molten Salt Reactors. The key driving forces of SMR development are fulfilling the need for flexible power generation for a wider range of users and applications, replacing aging fossil-fired units, enhancing safety performance, and offering better economic affordability.

Many SMRs are envisioned for niche electricity or energy markets where large reactors would not be viable. SMRs could fulfill the need of flexible power generation for a wider range of users and applications, including replacing aging fossil power plants, providing cogeneration for developing countries with small electricity grids and remote and off grid areas, and enabling hybrid nuclear/renewables energy systems. Through modularization technology, SMRs target the economics of serial production with shorter construction time. Near-term deployable SMRs will have safety performance comparable or better to that of evolutionary reactor designs.

Furthermore, *small-* and *medium*-sized reactors include a large variety of designs and technologies and, in general, consist of:

- Advanced SMRs, including modular reactors and integrated PWRs
- Innovative SMRs, including small-sized Gen IV reactors with non-water coolant/moderator
- Converted or modified SMRs, including barge-mounted floating NPP and seabed-based reactors
- Conventional SMRs, those of Gen-II technologies and still being deployed

Advanced SMRs having an equivalent electric power of less than 700 MW(e) or even less than 300 MW(e) are part of a new generation of nuclear power plant designs being developed to provide a flexible, cost-effective energy for various applications. Advanced SMR designs include water-cooled reactors, high-temperature gas-cooled reactors, as well as liquid-metal-cooled reactors with fast-neutron spectrum. The trend of development has been toward design certification of Small Modular Reactors, which are defined as advanced reactors that produce equivalent electric power less than 300 MW(e) designed to be built in factories and shipped to utilities for installation as demand arises. The SMR systems adopt modularization, by which the structures, systems, and components are shop-fabricated then shipped and assembled on-site; thus, the construction time for SMRs can be substantially reduced. Some of the SMRs are to be deployed as multiple-module power plants allowing utilities to add additional reactor and power conversion modules as demand for local power increases.

Advanced SMRs will use different approaches from large reactors for achieving a high level of safety and reliability in their systems, structures, and components, and that will be the result of a complex interaction between design, operation, material, and human factors. Interest in SMRs continues to grow as an option for future power generation and energy security. However, the first phase of advanced SMR deployments will have to ultimately demonstrate high levels of plant safety and reliability and prove their economics in order for further commercialization to be feasible. This situation is an issue for the technology developers since building the first plants will be relatively expensive and the “nth of a kind” costs can only be confirmed after the first demonstration plants are deployed. As a result, new private-public partnership arrangements may be needed to support the first phase of advanced SMR deployments. These plants would have greater automation but will still rely on human interaction for supervision, system management, and operational decisions because operators are still regarded as the last line of defense if failures in automated protective measures occur.

2.17 Advantages of Small Modular Reactors (SMRs)

Associated with Smaller Modular Reactors (SMRs), there exist certain benefits, which these SMRs offer, and to start with is their cost-effectiveness form investment and total cost of ownership (TCO), thus the return on investment (ROI) point of view.

Small Modular Reactors offer a lower initial capital investment, greater scalability, and siting flexibility for locations unable to accommodate more traditional larger reactors. They also have the potential for enhanced safety and security compared to earlier designs. Deployment of advanced SMRs can help drive economic growth. These low initial costs are coming with certain associated benefits, that are listed here, and they are:

1. Modularity

   The term "modular" in the context of SMRs refers to the ability to fabricate major components of the nuclear steam supply system in a factory environment and ship to the point of use. Even though current large nuclear power plants incorporate factory-fabricated components (or modules) into their designs, a substantial amount of field work is still required to assemble components into an operational power plant. SMRs are envisioned to require limited on-site preparation and substantially reduce the lengthy construction times that are typical of the larger units. SMRs provide simplicity of design, enhanced safety features, the economics and quality afforded by factory production, and more flexibility (financing, siting, sizing, and end-use applications) compared to larger nuclear power plants. Additional modules can be added incrementally as demand for energy increases.
2. Lower Capital Investment

   SMRs can reduce a nuclear plant owner's capital investment due to the lower plant capital cost. Modular components and factory fabrication can reduce construction costs and duration.
3. Siting Flexibility

   SMRs can provide power for applications where large plants are not needed or sites lack the infrastructure to support a large unit. This would include smaller electrical markets, isolated areas, smaller grids, sites with limited water and acreage, or unique industrial applications. SMRs are expected to be attractive options for the replacement or repowering of aging/retiring fossil plants or to provide an option for complementing existing industrial processes or power plants with an energy source that does not emit greenhouse gases.
4. Greater Efficiency

   SMRs can be coupled with other energy sources, including renewables and fossil energy, to leverage resources and produce higher efficiencies and multiple energy end-products while increasing grid stability and security. Some advanced SMR designs can produce a higher-temperature process heat for either electricity generation or industrial applications.
5. Safeguards and Security/Nonproliferation

SMR designs have the distinct advantage of factoring in current safeguards and security requirements. Facility protection systems, including barriers that can withstand design basis aircraft crash scenarios and other specific threats, are part of the engineering process being applied to new SMR design. SMRs also provide safety and potential nonproliferation benefits to the United States and the wider international community. Most SMRs will be built below grade for safety and security enhancements, addressing vulnerabilities to both sabotage and natural phenomena hazard scenarios. Some SMRs will be designed to operate for extended periods without refueling. These SMRs could be fabricated and fueled in a factory, sealed and transported to sites for power generation or process heat, and then returned to the factory for defueling at the end of the life cycle. This approach could help to minimize the transportation and handling of nuclear material. Light water-based SMRs are expected to be fueled with low-enriched uranium, i.e., approximately 5% U-235, which is similar to existing large nuclear power plants. The "security by design" concepts being applied to these technologies are expected to increase SMR resistance to theft and diversion of nuclear material. Also, reactor cores for these light-water SMRs can be designed to burn plutonium as a mixed oxide (MOX) fuel. Further, SMRs based on non-Light Water Reactor coolants could be more effective at dispositioning plutonium while minimizing the wastes requiring disposal.

6. US Industry, Manufacturing, and Job Growth

   The case for SMR economic competitiveness is rooted in the concept that mass manufacture of modular parts and components will reduce the cost per kilowatt of electricity on par with current generating sources. There is both a domestic and international market for SMRs, and US industry is well positioned to compete for these markets. DOE hopes that the development of standardized SMR designs will also result in an increased presence of US companies in the global energy market. If a sufficient number of SMR units were ordered, it would provide the necessary incentive to develop the appropriate factory capacity to further grow domestic and international sales of SMR power plants.

7. Economic Development

   SMR deployment to replace retiring electricity generation assets and meet growing generating needs would result in significant growth in domestic manufacturing, tax base, and high-paying factory, construction, and operating jobs. A 2010 [8] study on economic and employment impacts of SMR deployment estimated that a prototypical 100 MW(e) SMR costing $500 million to manufacture and install would create nearly 7000 jobs and generate $1.3 billion in sales, $404 million in earnings (payroll), and $35 million in indirect business taxes. The report examines these impacts for multiple SMR deployment rates, i.e., low (1–2 units/year), moderate (30 units/year), high (40 units/year), and disruptive (85 units/year). The study indicates significant economic impact would be realized by developing an SMR manufacturing enterprise at even moderate deployment levels.

If we consider these above benefits that are coming from SMR-type nuclear power plants, we do see their cost-effectiveness and initial capital investment, in comparisons with today's analysis of cost of electivity from nuclear power plants.

2.17.1 Cost of Electricity from Nuclear Power

The newest estimates for the cost of nuclear power in the United States are (costs indicated are per delivered kWh):

11.2 US-cents per kWh nuclear power (MIT 2003)
14.1 US-cents per kWh nuclear power (Keystone June 2007)
18.4 US-cents per kWh nuclear power (Keystone midrange estimate)

In the United Kingdom, the cost of nuclear electricity was estimated to be 8.2 US-cents per kWh (for an interest rate of 10%) and US-cents 11.5 per kWh (for an interest rate of 15%). Here transmission and distribution of the electricity (usually about 3 US-cents per kWh) have to be added (Source: Thomas, Bradford, Froggatt, Milborrow, 2007) [9]. So, in the United Kingdom, the cost of nuclear energy is estimated to be in the same range as indicated above.

In comparison with other technologies, e.g., large wind power farms: 7 US-cents per kWh delivered.

Off course most of the environmentalists are arguing over wind and solar energy as a cheap way of producing electricity and call them as an alternative renewable source of energy, and they are getting less expensive almost daily while nuclear power is getting more and more expensive. Isn't this alone a very strong indication that nuclear power is a technology from the past?

However, one needs bear in mind that wind does not blow 24 × 7 nor sun will shine all the time during day and night, providing during cloudy day, we may have this issue, while at night there is now sun out.

2.17.2 Cost of Nuclear Technology Is Too High

The reactor types you are describing are not available. A lot of research and development would be needed to develop this technology for productive applications. We probably talk about a time frame of 30–50 years, if not even longer. But we need a solution now! It is urgent.

The costs for the development are very difficult to estimate, and therefore the costs indicated above might be (and probably are) wrong. In the history of nuclear power, the cost of development has always been tremendously under estimated.

We should put full effort into developing sustainable energy technologies further and phase-out nuclear technology. This will open many opportunities and have many positive side effects. It is time for change!

In addition, we will inevitably have to change our behavior: We should only use as much energy as we are able to produce with sustainable technologies. Demand of energy has to follow available supply of sustainable energy. This will have many positive side effects, too.

2.18 Conclusions

Small Modular Reactors communities advocate for public/private partnerships—similar to those that provide support for the introduction of other new energy technologies such as solar and wind—to help ensure the successful commercialization of SMRs. Public/private partnerships stimulate the private investments required to ensure that the technology continues to advance and is capable of competing in domestic and overseas markets without additional direct support once the technology matures.

Today's energy markets are characterized by low natural gas prices, heavily subsidized renewable generation, and low growth in electricity demand. America vendors also compete in the international markets for the deployment of SMRs and other advanced nuclear designs. Unfortunately, most countries like Russia and China directly subsidize their nation's nuclear technologies in order to penetrate global markets.

The new waves of SMR designs has been supported in many recent inside the DOE's and NRC initiatives, and a final SMR design is necessary for the first power plant to begin construction and to take advantage of factory fabrication. But even after an SMR design certification application has been submitted to the US Nuclear Regulatory Commission (NRC), there is still significant cost and uncertainty involved in the design certification approval process, as well as work to finalize the design. A cost-sharing program for design finalization activities would help reduce the risk for first reactor design companies and would increase the probability that a final SMR design is produced for factory fabrication.

The low price of natural gas poses a challenge to low-carbon technology deployment, and a production tax credit (PTC) would help to value the carbon-free benefits of SMRs and close this price gap as it has for renewables. Power purchase agreements (PPAs) would help to create demand for new, low-carbon technologies in an era of slower electrical demand growth.

PPAs between SMR operators and federal facilities would align with the directive to support low-carbon technologies codified in Executive Order 13693, as well as promote national security.

Note that, traditionally, a **power purchase agreement**, or PPA, is a contract between a government agency and a private utilities company. The private company agrees to produce electricity, or some other power source, for the government agency over a long period of time.

Investments in advanced manufacturing will reduce cost and construction time for SMRs and create manufacturing jobs for components that currently have to be made outside the United States for the large traditional nuclear reactors.

New reactor designs are pretty advanced and ready to be rolled out. In January, NuScale Power out of Oregon announced their submission to the Nuclear Regulatory Commission of the first design certification application for any SMR in the United States. It is expected to be built in the early 2020s.

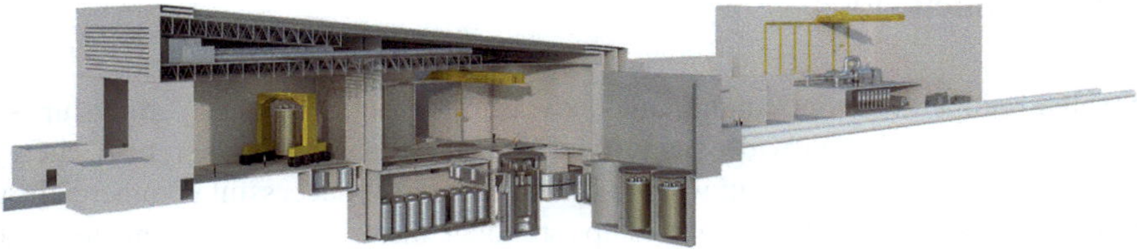

Fig. 2.14 Terrestrial Energy's IMSR Small Modular Reactors general depiction

Terrestrial Energy (Fig. 2.14), another innovative Small Modular Reactor company, has notified the Nuclear Regulatory Commission of its intentions to license their Integral Molten Salt Reactor (IMSR). They have plans to submit its licensing application in 2019.

IMSR® technology can be brought to market quickly. IMSR® power plants can be built in 4 years and produce electricity or industrial heat at prices competitive with fossil fuels while emitting no greenhouse gases.

The first IMSR® power plants are expected to come online in the 2020s. IMSR® technology is today's clean energy game changer. Learn more about the IMSR® and how Terrestrial Energy is leading the way to a bright energy future [22].

But what really matters to the public is safety. The small size and large surface area-to-volume ratio of SMRs, like NuScale Power, that sits below ground in a super seismic-resistant heat sink, allows natural processes to cool it indefinitely in the case of a complete power blackout, with no humans needed to intervene, no AC or DC power, no pumps, and no additional water for cooling.

References

1. Zohuri, B. (2015). *Combined cycle driven efficiency for next generation nuclear power plants: An innovative design approach* (1st ed.). New York: Springer.
2. Zohuri, B., & McDaniel, P. (2017). *Combined cycle driven efficiency for next generation nuclear power plants: An innovative design approach* (2nd ed.). New York: Springer.
3. Zohuri, B. (2018). *Small modular reactors as renewable energy sources*. New York: Springer.
4. Retrieved from http://www.ne.doe.gov/geniv/neGenIV1.html.
5. Sustainable Nuclear Energy Technology Platform. (2009, May). *Strategic research agenda*. Retrieved from www.SNETP.eu.
6. OECD. (2008). *Uranium 2007: Resources, production and demand*. OECD Nuclear Energy Agency and the International Atomic Energy Agency, NEA N 6345.
7. Nuclear Energy Agency. (2008). *Nuclear Energy Outlook 2008* (OECD/NEA Report No. 6348). Paris: Nuclear Energy Agency.
8. Massachusetts Institute of Technology. (2009, May). *2009 Update of the MIT 2003 future of nuclear power, an interdisciplinary MIT study, 2003*. Cambridge, MA: Massachusetts Institute of Technology. Retrieved from http://web.mit.edu/nuclearpower/pdf/nuclearpower-update2009.pdf.
9. Retrieved from http://www.ne.doe.gov/pdfFiles/NGNP_reporttoCongress.pdf.

10. *Updated emissions projections.* (2006, July). DTI. Retrieved from http://www.dti.gov.uk/files/file31861.pdf.
11. *DTI: Energy White Paper, Meeting the Energy Challenge.* Retrieved from http://www.dti.gov.uk/energy/whitepaper.
12. Retrieved from http://www.ne.doe.gov/geniv/documents/gen_iv_roadmap.pdf/.
13. Proposal for a COUNCIL DIRECTIVE (EURATOM) setting up a Community framework for nuclear safety COM(2008) 790/3. (2008, November).
14. COUNCIL OF THE EUROPEAN UNION Legislative Acts and Other Instruments 10667/09. (2009, June).
15. Yan, X. L. Japan Atomic Agency, Oarai-Machi, Ibaraki-ken, Japan.
16. Sato, H., Yan, X. L., Tachibana, Y., & Kunitomi, K. (2014). GTHTR300dA nuclear power plant design with 50% generating efficiency. *Nuclear Engineering and Design, 275*, 190–196.
17. Yan, X., Sato, H., Kamiji, Y., Imai, Y., Terada, A., Tachibana, Y., et al. (2014). GTHTR300 cost reduction through design upgrade and cogeneration. Paper HTR2014-21436. In *Proceedings of the HTR 2014, Weihai, China, October 27–31.*
18. Fu, J., Jiang, Y., Cheng, H., & Cheng, W. (2014). Overview and progress of high temperature reactor pebble-bed module demonstration project (HTR-PM), HTR2014-11125. In *Proceedings of the HTR 2014, Weihai, China, October 27–31.*
19. Petti, D. (2014). Implications of results from the advanced gas reactor fuel development and qualification program on licensing of modular HTGRs. Paper HTR2014-31252. In *Proceedings of the HTR 2014, Weihai, China, October 27–31.*
20. Shahrokhi, F., Lommers, L., Mayer, J., III, & Southworth, F. (2014). US HTGR deployment challenges and strategies. Paper HTR2014-11309. In *Proceedings of the HTR 2014, Weihai, China, October 27–31.*
21. Zohuri, B. (2018). *Hydrogen energy: Challenges and solutions for a cleaner future* (1st ed.). New York: Springer.
22. Retrieved from https://www.terrestrialenergy.com/technology/.

Chapter 3
Design and Analysis of Core Design for Small Modular Reactors

The pronuclear energy and advocates are lobbying that the sustainable development of the world's energy sector cannot be achieved without extensive use of nuclear energy and the advantages of nuclear-related technologies, including upcoming new generation of the Small Modular Reactors in near future horizon. The dawn of these SMRs requires new design and analysis no matter if they are falling into Light Water Reactor (LWR), pressurized water reactor (PWR), or even multi-application small light water reactor (MASLWR) categories, depending on the vendor involved with these new technologies and consequently safety standards and their nonproliferation requirements as well. This chapter visits these standards for core design and generally elaborated on them with understanding that readers need to refer just beyond this book and this chapter for more details.

3.1 Introduction

In order to meet national and international requirements to achieve maximum safety and efficiency levels and address worldwide proliferation concerns, a variety of innovative new technologies implemented into the new generation of nuclear reactors will need to be developed.

There are manufacture vendors that are looking into design and development of Small Modular Reactor (SMR), such as NuScale using existing Light Water Reactor (LWR) for their NuScale Power Module as illustrated in Fig. 3.1 and Holtec International with their SMR-160 using similar technologies that exist around pressurized water reactor (PWR) and their scaling down that is used in nuclear vessels such as submarine as illustrated in Fig. 3.2.

Then, of course, there are other nuclear industrial vendors such as mPower as part of Babcock & Wilcox (B&W) that in 2009 announced its mPower reactor as depicted in Fig. 3.3, a 500 Mew integral PWR designed to be factory-made and railed to site. It was a deliberately conservative design to more readily gain

B. Zohuri, *Heat Pipe Applications in Fission Driven Nuclear Power Plants*,
https://doi.org/10.1007/978-3-030-05882-1_3

Fig. 3.1 NuScale Power Module

acceptance and licensing. In November 2012 the US Department of Energy (DOE) announced that it would support accelerated development of the design for early deployment, with up to $226 million, and it paid $111 million of this.

On the other hand, universities such as Oregon State University (OSU) is involved with the design study of the multi-application small light water reactor (MASLWR) as illustrated in Fig 3.4 with the following requirements in mind such as the passive safety systems, natural circulation, long core lifetime, off-site refueling capability, and the use of standard equipment is the basic design features of this reactor.

Then, globally, the country with the most advanced Small Modular Reactor (SMR) project is China, where Chinergy is starting to build the 210 MW(e) high-temperature gas-cooled reactor pebble-bed module (HTR-PM), which consists of twin 250 MWt high-temperature gas-cooled reactors (HTGRs) which build on the experience of several innovative reactors in the 1960s to 1980s, as illustrated in Fig. 3.5.

The HTR-PM reactor uses helium as a coolant instead of water, where after the helium is heated to 750 °C (1382 °F), it is sent to a steam generator where it heats water until it becomes high-temperature steam, and that steam then flows into a steam turbine to generate electricity. The process uses a graphite-moderated nuclear reactor and a once-through uranium fuel cycle.

SNPTC's project as part of China Advanced HTGR Technology is consist of two 250 Mwt high-temperature reactor pebble-bed modules as illustrated in Fig. 3.5, which is located in Shandong province. Tests at the project are expected to end in April 2018, at which time the reactor will go into commercial operation as part of new generation of Advanced Small Modular Reactor (AdvSMR) series.

The success of this project will establish a milestone for the nuclear industry and its advocates to push this industry as an option for a renewable source of energy, where the SMRs are going to be utilized while helping decarbonization of our

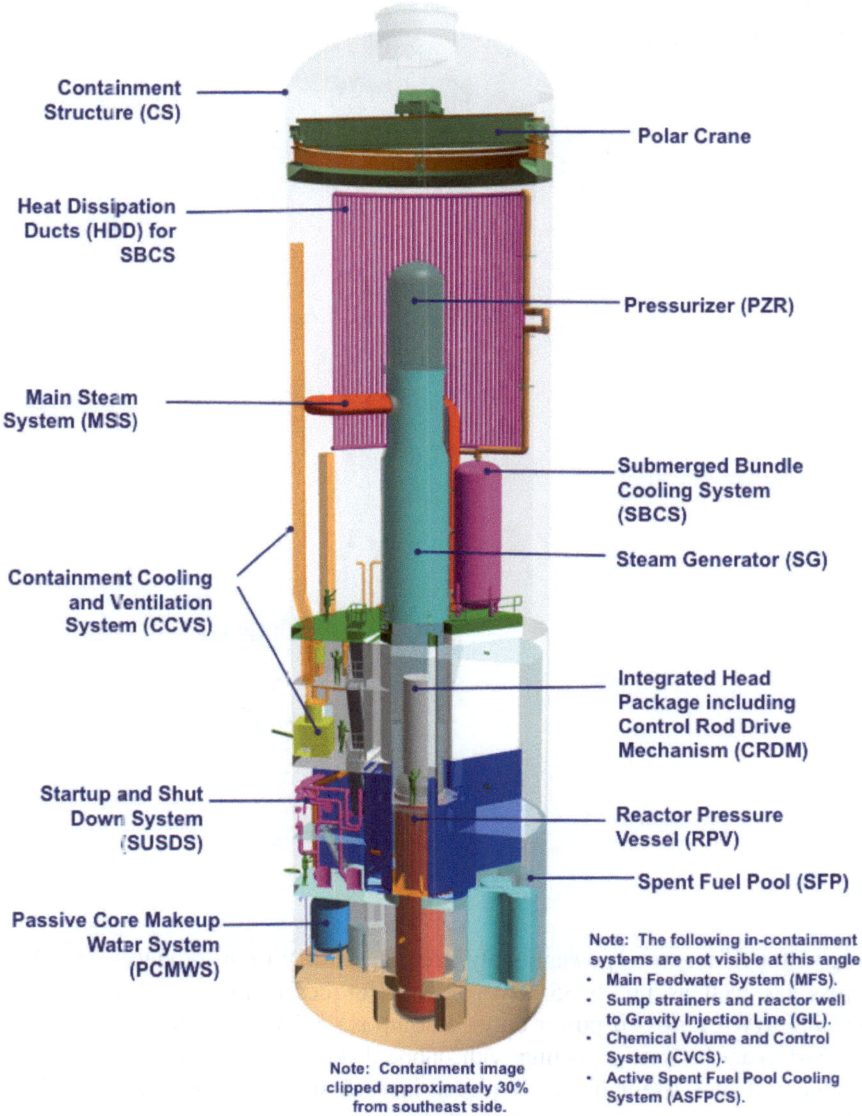

Fig. 3.2 Holtec International SMR-160

environment and pave the way to meet the demand on electricity due to global population growth and related industries to meet this growth.

Others such as Russian design KLT-40S (Fig. 3.6) from OKBM (full name for OAO I. I. Afrikantov OKB Mechanical Engineering. The company was founded in 1945. In 1998, it was named after its former chief designer and director I. I. Afrikantov). Afrikantov is derived from the KLT-40 reactor well in icebreakers and now—with low-enriched fuel—proposed for wider use in desalination and, on

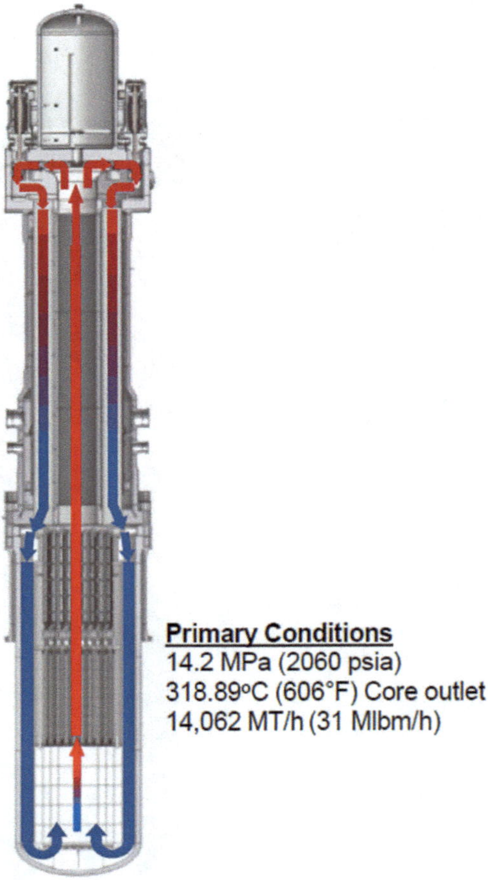

Fig. 3.3 mPower nuclear reactor depiction

barges, for remote area power supply. Here a 150 MWt unit produces 35 MW(e) (gross) as well as up to 35 MW of heat for desalination or district heating (or 38.5 MW(e) gross if power only). Burnup is 45 GWd/t. Units are designed to run 3–4 years between refueling with onboard refueling capability and used fuel storage. All fuel assemblies are replaced in each such refueling. At the end of a 12-year operating cycle, the whole plant is taken to a central facility for overhaul and storage of used fuel. Operating plant lifetime is 40 years. Two units will be mounted on a 20,000 ton barge to allow for outages (70% capacity factor). It may also be used in Kaliningrad.

Although the reactor core is normally cooled by forced circulation (four-loop), the design relies on convection for emergency cooling. Fuel is aluminum-uranium (Al_2U) silicide with enrichment levels of up to 20%, giving up to 4-year refueling intervals. A variant of this is the KLT-20, specifically designed for floating nuclear plants. It is a two-loop version with the same enrichment but with a 10-year refueling interval.

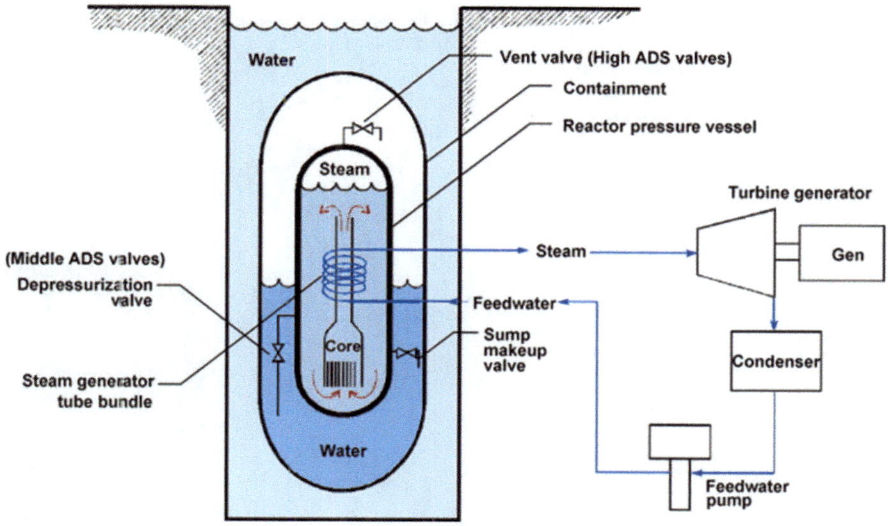

Fig. 3.4 Overview of multi-application small light water reactor

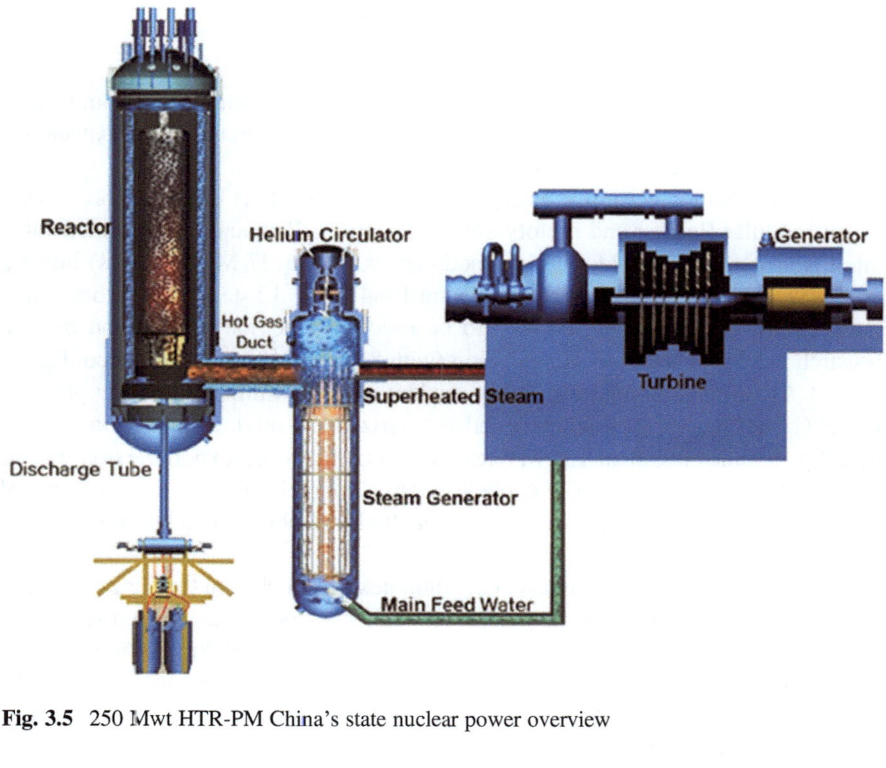

Fig. 3.5 250 Mwt HTR-PM China's state nuclear power overview

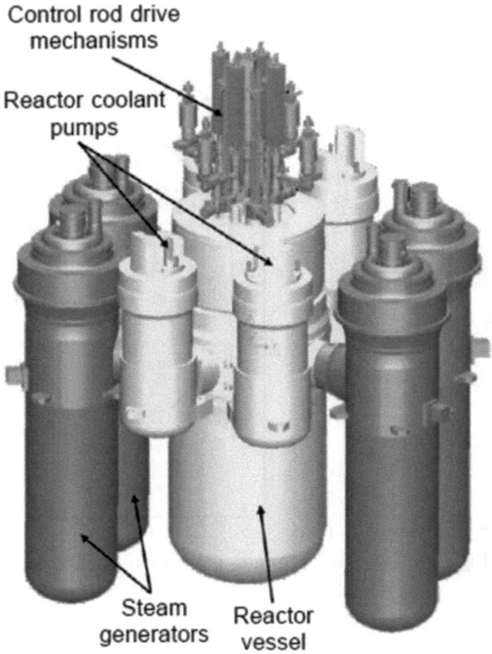

Fig. 3.6 KLT-40S, Afrikantov OKBM depiction

The first floating nuclear power plant, the *Akademik Lomonosov*, commenced construction in 2007. Due to insolvency of the shipyard, the plant is now expected to be completed in late 2018 and operational in 2019.

Argentina National Atomic Energy Commission (CNEA) is also involved with Small Modular Reactor and prototyping the CAREM-25 reactor, with considerable input from INVAP, an older design modular 100 MWt (27 MW(e) gross) integral pressurized water reactor, first announced in 1984. It has 12 steam generators within the pressure vessel and is designed to be used for electricity generation or as a research reactor or for water desalination (with 8 MW(e) in cogeneration configuration). CAREM has its entire primary coolant system within the reactor pressure vessel (11 m high, 3.5 m diameter), self-pressurized and relying entirely on convection (for modules less than 150 MW(e)). The final full-sized export version will be 100 MW(e) or more, with axial coolant pumps driven electrically. Fuel is standard 3.1- or 3.4%-enriched PWR fuel in hexagonal fuel assemblies, with burnable poison, and is refueled annually.

The 25 MW(e) prototype unit is being built next to Atucha, on the Parana River in Lima, 110 km northwest of Buenos Aires, and the first larger version (probably 100 MW(e)) is planned in the northern Formosa province, 500 km north of Buenos Aires, once the design is proven. Some 70% of CAREM-25 components will be locally manufactured. The pressure vessel is being manufactured by Industrias Metalurgicas Pescarmona S.A.I.CyF. (IMPSA). See Fig. 3.7.

The CAREM-25 is an integral-type PWR to produce 27 MW(e) developed by CNEA, and the reactor is adopting advanced features that include an integral primary

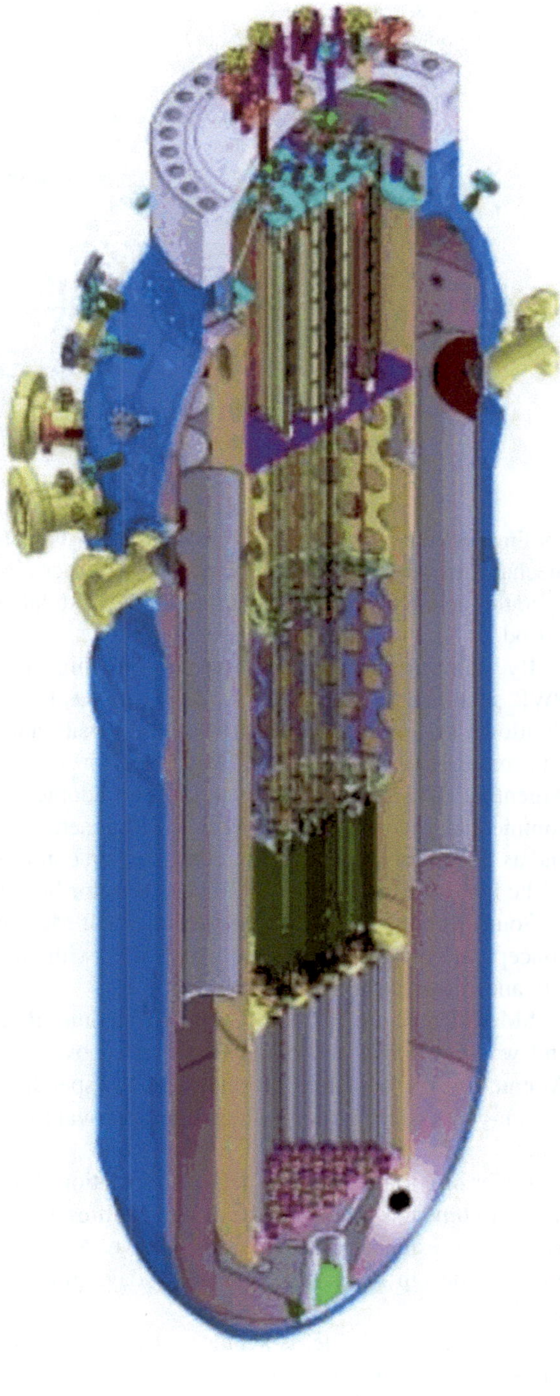

Fig. 3.7 Argentina CAREM-25 small modular reactor image

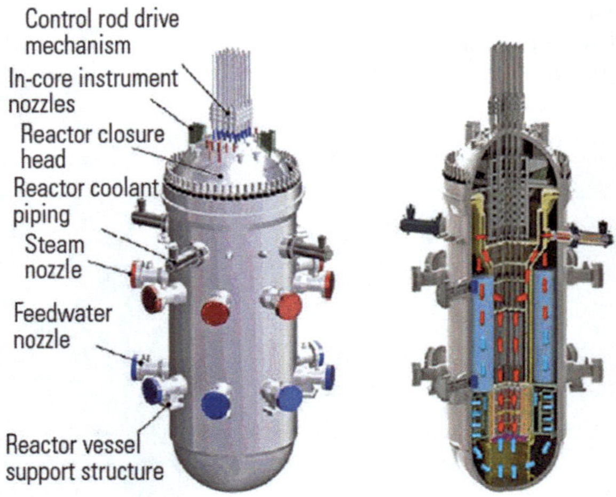

Fig. 3.8 SMART reactor depiction by South Korea

cooling system using natural circulation, in-vessel hydraulic control rod drive mechanisms, and passive safety systems. Neither air-conditioning (AC) power nor operator actions are required to mitigate postulated design events during grace period.

By adopting the integrated primary cooling system, the possibility of typical PWR-postulated design basis accident, such as large loss of coolant accidents and shutdown systems, are designated with diversity and redundancy that meet regulatory requirements. The CAREM-25 reactor is currently under construction in Argentina as a prototype to validate the adopted innovative features for future commercial version of CAREM that will generate up to 300 Mw(e) output power and as the prototype plant is targeting start-up commissioning in 2018, which is one of the many Advanced Small Modular Reactor built by other forging countries.

South Korea's SMART approach to Small Modular Reactor is beginning of the conceptual design is another AdvSMR with its System-Integrated Modular Advanced reactor as illustrated in Fig. 3.8.

SMART reactor is a 300 MWt Small Modular Reactor design that was firmed up and was developed and components tested over the next decade with the Korea Atomic Energy Research Institute (KAERI) spending some $300 million and 1500 person-years on the project, which won approval from the Korea Institute of Nuclear Safety.

According to the World Nuclear Association, SMART's design life is 60 years, fuel enrichment is 4.8%, and the design features a 3-year refueling cycle. As in many SMRs, the residual heat removal is passive. According to KAERI, the passive heat recovery design gives the plant a "20 days grace period against Fukushima-type accidents."

In the design, all fuel is submerged in water, and the containment building can withstand a crash from a Boeing 767. The containment also includes a passive hydrogen removal system to prevent hydrogen explosions.

Many of today's SMR plans have their roots in naval reactor technology, as did Shippingport. Its technology was based on Westinghouse reactors that powered the first US nuclear submarines. Argentina's CAREM-25 reactor design came from the Argentine navy. The country unveiled the design at a 1984 IAEA conference. The project then got shelved but was revived in 2006 as Argentina moved to revitalize its nuclear power program in the face of limited supplies and high prices for imported natural gas. Argentina has few easily accessible indigenous energy resources.

Russia's floating nukes also rely on maritime technology, reactors developed for its successful fleet of nuclear icebreakers, dating back well into the days of the Soviet Union. The nation's first nuclear icebreaker, the NS Lenin, was launched in 1957, the same year that Shippingport went into commercial service.

In the USA, two of the major SMR industrial developers, Babcock & Wilcox and Westinghouse, both have extensive experience with naval reactors. But that technology advantage has not provided commercial leverage, as both companies have scaled back their SMR programs in the face of a lack of demand for their product (see "What Went Wrong with SMRs?" in the September 2014 issue).

The Japan Atomic Energy Research Institute (JAERI) designed the MRX, a small (50–300 MWt) integral PWR reactor for marine propulsion or local energy supply (30 MW(e)). The entire plant would be factory-built. It has conventional 4.3%-enriched PWR uranium oxide fuel with a 3.5-year refueling interval and has a water-filled containment to enhance safety. Little has been heard of it since the start of the Millennium.

In summary, what accounts for the inability of the US (and European, for that matter) market to embrace SMR technology when less-developed and less-financially muscular countries and utilities are moving ahead?

Giorgio Locatelli of the UK's University of Lincoln, who published a recent paper on the economics of SMRs, "Small Modular Reactors: A Comprehensive Overview of Their Economics and Strategic Aspects," argues that the smaller reactors make sense in developing countries, where "it can be very tricky to get equity to make the investments. But with a small modular reactor, you build the first one, which comes cheaper, and then when you've raised more money you create the second, and then you start to sell electricity with the first and the second, and by selling electricity you can finance the construction of the third and then the fourth."

3.2 Heat Pipe Micro-reactor

It is worth mentioning here that there exists another distinct form of reactor that is different than other small reactor designs, namely, heat pipe reactor that is known as eVinci by Westinghouse.

Distinct from other small reactor designs, the eVinci is a heat pipe reactor, using a fluid in numerous sealed horizontal steel heat pipes to conduct heat from the hot fuel (where the fluid vaporizes) to the external condenser (where the fluid releases latent heat of vaporization) with heat exchanger. No pumps are needed to effect continuous

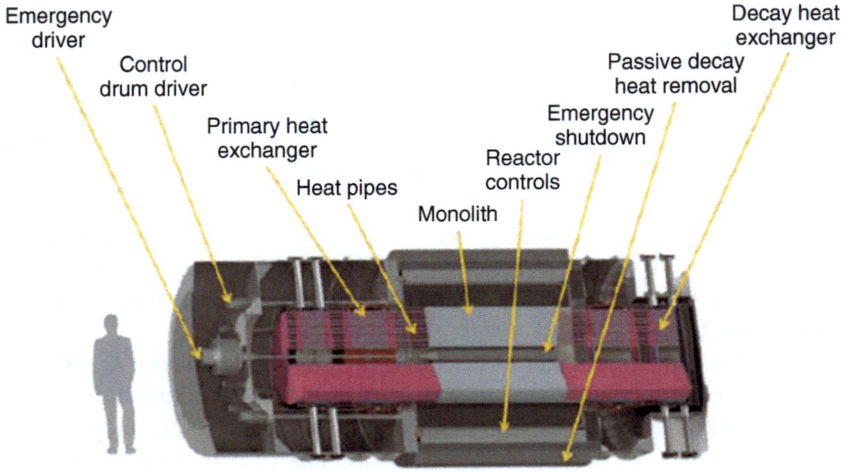

Fig. 3.9 Westinghouse eVinci heat pipe reactor

isothermal vapor/liquid internal flow at low pressure. The principle is well established on a small scale, but here a liquid metal is used as the fluid and reactor sizes up to several megawatts are envisaged. Experimental work on heat pipe reactors for space has been with much smaller units (about 100 kW(e)), using sodium as the fluid. They have been developed since 1994 as a robust and low technical risk system for space exploration with an emphasis on high reliability and safety.

The eVinci reactors would be fully factory built and fueled. As well as power generation, process heat to 600 °C would be available. Units would have 5- to 10-year operational lifetime, with walk-away safety due to inherent feedback diminishing the nuclear reaction with excess heat, also effecting load-following. Although eVinci is not considered as an Advanced Small Modular Reactor (AdvSMR), due to its small-scale size, it is a good candidate for space-based source of energy as mentioned in above and falls into the category of micro-reactor for lack of better classification choice. As illustrated in Fig. 3.9, the eVinci micro-reactor's innovative design is a combination of nuclear fission, space reactor technologies, and 50+ years of commercial nuclear system design, engineering, and innovation. The eVinci micro-reactor aims to create affordable and sustainable power with improved reliability and minimal maintenance, particularly for energy consumers in remote locations. The small size of the generator allows for easier transportation and rapid, on-site installation in contrast to large, centralized stations. The reactor core is designed to run for more than 10 years, eliminating the need for frequent refueling.

The key benefits of the eVinci micro-reactor are attributed to its solid core and advanced heat pipes. The core encapsulates fuel to significantly reduce proliferation risk and enhances overall safety for the user. The heat pipes enable passive core heat

Fig. 3.10 Various application of eVinci in off-grid mode

extraction and inherent power regulation, allowing autonomous operation and inherent load-following capabilities. These advanced technologies together make the eVinci micro-reactor a pseudo "solid-state" reactor with minimal moving parts.

The key attributes of eVinci micro-reactor are:

- Transportable energy generator
- Fully factory-built, fueled, and assembled
- Combined heat and power—200 kW(e) to 25 MW(e)
- Up to 600 °C process heat
- 5- to 10-year life with walkaway inherent safety
- Target less than 30 days on-site installation
- Autonomous load management capability
- Unparalleled proliferation resistance
- High reliability and minimal moving parts
- Green field decommissioning and remediation

As part of the feature of eVinci as a micro-reactor design, it could represent itself as an ultimate energy solution for the off-grid customer as depicted in Fig. 3.10.

The use of the heat pipes in a reactor system addresses some of the most difficult reactor safety issues and reliability concerns present in current Generations II and III and, to some extent, Generation IV concept commercial nuclear reactors—in particular, loss of primary coolant. Heat pipes operate in a passive mode at relatively low pressures, less than an atmosphere. Each individual heat pipe contains only a small amount of working fluid, which is fully encapsulated in a sealed steel pipe. There is no primary cooling loop, hence no mechanical pumps, valves, or large-diameter primary loop piping typically found in all commercial reactors today. Heat pipes simply transport heat from the in-core evaporator section to the ex-core condenser in continuous isothermal vapor/liquid internal flow. Heat pipes offer a new and unique means to remove heat from a reactor core.

3.3 High-Temperature Gas-Cooled Reactors/Advanced Small Modular Reactor

The high-temperature gas-cooled reactors (HTGRs) as illustrated in Fig. 3.11 design feature and operational temperature requirements are considered as family of an Advanced Modular Reactor (AdvSMR), which fit them into category where the innovative open-air Brayton or closed-CO_2 combined Brayton cycle takes advantages of the high temperature into account in order to boost thermal efficiency of these HTGRs to its maximum possible [1–3].

These types of Advanced Small Modular Reactors use graphite as moderator unless they are fast neutron-type AdvSMRs and either helium, carbon dioxide, or

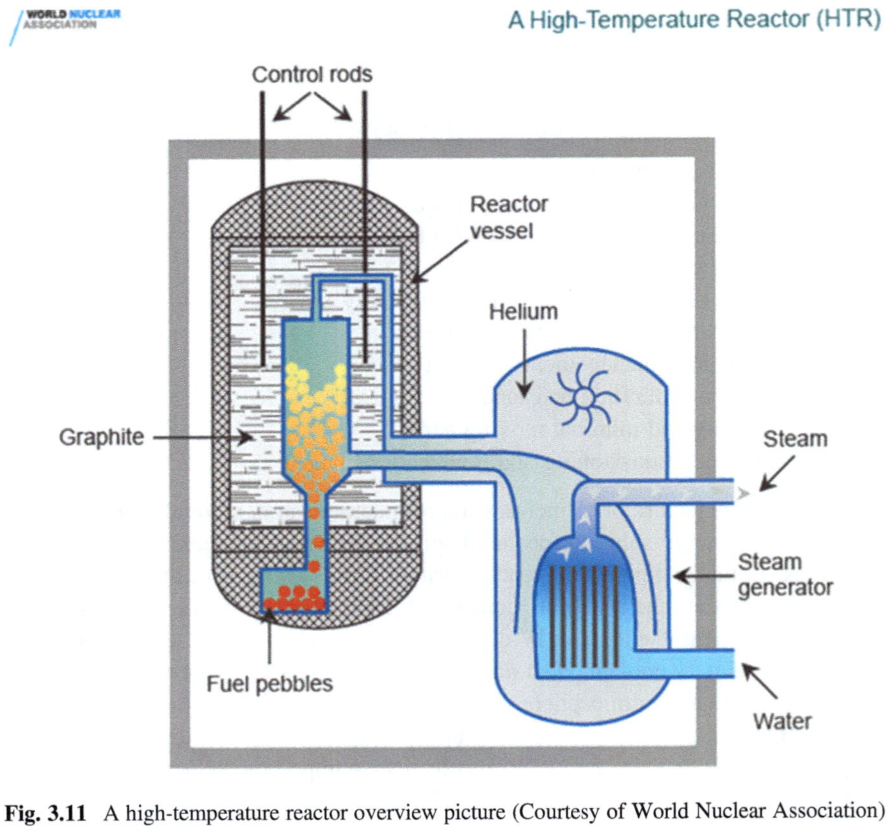

Fig. 3.11 A high-temperature reactor overview picture (Courtesy of World Nuclear Association)

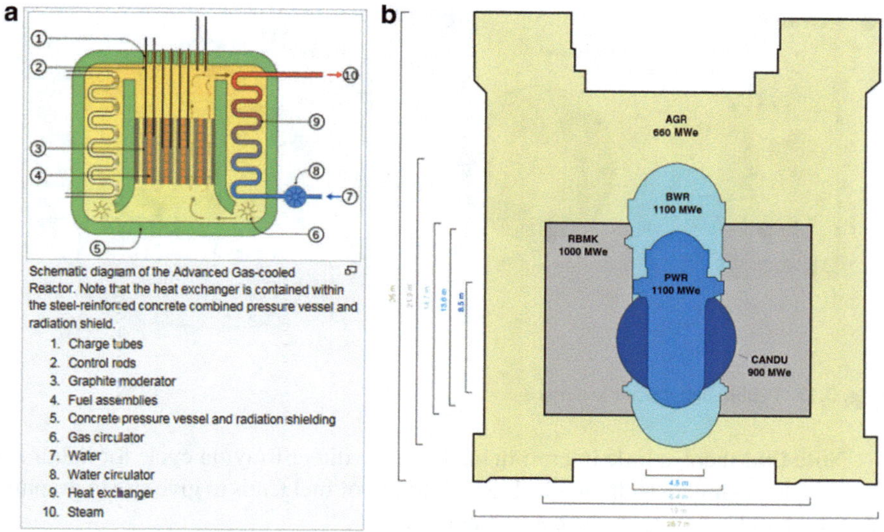

Fig. 3.12 Advanced gas-cooled reactor (AGR). (**a**) AGR components overview. (**b**) AGR size compared to other technologies

nitrogen as primary coolant. The experience with several innovative reactors built in the 1960s and 1970s, notably those in Germany, has been analyzed, especially in light of the US plans for its Next Generation Nuclear Plant (NGNP) and China launching its high-temperature reactor pebble module (HTR-PM) project in 2011. Lessons learned and documented for NGNP include the use of TRISO fuel, use of a reactor pressure vessel, and use of helium cooling (UK advanced gas-cooled reactors (AGRs) are the only high-temperature reactors (HTRs) to use CO_2 as primary coolant); see Fig. 3.12a and b. However, US government funding for Next Generation Nuclear Plant (NGNP) has now virtually ceased, and the technology lead has been passed to China these days.

New high-temperature gas-cooled reactors (HTGRs) are being developed which will be capable of delivering high-temperature (700–950 °C and eventually up to about 1000 °C) helium either for industrial application via a heat exchanger or to make steam conventionally in a secondary circuit via a steam generator or directly to drive a Brayton cycle gas turbine for electricity with almost 50% thermal efficiency possible (efficiency increases around 1.5% with each 50 °C increment) [1–4]. One design uses the helium to drive an air compressor to supercharge a Combined Cycle Gas Turbine (CCGT) unit. Improved metallurgy and technology developed in the last decade make HTRs more practical than in the past, though the direct cycle means that there must be high integrity of fuel and reactor components [4]. All but one of those described below have neutron moderation by graphite, one is a fast neutron reactor.

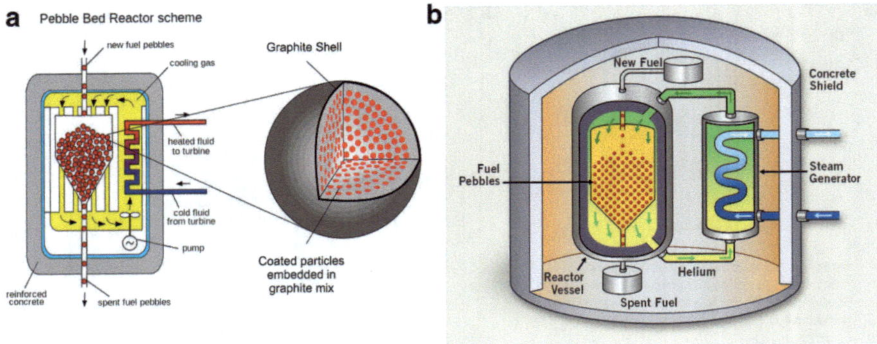

Fig. 3.13 Pebble-bed reactor schema

Note that there is little interest in pursuing the direct Brayton cycle for helium at present due to higher technological risk. Attrition of fuel tends to give rise to graphite dust with radioactivity in the coolant circuit.

Fuel for these reactors is in the form of TRISO (tristructural-isotropic) particles less than a millimeter in diameter. Each has a kernel (*ca.* 0.5 mm) of uranium oxycarbide (or uranium dioxide), with the uranium enriched up to 20% U-235, though normally less. This is surrounded by layers of carbon and silicon carbide, giving a containment for fission products which is stable to over 1600 °C.

There are two ways in which these particles are arranged: in blocks—hexagonal "prisms" of graphite—or in billiard ball-sized pebbles of graphite, each with about 15,000 fuel particles and 9g uranium. There is a greater volume of used fuel (20 times) than from the same capacity in a Light Water Reactor, due to the fact that the fuel pebbles are mainly graphite—less than one percent is uranium. However, the used fuel is overall less radiotoxic and produces less decay heat due to higher burnup. The HTR moderator is graphite.

HTRs can potentially use thorium-based fuels, such as highly-enriched or low-enriched uranium with Th, U-233 with Th, and Pu with Th. Most of the experience with thorium fuels has been in HTRs.

With negative temperature coefficient of reactivity (the fission reaction slows as temperature increases) and passive decay heat removal, the reactors are inherently safe. HTRs therefore are put forward as not requiring any containment building for safety. They are sufficiently small to allow factory fabrication and will usually be installed below ground level.

Three HTR designs in particular—pebble-bed modular reactor (PBMR) (Fig. 3.13a, b), gas turbine-modular helium reactor (GT-MHR) (Figs. 3.14 and 3.15), and AREVA's Steam Cycle High-Temperature Gas Reactor (SC-HTGR) (Fig. 3.16a, b)—were contenders for the Next Generation Nuclear Plant (NGNP) project in the USA. In 2012 AREVA's HTR was chosen. However, the only HTR project currently proceeding is the Chinese HTR-PM.

Hybrid power technologies have a hybrid-nuclear Small Modular Reactor (SMR) coupled to a fossil fuel-powered gas turbine.

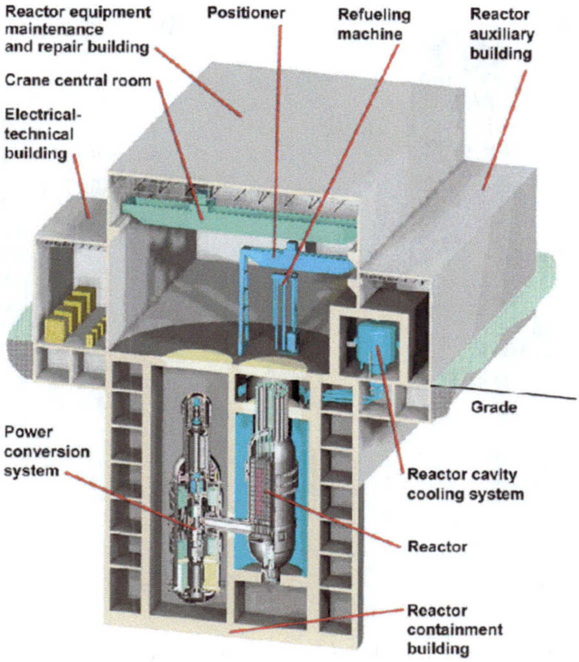

Fig. 3.14 Schematic of GT-MHR overview. M.P. LaBar, A.S. Shenoy, W.A. Simon, and E.M. Cambel, "Status of the GT-MHR for Electricity Production", World Nuclear Association Annual Symposium 2003, http://www.world-nuclear.org/sym./2003/fig-htm/labf2-h.htm

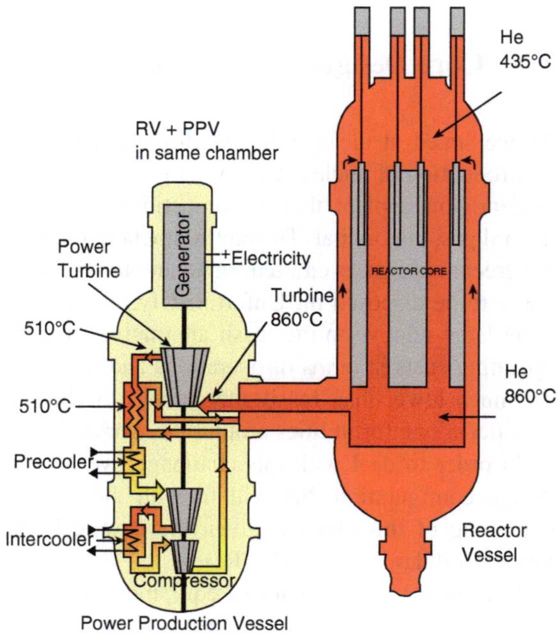

Fig. 3.15 GT-MHR layout

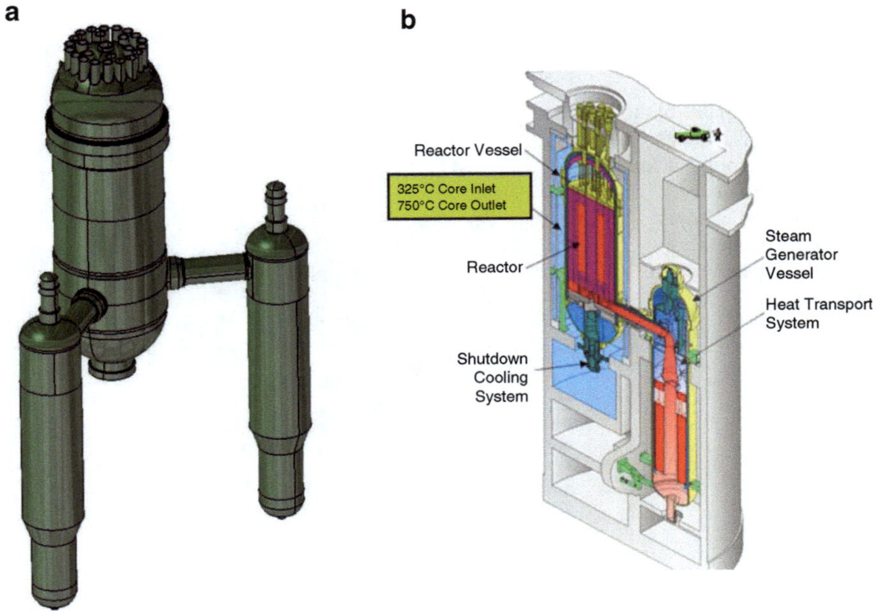

Fig. 3.16 AREVA's design layout of SC-HTGR. (**a**) Design certificate boundaries. (**b**) Reactor and steam generator general arrangement

3.4 Core Design and Analysis

As we stated at the opening of this chapter, one of the major challenges of the reintroduction of nuclear energy into the world energy mix is the development of a nuclear power plant that is competitive with other energy alternatives, such as natural gas, oil, or coal. The environmental imperative of nuclear energy is obvious. No greenhouse gases emitted, small amounts of fuel required, and small quantities of waste to be disposed of. Unfortunately, the capital costs of new nuclear plants are quite large relative to the fossil alternatives. Despite the fact that nuclear energy's operating costs in terms of operations and maintenance and, most importantly, fuel are much lower than fossil alternatives, the barrier of high initial investment is a significant one for utilities around the world.

In order to deal with this challenge, we need proper and suitable reactor core design configuration that will fit with these complex requirements stated at the beginning of this chapter, given the new technology that goes along with these Small Modular Reactors (SMRs).

Despite the use of standard equipment such as standard fuel geometry, in-core internals, the combination of features listed at the beginning of this chapter will require fuel characteristics for these SMRs and AdvSMRs that are different from traditional Light Water Reactor (LWR) fuel assemblies. The design of small, natural circulation LWR cores requires an understanding of the Reactor Physics, Thermal-hydraulic [5], *Neutronic Analysis for Nuclear Reactor* (i.e., this book) and an accurate and efficient set of computational tools and software and a well-developed design methodology.

The purpose of this section along with this chapter as a whole is to evaluate the effects of increased enrichment, burnup rates, increased leakage, and burnable poison distribution on reactor performance and to design in general a prototype core for the use in a small light water reactor from top level point of view; however, readers can find more detail-oriented information in a book by Walter [6].

However, if we use the state-of-the-art tools based on modeling of traditional Light Water Reactors, for example, vendors such as NuScale using their design approach to Small Modular Reactor technologies, it could be helpful for obtaining licensing operational process. Codes like RELAP which is used by Nuclear Regulatory Commission (NRC) for validation and verification of such design has been developed around traditional LWRs and can be easily applied with toward these types of SMRs, when it comes to safety analyses and shows that the core will remain covered during the entire transient procedure of design-basis from loss of coolant accident's point of view. It is important to notice that LWRs are confined under vapor dome as pointed out by Zohuri and McDaniel [1, 2].

Furthermore the above design approach and data generated through these simulations cover a new range of fuel enrichments, temperatures, pressure, power, and flow rates that are very important for an improved core design of new generation of nuclear power plants of GEN-IV from top point of view. An evaluation of the reactor control system options can also be performed in order to meet the stringent design criteria of these types Light Water Reactors (LWRs).

In such situation one can concentrate on the increased enrichment fuel, and analysis of phenomena related with the use of this fuel rate at a specific operational conditions from top point of view, where, for example, multi-application small light water reactor (MASLWR) may be considered. As part of first level of analysis, it can be done via neutron transport burnup code such as CASMO-4 or CASMO-5 for a fuel assembly with reflected boundary conditions. This analysis is able to evaluate characteristics of the MASLWR fuel and compare them with characteristics of conventional pressurized water reactor (PWR) fuel, and results are shown by Soldatov [7].

Note that Light Water Reactor Core Simulations using CASMO codes are able to the following analyses:

- Lattice physics codes (e.g., CASMO) are used to prepare problem-dependent few group cross sections based on a smaller region (e.g., fuel assembly) for input to a whole-core program (e.g., SIMULATE).
- The CASMO few group cross sections are curve fitted to functions of important variables such as burnup, moderator temperature and void content, fuel temperature, and soluble poison concentration.
- These workhorse codes (CASMO) are all deterministic. There is also growing use of stochastic "Monte Carlo" programs for these applications (MCNP will be

introduced in later lab sessions), but code running times are still excessive for many purposes.

The results of such above simulations will effects of the increased fuel enrichment and operational conditions onto:

1. Infinite lattice multiplication factor
2. The fuel isotopic content
3. Neutron energy spectrum
4. Reactor kinetic parameters
5. Reactivity feedbacks
6. Worth of the control rods and soluble boron

In addition the series of COSMO computer code modules will have the following abilities as well, and they are:

- Transport-burnup codes:
  - Infinite lattice, fuel cell, or assembly
  - Many energy groups
  - Region average cross-sectional data
  - Time (burnup) dependent
  - Operation condition dependent (T_f, T_m, Power, Xe, boron concentrations)
- Nodal diffusion codes:
  - 3D core power distribution
  - Two energy groups
  - Account for thermal feedback

Furthermore, former analyses mentioned above that, by use of the increased enrichment fuel in a reactor such as MASLWR core, it can lead to the high reserve of the reactivity for a fuel burnup. This reactivity reserve is compensating with a gadolinium burnable absorbers.

The self-shielding of the gadolinia in a MASLWR fuel, fuel shadowing, and control rod shadowing effects was evaluated for such type of reactor fuel and discussed by Soldatov [7].

Taking into consideration the core design of Small Modular Reactors, the small reactor core have a high neutron leakage based of traditional LWR core design approach. The evaluation of such neutron leakage on the multiplication factor and fuel utilization and discussion of the core reflector issues are also discussed in the reference by Soldatov [7].

In this case, the neutron leakage, use of the burnable absorbs and coupled neutronic, and thermal hydraulic analysis (i.e., Zohuri [5]) should also be taken into account during a core design as it was mentioned before.

Additionally, reactor theory uses will reveal the following technical points as well, and they are:

Table 3.1 Current code sets used for LWR reload analysis

Code supplier	Code set[a]	Users		
		Vendor	Licensee	Consultant
ABB-CE	PHOENIX/POLKA	✓	✓	
B&W/Framatome	CASMO[b]/NEMO	✓	✓	
General Electric	TGEBLA/PANACEA	✓	✓	
Siemens	CASMO[b]/MICROBURN	✓	✓	
Westinghouse	PHOENIX/ANC	✓	✓	
EPRI	CPM-2/NODE-P		✓	✓
EPRI	CPM-2/SIMULATE-E		✓	✓
EPRI	CPM-3/CORETRAN			✓
Studsvik	CASMO[b]/SIMULATE-3[b]		✓	✓
Scanpower[c]	HELIOS/RAMONA		✓	

[a]Many different versions of these codes are in use today
[b]Codes developed by Studsvik of America
[c]Scanpower has merged with Studsvik

- Slowing down:
  - Energy spectrum calculation
  - Homogenization of heterogeneous regions
- Transport:
  - Spatial power distribution where diffusion does not work
  - Preparation of region-averaged cross sections
- Diffusion:
  - Nodal methods for 3D core power distribution
- Burnup:
  - Evolution of nuclide concentrations with time

All these technical features possibly could be handled with computer codes such as CASMO code family.

Current computer code sets used for Light Water Reactor reload analysis are presented in Table 3.1.

However, bear in your mind that several applications in nuclear reactor physics and dynamics require three-dimensional modeling of the core neutronics. Such calculations are presently carried out using few-group nodal diffusion codes. Thus, the geometry in the diffusion calculation should be consisted of homogenized material regions or nodes. The input data for the calculation may include spectrum-averaged cross sections, kinetic parameters, and other group constants, which are generated in such a way that the integral reaction rate balance is preserved within each node [8].

In summary, CASMO-5 is the latest code version in the Studsvik Scandpower series of lattice physics codes for general LWR analysis and has capabilities, features, and numerical models not previously available in CASMO-4. Among the new capabilities of CASMO-5 are a "quadratic gadolinium depletion model" and the ability to perform 2D transport calculations in hundreds of energy groups. The description of CASMO-5 capabilities and the role of CASMO-5 in generating cross section and discontinuity factor data for SIMULATE-4 can be found in the web site of the vendor of the code, where also it describes the new CASMO-5 data libraries (586 group neutron and 18 group gamma), and interested readers should be refereeing themselves to that site.

3.5 Small Modular Reactors' General Concepts

The driving forces in the development of Small Modular Reactors (SMRs) are their specific characteristics. They can be deployed incrementally to closely match increasing energy demand resulting in a moderate financial commitment for countries or regions with smaller electricity grids. SMRs show the promise of significant cost reduction through modularization and factory construction which should further improve the construction schedule and reduce costs. In the area of wider applicability SMR designs and sizes are better suited for partial or dedicated use in non-electrical applications such as providing heat for industrial processes, hydrogen production, or sea-water desalination. Process heat or cogeneration results in significantly improved thermal efficiencies leading to a better return on investment. Some SMR designs may also serve niche markets, for example, to burn nuclear waste.

As it was stated in former few sections of this chapter, Small Modular Reactors (SMRs) are designed with a potential of providing clean and cost-effective energy. As per the International Atomic Energy Association (IAEA) classification, SMRs are defined as reactors which produce power output of less than or equal to 300 MW(e); but in general any reactor with an electrical output less than 700 MW is considered as a SMR. Based on the characteristics, they are further categorized into three types:

1. Light Water Reactors (LWRs)
2. High-temperature gas-cooled reactors (HTGRs), and
3. Liquid metal (i.e., Fast Sodium Reactors (FSRs) and gas-cooled reactors (GCRs)

As shown in Fig. 3.17, the scalability, modularity, robust design, and enhanced safety features of the SMR offer great advantages over large commercial reactors' modularity and flexibility: In contrast to large reactors, SMRs can be fabricated and assembled in a factory environment and then transported to the nuclear power site. This will help limit the on-site preparation and also reduce the lengthy construction duration further reducing the construction cost and making the SMRs 20–30% less expensive [9]. In addition, the scalability and flexibility features of SMR also make it more suitable for small isolated areas with low-energy demands, limited infrastructure, and smaller and less established grid system.

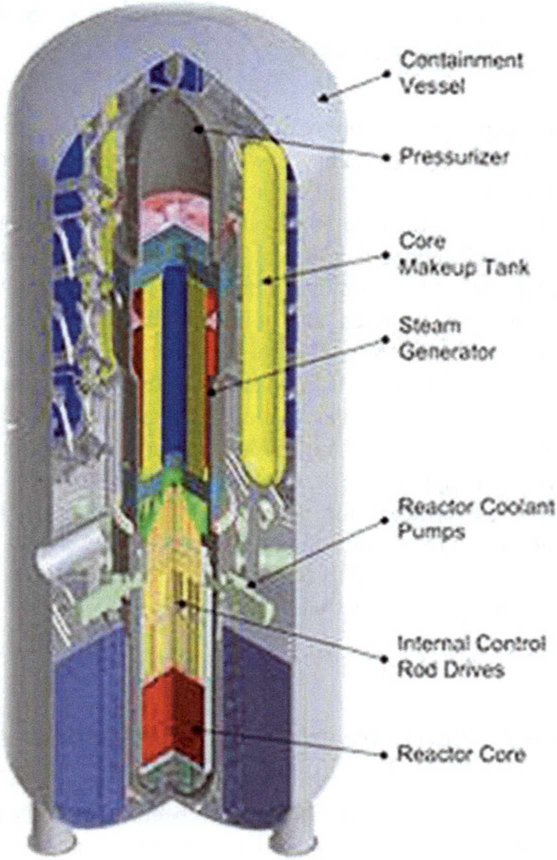

Fig. 3.17 Isometric section of Small Modular Reactor (Courtesy of Nuclear Regulatory Commission)

3.5.1 Modularity and Flexibility

In contrast to large reactors, SMRs can be fabricated and assembled in a factory environment and then transported to the nuclear power site. This will help limit the on-site preparation and also reduce the lengthy construction duration further reducing the construction cost and making the SMRs 20–30% less expensive [9]. In addition, the scalability and flexibility features of SMR also makes it more suitable for small isolated areas with low-energy demands, limited infrastructure, and smaller and less established grid system.

3.6 Safety Features and Licensing of Small Modular Reactors (SMRs)

Nuclear energy can play a very significant long-term role for meeting the world's increasing energy demands while simultaneously addressing challenges associated with global climate and environmental impact. Many nations of the world, particularly the Asia/Pacific Rim countries, are actively engaged in a major expansion of their nuclear energy complex. The degree to which nuclear energy can address long-term energy needs, either globally or regionally, will be dictated by the pace and adequacy of technical and policy solutions for waste, safety, security, and nonproliferation issues, as well as the capital cost of construction. Small Modular Reactors (SMRs) could successfully address several of these issues. SMRs offer simpler, standardized, and safer modular design by being factory-built, requiring smaller initial capital investment and having shorter construction times. The SMRs could be small enough to be transportable, could be used in isolated locations without advanced infrastructure and without power grid, or could be clustered in a single site to provide a multi-module, large capacity power plant.

Both topics of safety and licensing, in general, are described as below and divided into two general summaries as follows.

3.6.1 Safety Features of Small Modular Reactors

The safety aspects of SMRs have been intensively in several recent publications, mostly originating from the International Atomic Energy Agency (IAEA), and they all can be found in IAEA web site, which are summarized below. However, one should keep in mind that the safety features of SMRs will be reanalyzed following the Fukushima Daiichi accident in order to take into account the lessons learned from it.

The major findings regarding Small Modular Reactor safety are as follows:

- The design manufactures of Advanced Small Modular Reactors (AdvSMRs) aim to implement safety design options with maximum use of inherent and passive safety feature, which is also referred to as "by design" safety features.
- On their own, the "by design" safety features used in SMRs are in most cases not size-dependent and could be applied in the reactors of large capacity. However, SMRs offer broader possibilities to incorporate such features with higher efficiency.
- In the case of some technologies such as high-temperature gas reactors, the incorporation of passive safety features limits the reactor capacity.
- All of the SMR designs considered with an aim to meet international safety norms and standard set by IAEA, such as those formulated in the IAEA Safety Standard NS-R-1 Safety of the Nuclear Power Plants, which includes design requirements,

regarding implementation of the defense-in-depth strategy and provision of redundant and diverse *Active and Passive Safety Systems*.

- The available information on safety features of Advanced SMRs for plant protection against the impacts of natural and human-induced events is generally sparser compared to that on internal events.
- The core damage frequencies (CDFs) indicated by the designers of Advanced SMRs are comparable to, or even lower than the ones indicated for the state-of-the-art, large water-cooled reactors (LWCRs).

Note that in case of *Passive Safety System*, the integral design of Small Modular Reactor makes it safer in case of any severe accidents preventing any radiation leak into the environment. The passive safety system is another important safety feature in the SMRs. In case of loss of coolant accident, this system shuts down the reactor and cools it without any human intervention or air-conditioning (AC) power for a period of 7 days. The safety system incorporates an on-site water inventory, which operates on natural forces (i.e., natural circulation or gravity or compressed gas) [10].

3.6.2 Licensing of Small Modular Reactors

The licensing of SMRs will be affected by the Fukushima accident in the same way as for large reactors that are in place today. Regarding licensing status and regulatory issues relevant to SMRs, the analysis of recent publications leads to the following observations as listed here:

- According to the vendors and designers of the Advanced Small Modular Reactors (AdvSMRs) listed in Table 3.2, they have been designed or are being designed in compliance with their current national (i.e., imposed by NRC) and international (i.e., imposed by IAEA) regulations.
- The SMRs available for deployment, which are the CANDU-6, the PHWR, the QP-300, the CNNP-600, and the KLT-40S, have already completed the licensing procedures in the countries of origin. The CANDU-6 and the QP-300 have also been licensed and deployed in countries other than the country of origin.
- For Advanced SMR designs, three of them are in a formal licensing process in Argentina, China, and the Republic of Korea (i.e., South Korea) and several others are in pre-licensing negotiations in the United States and India as indicated in Table 3.2.

However, government support for licensing of selected, Advanced SMRs (AdvSMRs) could help overcome the corresponding delays.

Another important set of regulatory requirements concern the ability of SMRs to resist nuclear proliferation. All advanced light water PWR SMRs use conventional low-enriched uranium (LEU) fuel, and most of the PWR SMR designs use the *same* fuel as large PWRs. However, particular attention should be paid to the

Table 3.2 Design status and potential timeframes for deployment of advanced SMRs

SMR	Technology family	Electric output, MW(e)	Plant configuration	Design status	Licensing status/ completion (application) date	Targeted deployment date
KLT-40S, Russia	PWR	2 × 35	Twin-unit barge-mounted plant	Detailed design completed	Licensed/Under construction	2013
VBER-300, Kazakhstan, Russia	PWR	302	Single module or twin-unit, land-based or barge-mounted plant	Detailed design nearly completed.	n/a	>2020
ABV, Russia	PWR	2 × 7.9	Twin-unit barge-mounted or land-based plant	**Barge-mounted plant:** detailed design completed **Land-based plant:** detailed design for plant modification in progress	Part of design licensed	2014–2015
CAREM-25, Argentina	PWR	27	Single module land-based plant	Detailed design being finalised	Licensing in progress/ 2011	Prototype: 2015
SMART, Republic of Korea	PWR	90	Single module land-based plant	Detailed design in progress	Licensing in progress/ 2011	~2015
NuScale, USA	PWR	12 × 45	Twelve-module land-based plant	Detailed design being finalised	Licensing pre-application/ (Application: 2011)	FOAK in 2018
mPower, USA	PWR	×125	Multi-module land-based plant	Detailed design in progress	Licensing pre-application/ (Application: 2011)	~2018
IRIS[a], USA	PWR	335	Single module or twin-unit land-based plant	Basic design completed and is under review by the vendor		
Westinghouse SMR	PWR	>225				
HTR-PM, China	HTGR	2 × 105	Two-module land-based plant	Detailed design completed	Licensing in progress/ 2010 or 2011	FOAK in 2013

AHWR, India	Advanced heavy water reactor	300	Single module land-based plant	Detailed design being finalised	Licensing pre-application/ (Application: 2011)	~2018
SVBR-100, Russia	Pb-Bi cooled fast reactor	×101.5	Single module or multi-module land-based or barge-mounted plant	Detailed design in progress.	n/a/Prototypes have operated in Russian submarines	Prototype: 2017
New Hyperion power Module, USA	Pb-Bi cooled fast reactor	×25	Single module or multi-module land-based plant	n/a	Licensing pre-application/ (Application: not known)	FOAK by 2018
4S, Japan	Na cooled fast reactor	10	Single module land-based plant	Detailed design in progress.	Licensing pre-application/ (Application: 2012)	FOAK after 2014

[a]Late in 2010 the Westinghouse Electric Company stopped the development of the IRIS project and announced it would go with an alternative integral design PWR of a 200 MW(e) class. Very few technical details of this new SMR were available as of 2011

nonproliferation potential of some heavy-water or liquid metal-cooled designs, especially if they are intended to be deployed in politically unstable areas. The IAEA has an on-going activity on the options of incorporation of intrinsic proliferation resistance features in nuclear power plants (NPPs) with innovative Small Modular Reactors (SMRs).

3.6.3 Nonproliferation Resistant and Security

The SMR is a sealed unit built below grade thus making it safer against any terrorist activities or aircraft impact or any vulnerabilities due to natural phenomenon. They are also designed to operate for longer periods without refueling (i.e., approximately 18–24 months); and the reactor can be refueled in a factory environment and then transported back to site thus securing it against any proliferation issues.

3.7 Small Reactor Designs in Market

Carbon emission concerns and volatility in fossil fuel resources have renewed worldwide interest in nuclear energy as a solution to growing energy demands. Several large nuclear reactors are currently under construction in the United States, representing the first new construction in over 30 years. Small Modular Reactors (SMRs) have been in design for many years and offer potential technical and economic advantages compared with traditionally larger reactors. Current SMR capital and operational expenses have a wide range of uncertainty.

According to the Organisation for Economic Co-operation and Development/ Nuclear Energy Agency (OECD/NEA) estimates, nuclear power plants (NPPs) whether with a large reactor or with Small Modular Reactors (SMRs), with electric output of less than 300 MW(e), using light-water technology, are competitive with many other electricity generation technologies in the large majority of cases, the exceptions being natural gas in the United States with the current level of prices and large hydro.

However, SMRs, including multi-module plants, may have higher values of levelized cost of electricity (LCOE) than NPPs with large reactors. The LCOE for an SMR should decrease with large-scale serial production, which is the key element for proving the competitiveness of SMRs. However, large initial orders of SMRs are needed to launch the serial production process, and it is important to know who could be the first customers, and how many SMR designs will really be deployed in the near future.

Today, the world's major reactor vendors have targeted markets in developed countries and currently offer designs that have large power outputs (1000–1700 MW(e)) [11]. However, these large reactors are unsuitable for many developing countries for several reasons [12].

As with large reactors, the market for SMRs is difficult to estimate. However, using the commonly accepted assumption that SMRs could be competitive with many nonnuclear technologies for generating electricity, in the cases when NPPs with large reactors are, for whatever reason, unable to compete, one can try to identify the key market opportunities.

However, many developing countries have limited electric grid capacity that cannot accommodate a single power plant with output approaching or exceeding 1000 MW(e). Also, the grid in some countries is localized in a few isolated population centers with minimal interconnections. This situation favors the use of smaller power plants sited at geographically separated locations.

As we know about the nuclear power industry so far, nuclear power plants traditionally have a large capital cost relative to fossil power plants, which creates an additional barrier to choosing nuclear power. By virtue of their reduced size and complexity, smaller-sized nuclear plants will have a lower capital cost per plant and shorter construction time. Thus, the initial power unit can be generating revenue before the second and third units are constructed, reducing the maximum capital outlay for the combined generating capacity. This is especially important for developing economies, which typically have limited availability of capital funds.

Because of the lower power levels of small- or medium-sized nuclear plants, countries have more flexibility to install generating capacity in smaller increments that better match their rate of power demand and economic growth. The reduced power levels allow greater use of passive safety systems and plant simplifications, such as natural circulation of the primary coolant. These features enhance the safety and reliability of the nuclear power facility, which is especially advantageous in countries that have limited nuclear experience and trained workforces.

Current, worldwide nuclear energy is based on medium and high-capacity level reactors [13] (IAEA classification of capacity levels 300–900 and 900–1600 MW(e), respectively [14]). The primary technology for operational nuclear reactors worldwide is Light Water Reactor (LWR) technology [14–16]. The world "technology" here does not specifically refer to the nuclear power plant design but also the LWR-oriented fuel cycle facilities and comprehensive supply chain. LWR technology is well studied and well reflected in regulatory practice [17–21]. This makes LWR technology most attractive for innovative reactor design in the short-term and mid-term timeframes.

Several countries are currently working on the design of innovative nuclear power plants in the low-power range (10–300 MW(e)). There are a variety of design concepts being considered worldwide (such as high-temperature gas-cooled reactors and liquid metal fast breeder reactors) [14, 22–24]. However, LWRs are considered the most economically feasible and deployable technology for innovative reactors in the near future [25].

In order to design a competitive reactor and successfully promote into international markets, the overview of a close competitor is necessary. In order to perform this analysis, first of all, it is necessary to recognize and list all major reactor design concepts (projects, designs) being proposed during the last decade, identify major advantages and disadvantages of the proposed designs, identify designs which has similarities with multi-application small light water reactor (MASLWR), and try to understand the ideas and motivation behind these designs. Learning the difficulties and unsuccessful designs of small reactors (particularly LWRs) is also important in order to avoid the repetition of the other's mistakes and potential challenges in the early design stage. The general overview of the competing designs is given for large-scale innovative LWRS [26–30] and for different designs of small reactors [25] including innovative LWR. Detailed discussions of the innovative reactor designs are given in the sources [31–39].

3.8 Conclusions

A principal conclusion of this chapter is that Small Modular Reactors (SMRs) have a significant potential to expand the peaceful applications of nuclear power by catering to the energy needs of those market segments that cannot be served by conventional nuclear power plants (NPPs) with large reactors. Such segments could be:

- Niche applications in remote or isolated areas, where large generating capacities are not needed, the electrical grids are poorly developed or absent, and where the non-electrical products such as heat or desalinated water are as important as the electricity.
- Replacement for the decommissioned small- and medium-sized fossil fuel plants, as well as an alternative to newly planned such plants, in the cases when certain siting restrictions exist, such as limited free capacity of the grid, limited spinning reserve, and/or limited supply of water for cooling towers of a power plant.
- Replacement for those decommissioned fossil-fueled combined heat and power plants, where the SMR power range seems to better fit the requirements of the currently existing heat distribution infrastructure.
- Power plants in liberalized energy markets or those owned by private investors or utilities for whom small upfront capital investments, short on-site construction time with the accordingly reduced cost of financing, and flexibility in plant configuration and applications matter more than the levelized unit electricity cost.

It should be noted, however, that none of the smaller reactors have yet been licensed for these applications, and there remain both development challenges to overcome and regulatory approvals to obtain before deployment, especially in light of the recent accident at Fukushima.

These observations in this chapter have found no situations, where NPPs with SMRs could compete with the NPPs with state-of-the-art large reactors, on low-enriched uranium (LEU) basis. However, it also found that **SMRs could be competitive with many nonnuclear technologies in the cases, when NPPs with large reactors are, for whatever reason, unable to compete.**

References

1. Zohuri, B. (2015). *Combined cycle driven efficiency for next generation nuclear power plants: An innovative design approach* (1st ed.). New York: Springer.
2. Zohuri, B., & McDaniel, P. (2017). *Combined cycle driven efficiency for next generation nuclear power plants: An innovative design approach* (2nd ed.). New York: Springer.
3. Zohuri, B. (2018). *Small modular reactors as renewable energy sources*. New York: Springer.
4. Zohuri, B., McDaniel, P., & De Oliveira, C. (2015). Advanced nuclear open-air-Brayton cycles for highly efficient power conversion. *Nuclear Technology Journal*.
5. Zohuri, B. (2017). *Thermal-hydraulic analysis of nuclear reactors* (2nd ed.). New York: Springer.
6. Walter, A. E. (1981). *Fast breeder reactors*. Oxford: Pergamon Press.
7. Soldatov, A. (2009). *Design and analysis of a nuclear reactor core for innovative small light water reactors*. A Dissertation Submitted to Oregon State University, March 9, 2009.
8. Leppänen, J. (2005). NEW ASSEMBLY-LEVEL MONTE CARLO NEUTRON TRANSPORT CODE FOR REACTOR PHYSICS CALCULATIONS. In *Mathematics and Computation, Supercomputing, Reactor Physics and Nuclear and Biological Applications Palais des Papes, Avignon, France, September 12–15, 2005, on CD-ROM, American Nuclear Society, LaGrange Park, IL*.
9. Nuclear Energy Association (NEA). (2011, June). *Current status, technical feasibility and economics of small modular reactors*.
10. Kozlowski, T., & Downar, T. J. (2003, December). *OECD/NEA AND U.S. NRC PWR MOX/UO2 Core Transient Benchmark*.
11. International Atomic Energy Agency. (1999). IAEA-TECDOC-1117. Evolutionary water-cooled reactors: Strategic issues, technologies and economic viability. In *Proceedings of a Symposium, Seoul, 30 November–4 December 1998*. Vienna: IAEA. ISSN 1011-4289.
12. Ingersoll, D. T., & Poore, W. P., III. (2007). *Reactor technology options study for near-term deployment of GNEP grid-appropriate reactors*. ORNL/TM-2007/157, ORNL, Oak Ridge, TN, September 26, 2007.
13. *2005 World nuclear industry handbook*. (2005). Nuclear Engineering International.
14. IAEA CSP-14 (Parts 1–5). (2002). Small and medium sized reactors: Status and prospects. In *Proceeding of International Seminar, Cairo, Egypt, May 27–31, 2001*. Vienna: IAEA.
15. IAEA PUB 1280. (2006). *Operating experience with nuclear power stations in member states in 2005*. Vienna: IAEA.
16. *Nuclear energy today. Nuclear development*. NEA/OECD. (2003).
17. *European utility requirements for LWR nuclear power plants* (Vol. 1: Main policies and objectives. Revision C, EUR-Club, 2001). Chapter 1-6 with list of acronyms and definitions section.
18. *European utility requirements for LWR nuclear power plants* (Vol. 2: Generic nuclear island requirements. Revision C, EUR-Club, 2001). Chapter 0-19 with list of acronyms and definitions section annexes.
19. *European utility requirements for LWR nuclear power plants* (Vol. 3: Application of EUR to the specific projects. Revision C, EUR-Club, 2001). Chapter 0-19 with list of acronyms and definitions section, annexes.
20. *European utility requirements for LWR nuclear power plants* (Vol. 4: Generic conventional island requirements. Revision B, EUR-Club, 2000). Chapter 0-19 with list of acronyms and definitions section annexes.
21. 10 CFR Part 52. *Domestic licenses, certifications, and approvals for nuclear power plants*. US Nuclear Regulatory Commission, Electronic Publication. Retrieved from http://www.nrc.gov/reading-rm/doc-collections/cfr/part052/.
22. International Atomic Energy Agency. (2006). IAEA-TECDOC-1485. *Status of innovative small and medium sized reactor designs 2005. Reactors with conventional refueling schemes*. Vienna: IAEA. ISSN 1011-4289.

23. IAEA. (2007). Progress in design and technology development for innovative small and medium sized reactors. In *Proceeding of IAEA 51st General Conference, September 17, 2007*. Vienna: IAEA.
24. IAEA. (2007). Current trends in nuclear fuel for power reactors. In *Proceeding of IAEA 51st General Conference, September 17, 2007*. Vienna: IAEA.
25. International Atomic Energy Agency. (2004). IAEA-TECDOC-1391. *Status of advanced light water reactor designs 2004*. Vienna: IAEA.
26. International Atomic Energy Agency. (2002). IAEA-TECDOC-1290. *Improving economics and safety of water-cooled reactors: Proven means and new approaches*. Vienna: IAEA. ISSN 1011-4289.
27. Kupitz, J., Bussurin, Y., & Gowin, P. *Considerations related to specific concepts: Water-cooled reactors, gas-cooled reactors, metal-cooled reactors, non-conventional reactors*. Retrieved from http://www.mi.infn.it/~landnet/Doc/Reactors/kupitz.pdf.
28. IAEA. (2003). *Overview of global development of advanced nuclear power plants*. Annex 1 (pp. 1094–1110). Vienna: IAEA.
29. IAEA. (2006). *IAEA overview of global development of advanced nuclear power plants. Information NPTDS brochure*. Vienna: IAEA.
30. Guerrini, B., & Paci, S. (1998). *Lessons on nuclear plants part IIB: Advanced reactors*. Training materials, RL (811) 99, University of Pisa.
31. Fisher, J. E., Modro, S. M., Weaver, K. D., Reyes, J. N., Jr., Groome, J. T., & Bapka, P. (2001). Performance and safety studies for Multi-Application, Small, Light Water Reactor (MASLWR). In *Proceedings of RELAP5 International Users Seminar, Sun Valley, ID, September 5–7, 2001*.
32. Fisher, J. E., Modro, S. M., Weaver, K. D., Reyes, J. N., Jr., Groome, J. T., & Bapka, P. (2002). Performance and safety studies for Multi-Application, Small, Light Water Reactor (MASLWR). In *Proceedings of RELAP5 International Users Seminar, Park City, UT, September 4–6, 2002*.
33. Modro, S. M., Fisher, J. E., Weaver, K. D., Reyes, J. N., Jr., Groome, J. T., Bapka, P., & Wilson, G. (2002). Generation-IV Multi-Application Small Light Water Reactor (MASLWR). In *10th International Conference on Nuclear Energy, INEEL/CON-02-00017, April 14, 2002*.
34. Modro, S. M., Fisher, J. E., Weaver, K. D., Reyes, J. N., Jr., Groome, J. T., Bapka, P., & Carlson, T. M. (2003, December). *Multi-application small light water reactor final report*. INEEL/EXT-04-01626.
35. Reyes, J. N., Groome, J., Woods, B. G., Young, E., Abel, K., Yao, Y., & Yoo, Y. J. (2007). Testing of the Multi-Application Small Light Water Reactor (MASLWR) passive safety systems. *Nuclear Engineering and Design, 237*, 1999–2005.
36. Woods, B. G., Reyes, J. N., & Wu, Q. (2005). *Flow stability testing under natural circulation conditions in integral type reactors*. Presentation at IAEA Natural Circulation Research Coordination Meeting, Corvallis, OR, September 1, 2005.
37. Soldatov, A., Marcum, W., Magedanz, J., Nelson, K., & Dahl, J. (2007, March). *Advanced thermal accessible reactor final report*. Course NE 574—Design Project Report, OSU.
38. Soldatov, A., Magedanz, J., & Dahl, J. (2007, June). *MASLWR extended core live design and preliminary instrumentation*. Course NE 575—Design Project Report, OSU.
39. Galvin, M., & Soldatov, A. (2007, March). *Multi-Application Small Light Water Reactor (MASLWR), Global Viability*. Course ECE599—Design Project Report, OSU.

Chapter 4
Thermodynamic Cycles

This chapter focuses on the turbine cycle, thermodynamics, and heat engines, where briefly are presented to remind readers about basic knowledge of these subjects. Further studies resources are provided in the reference section of this chapter [1, 2].

4.1 Introduction

Thermodynamic cycle is a process that series of them causes changing in the volume, temperature, and pressure of a gas at different stage of these process and series but which at the end of the cycle return to conditions as at the start. Understanding of these thermodynamic cycles is important for purpose of combined cycle's application with new generation of nuclear power plants and how they impact to improve the thermal efficiencies turbo machinery, and they can be repeated indefinitely. Understanding of thermodynamic cycles was extremely important to the industrial revolution, and they remain keys to most large-scale manufacturing processes, most engines, refrigerators, and air conditioners.

4.2 Work

The infinitesimal amount of work done by a system as shown in Fig. 4.1 below is

$$\begin{aligned} dW &= Fdx \\ &= PAdx \\ &= PV \end{aligned} \tag{4.1}$$

In Fig. 4.2 below, the shaded area represents the work done in moving from state **a** to state **b** as it can be seen in that figure as a simple *P-V* diagram , i.e.,

B. Zohuri, *Heat Pipe Applications in Fission Driven Nuclear Power Plants*,
https://doi.org/10.1007/978-3-030-05882-1_4

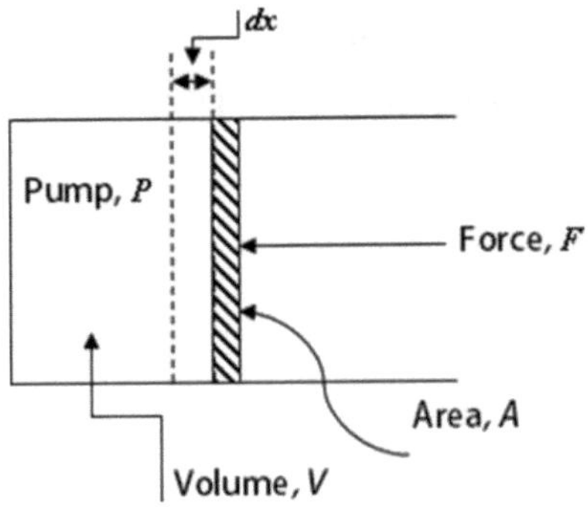

Fig. 4.1 A simple system for doing work

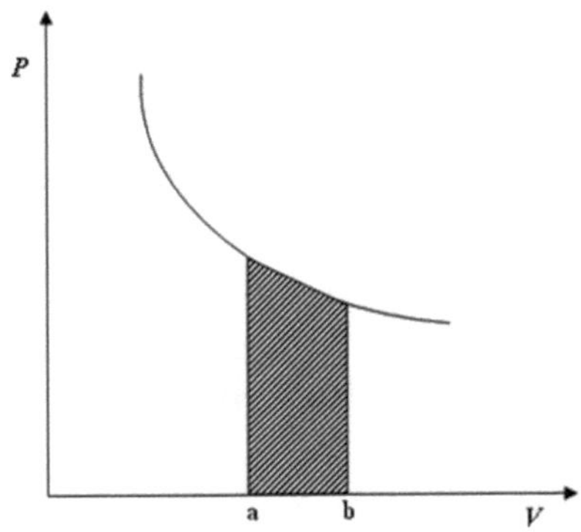

Fig. 4.2 A simple *P-V* diagram

$$W_{ab} = \int_{V_a}^{V_b} PdV \tag{4.2}$$

In the MKS unit system, pressure, volume, and work units can be identified as

$$\begin{array}{l}\text{Pressure, } P = \text{N/m}^2 \\ \text{Volume, } V = \text{m}^3\end{array} \Rightarrow \text{Work, } W = \text{N} \cdot \text{m} = \text{J}$$

4.3 First Law of Thermodynamics

The first law of thermodynamics is one of the important laws that deals with combined cycles and heat of turbine machinery, and its understanding relies on it. We briefly touch up on it here, but encouraging readers to refer to references that are out in open market weather is traditional thermodynamics text book or specifically designed around thermodynamics of cycles (i.e., book by Zohuri and McDaniel) [1].

"The total work is the same in all adiabatic processes between any two equilibrium states having the same kinetic and potential energies." We generalize this to include the internal energy. This is just conservation of energy:

$$dE = dQ - dW \tag{4.3a}$$

where

$$\begin{aligned} E &= U + E_k + E_p \\ &= \text{internal energy} + \text{kinetic energy} + \text{potential energy} \end{aligned} \tag{4.3b}$$

4.4 Enthalpy, *h*

Consider a substance undergoing a phase change (state 1 → state 2) at constant temperature. Since some change in volume generally occurs, the total change resulting from the energy input is

$$\int_1^2 dQ = \int_1^2 dU + \int_1^2 PdV \tag{4.4}$$

On a unit mass basis, $q = Q/M$, $u = U/M$, and $v = V/M$, where $M =$ mass. Thus

$$q_2 - q_1 = u_2 - u_1 + P(v_2 - v_1) \tag{4.5}$$

(assuming that $P =$ constant for this process)

The combination $u+Pv$ occurs frequently and is called the specific enthalpy, h:

$$h = u + Pv \tag{4.6}$$

Example 4.1 Consider the change in phase from liquid water to water vapor at 100 °C. The latent heat of vaporization is 22.6×10^5 J/kg. The vapor pressure at 100 °C is 1 atm $= 1.01 \times 10^5$ N/m^2 and $v_g = 1.8$ m^3/kg and $v_f = 10^{-3}$ m^3/kg.

Solution The work done (in pushing back the atmosphere to make room for vapor) is

$$w = P(v_g - v_f) = 1.7 \times 10^5 \text{ J/kg} \tag{4.7}$$

Thus $(u_g - u_f) = \{22.6 \times 10^5 - 1.7 \times 10^5\} = 20.9 \times 10^5$ J/kg

Therefore, 92% of the energy of transformation is used in increasing the internal energy, and 8% is used in Pv work.

4.5 Energy Equation

The law of conservation of energy states:

4.5.1 Total Energy of an Isolated System Is Constant

Consider the case, as shown in Fig. 4.3, where a mass is added at point 1 and leaves at point 2. The system is at steady state. Thus, we can compute for net energy as

$$\begin{aligned}\Delta E = 0 &= \sum \text{energy inflow} - \sum \text{energy outflow} \\ &= \frac{1}{2}mv_1^2 - mu_1 + Q + P_1V_1 + mgZ_1 \\ &-\frac{1}{2}mv_2^2 - mu_2 + W - P_1V_1 - mgZ_2\end{aligned} \tag{4.8}$$

On a per unit mass basis, where $q = Q/m$ and $w = W/m$, we find

$$\begin{aligned}&\left(u_2 + Pv_2 + \frac{1}{2}mv_2^2 + gZ_2\right) - \left(u_1 + Pv_2 + \frac{1}{2}mv_1^2 + gZ_1\right) = q - \mathrm{w} \\ &\text{or} \\ &\left(h_2 + \frac{1}{2}mv_2^2 + gZ_2\right) - \left(h_1 + \frac{1}{2}mv_1^2 + gZ_1\right) = q - w\end{aligned} \tag{4.9}$$

Special case: Turbine $q \approx 0$ $Z_1 \approx Z_2$

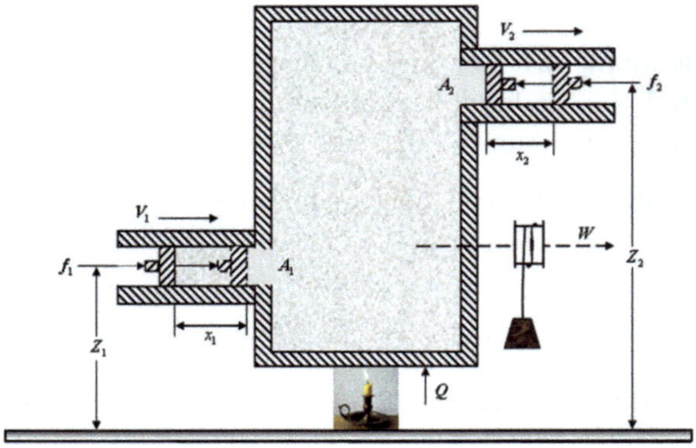

Fig. 4.3 Steady flow process

$$w = (h_1 - h_2) + \frac{1}{2}\left(v_1^2 - v_2^2\right) \tag{4.10}$$

Special case: Flow through a nozzle $q = 0 = w$

$$v_2^2 = v_1^2 + 2(h_1 - h_2) \tag{4.11}$$

Special case: Bernoulli equation $q = 0 = w$

$$\begin{aligned} h_1 + \frac{1}{2}v_1^2 + gZ_1 &= h_2 + \frac{1}{2}v_2^2 + gZ_2 = \text{constant} \\ u + Pv + \frac{1}{2}v^2 + gZ &= \text{constant} \end{aligned} \tag{4.12}$$

If there is no friction, $u =$ constant. In addition, if the fluid is incompressible, then we have

$$\begin{aligned} Pv + \frac{1}{2}v^2 + gZ &= \text{constant} \\ P + \frac{1}{2}v^2 + \rho gZ &= \text{constant} \end{aligned} \tag{4.13}$$

4.6 Open Cycle

When internal combustion engine operation is examined, it is seen to differ in the process of heat supply for a typical heat engine cycle because there is a permanent change in the working fluid during combustion. Therefore, the fluid does not pass through a cycle, so the internal combustion engine is often referred to as an "open cycle" device, not a cyclic thermodynamic heat engine.

The term "open cycle," while meaningless from a thermodynamic perspective, refers to the fact that energy is supplied to the engine from outside in the form of petroleum fuel, and the unconverted portion of energy remaining in the spent combustion mixture is exhausted to the environment. "Closing the cycle", i.e., returning the rejected products to the starting point where they can be reused, is left for nature to accomplish—hence the term "open cycle" is coming to play.

An internal combustion engine is therefore a device for releasing mechanical energy from petroleum fuel using air as the working medium rather than a heat engine for processing air in a thermodynamic cycle. Heat, as such, is not supplied to the internal combustion engine, so it cannot be a heat engine in the sense described in most thermodynamic references.

A simulated heat engine cycle can be constructed to correspond approximately to the operation of an internal combustion engine by substitution of analogous heat

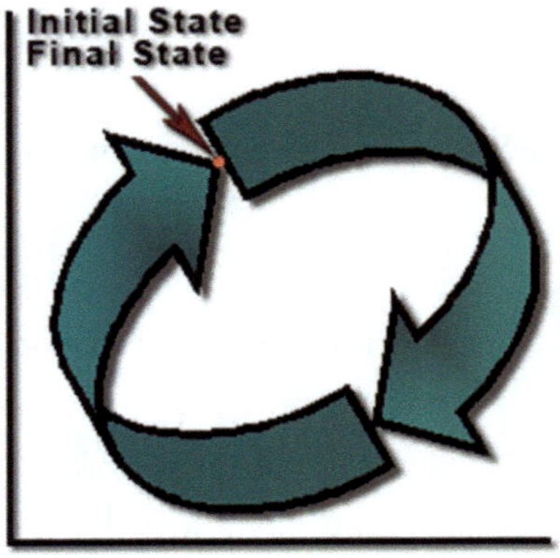

Fig. 4.4 Illustration of a thermodynamic closed cycle

transfer processes for some of the actual engine processes. The specific mechanism of such heat transfer is neglected because the simulation is only a theoretical model of the engine, not an actual device. Such cycles, called *air-standard cycles*, which are subject of study in Sect. 13.3 of the book by Zohuri and McDaniel [1], are useful in the elementary study of internal combustion engines.

4.7 Closed Cycle

Thermodynamic cycles can be categorized yet another way closed, and open cycles and open cycle is described in Sect. 13.1.1 of book by Zohuri and McDaniel [1]. In closed cycles as demonstrated in Fig. 4.4, the working fluid is returning to the initial state at the end of the cycle and is recirculated. By the same talking, in open cycles, the working fluid is renewed at the end of each cycle instead of being recirculated. For example, in automobile engines, the combustion gases are exhausted and replaced by fresh air-fuel mixture at the end of each cycle. The engine operates on a mechanical cycle, but the working fluid does not go through a complete thermodynamic cycle [2].

As we said before, any thermodynamic cycle is essentially a *closed cycle* in which the working substance undergoes a series of processes and is always brought back to the initial state.

4.8 Gas Compressors and Brayton Cycle

The work in a gas compressor is calculated by [1]

$$\dot{W}_{\text{comp}} = \dot{m}\,(h_e - h_i) \tag{4.14}$$

If we assume that the gas in the compressor is calorically perfect, then we have

$$\dot{W}_{\text{comp}} = \dot{m}\,C_p(T_e - T_i) \tag{4.15}$$

In many cases, this is a reasonable approximation. For noble gases, it is very accurate because they are calorically perfect. For air and similar working fluids, it is reasonable because the temperature rise is not that great and an average value of C_p is usually adequate. However, the average value of C_p should be chosen based on a temperature between T_e and T, not one at 300 K.

If then we assume that a compressor operates isentropically (adiabatic and reversible), the exit temperature can be related to the pressure rise in the compressor as shown below in 4.16:

$$\begin{aligned} T_e &= T_i\left(\frac{p_e}{p_i}\right)^{\frac{\gamma-1}{\gamma}} \\ \dot{W} &= \dot{m}\,C_p[T_e - T_i] = \dot{m}\,C_pT_i\left[\frac{T_e}{T_i} - 1\right] = \dot{m}\frac{\gamma R}{\gamma - 1}T_i\left[\left(\frac{p_e}{p_i}\right)^{\frac{\gamma-1}{\gamma}} - 1\right] \end{aligned} \tag{4.16}$$

There are basically three types of compressors, reciprocating, centrifugal flow, and axial flow. In a reciprocating or positive displacement compressor, a piston slides in a cylinder, and valves open and close to admit low-pressure fluid and exhaust high-pressure fluid. In centrifugal flow and axial flow compressors, the fluid enters at one end and is compressed by rotating blades and exits at the opposite end of the compressor. In the centrifugal flow compressor, the flow is in a radially outward direction, and the compression is achieved by forcing the flow against the outer annulus of the compressor. In an axial flow compressor, a set of rotating blades move the flow through the compressor, acting as airfoils. They force the flow through an increasingly narrower channel, thus increasing the density and pressure. Gasoline and diesel engines are examples of reciprocating compressors, as are positive displacement pumps. Water pumps are examples of centrifugal flow compressors, similar to the rotor in a washing machine. Jet engine compressors are typically axial flow compressors. Reciprocating compressors require no priming and can reach very high pressures but only moderate flow rates. Centrifugal flow and axial flow compressors usually require priming and can reach very high flow rates but moderate pressures.

In this section we discuss the *Brayton thermodynamic cycle* which is used in all gas turbine engines. Figure 4.5 shows a *T-s* diagram of the Brayton cycle. Using the turbine engine station numbering system, we begin with free stream conditions at *station 0*. In cruising flight, the inlet slows the air stream as it is brought to the compressor face at *station 2*. As the flow slows, some of the energy associated with the aircraft velocity increases the static pressure of the air, and the flow is

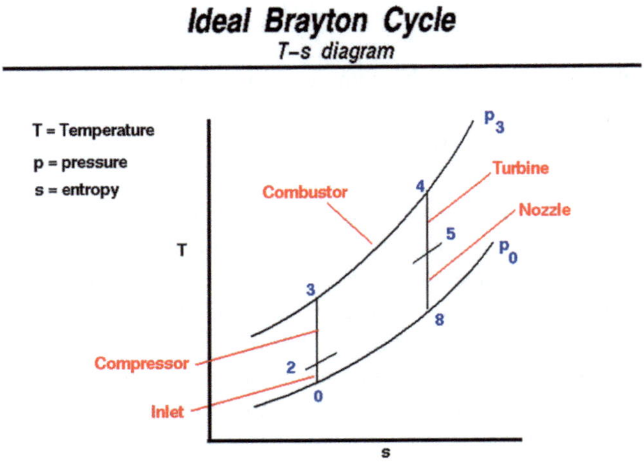

Fig. 4.5 Illustration of Brayton cycle (Courtesy of NASA)

compressed. Ideally, the compression is isentropic, and the static temperature is also increased as shown on the plot. The compressor does work on the gas and increases the pressure and temperature isentropically to *station 3* the compressor exit. Since the compression is ideally isentropic, a vertical line on the *T-s* diagram describes the process. *In reality, the compression is not isentropic, and the compression process line leans to the right because of the increase in entropy of the flow.* The combustion process in the burner occurs at constant pressure from *station 3* to *station 4*. The temperature increase depends on the type of fuel used and the fuel-air ratio. The hot exhaust is then passed through the power turbine in which work is done by the flow from *station 4* to *station 5*. Because the turbine and compressor are on the same shaft, the work done on the turbine is exactly equal to the work done by the compressor, and, ideally, the temperature change is the same. The nozzle then brings the flow isentropically (adiabatic and reversible) back to free stream pressure from *station 5* to *station 8*. Externally, the flow conditions return to free stream conditions, which complete the cycle. The area under the *T-s* diagram is proportional to the useful work and thrust generated by the engine. The *T-s* diagram for the ideal Brayton cycle is shown here:

The Brayton cycle analysis is used to predict the thermodynamic performance of gas turbine engines.

As we know the gas turbine is another mechanical system that produces power, and it may operate on a cycle when used as an automobile or truck engine or on a closed cycle when used in a nuclear power plant [1].

Usage of Brayton process in a simple gas turbine cycle can be described by first in an open cycle operation where air enters the compressor and passes through a constant pressure combustion chamber, then goes through turbine, and then exits as product of combustion to the atmosphere, as shown in Fig. 4.6a. Similar situation can be studied when the combustion chamber a heat exchanger gets add on to the

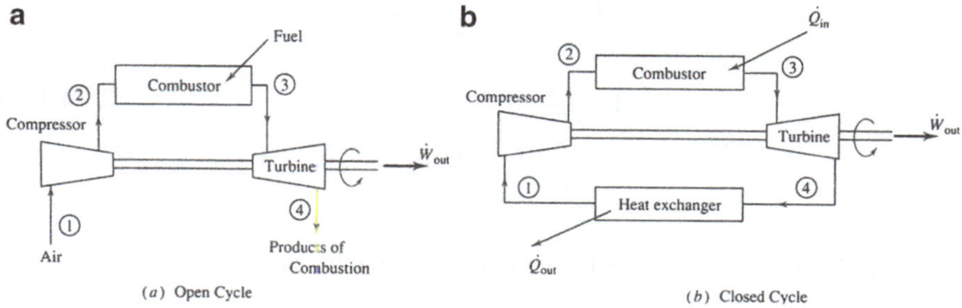

Fig. 4.6 Illustration of Brayton components for (**a**) open and (**b**) closed cycles

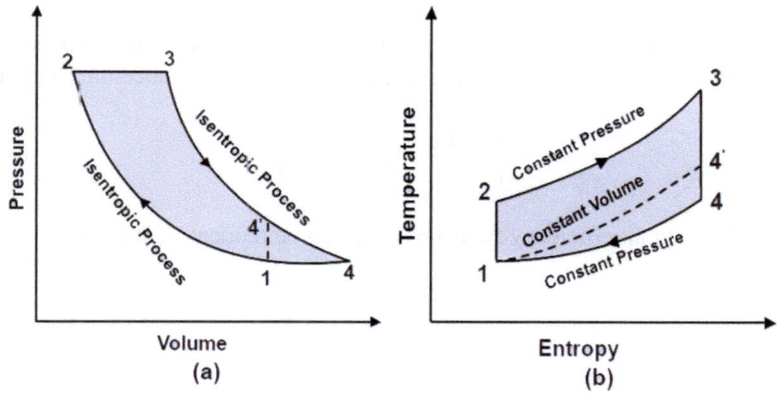

Fig. 4.7 Illustration of the Brayton cycle on *P-V* and *T-s* diagram

loop of Fig. 4.6a, in order to organize a closed cycle as it can be seen in Fig. 4.6b. Energy from some external source enters the cycle, and an additional heat exchanger that has been added on into the loop transfers heat from the cycle so that the air can be returned to its initial state, as clearly can be seen in Fig. 4.6b.

The Brayton cycle is a theoretical cycle for simple gas turbine. This cycle consists of two isentropic and two constant pressure processes. Figure 4.7 shows the Brayton cycle on *P-V* and *T-s* coordinates. The cycle is similar to the Diesel cycle in compression and heat addition. The isentropic expansion of the Diesel cycle is further extended followed by constant pressure heat rejection.

The following notation gives the thermal efficiency in mathematical format for ideal cycle used to model the gas turbine, which utilizing isentropic compression and expansion in Brayton process as

$$\eta_{\text{th}} = \frac{\text{Heat added-Heat rejected}}{\text{Heat added}} = \frac{\dot{Q}_{\text{out}}}{\dot{Q}_{\text{in}}} \tag{4.17a}$$

$$\begin{aligned}\eta_{\text{th}} &= \frac{mC_{\text{p}}(T_3 - T_1) - mC_{\text{p}}(T_4 - T_1)}{mC_{\text{p}}(T_3 - T_2)} \\ &= 1 - \frac{T_4 - T_1}{T_3 - T_2} \\ &= 1 - \frac{T_1\,(T_4/T_1) - 1}{T_2\,(T_3/T_2) - 1}\end{aligned} \tag{4.17b}$$

Using the following isentropic process and relations that we have

$$\frac{T_2}{T_1} = \left(\frac{P_2}{P_1}\right)^{\frac{\gamma-1}{\gamma}} \quad \text{and} \quad \frac{T_3}{T_4} = \left(\frac{P_3}{P_4}\right)^{\frac{\gamma-1}{\gamma}} \tag{4.17c}$$

For ideal gas and observation, *P*-*V* diagram of Fig. 4.7a obviously shows that we can state $P_2 = P_3$ and $P_1 = P_4$; as a result, using 4.17c will induce the following:

$$\frac{T_2}{T_1} = \frac{T_3}{T_4} \quad \text{or} \quad \frac{T_4}{T_1} = \frac{T_3}{T_2} \tag{4.17d}$$

Then the thermal efficiency η_{th} from 4.17a can be reduced to the following form as

$$\eta_{\text{th}} = 1 - \frac{T_4}{T_3} = 1 - \frac{T_1}{T_2} \tag{4.17e}$$

Now if we introduce a term of the pressure ration $r_{\text{p}} = P_2/P_1$, the thermal efficiency from 4.17e will take a form of the following:

$$\frac{T_4}{T_3} = \frac{T_4}{T_3} = \frac{V_2}{V_1} = \frac{1}{r_{\text{p}}^{\gamma-1}} \tag{4.17f}$$

$$\frac{1}{r_{\text{p}}^{\gamma-1}} = \left(\frac{V_2}{V_1}\right)^{\gamma-1} \left\{\left(\frac{P_1}{P_2}\right)\right\}^{(\gamma-1)} = (r_{\text{p}})^{\frac{\gamma-1}{\gamma}} \tag{4.17g}$$

$$\eta_{\text{th}} = 1 - \frac{T_1}{T_2} = 1 - \left(\frac{P_1}{P_2}\right)^{(\gamma-1)/\gamma} \tag{4.17h}$$

or

$$\eta_{\text{th}} = 1 - r_{\text{p}}^{(1-\gamma)/\gamma} \tag{4.17i}$$

Note that the above final expression for thermal efficiency η_{th} in both forms of Eqs. 4.17h and 4.17i was obtained based on assumption of using constant specific heats. For more accurate calculations, the gas tables should be utilized.

In actual gas turbine, the compressor and the turbine are not isentropic, and some losses are taking place. These losses, usually in the neighborhood of 85%, significantly reduce the efficiency of the gas turbine engine.

Considering all the above we can see that the back work ratio is defined for a Brayton system as W_{comp}/W_{turb}. This is an important feature of the gas turbine that limits thermal efficiency and that is required for compressor to have high work and is measured by this ration. This can actually be fairly large approaching 1.0. If the compressor is too inefficient, the Brayton cycle will not work. Only when efficient air compressors were developed was the jet engine feasible.

Example 4.2 Air enters the compressor of a gas turbine at 100 kPa and 25 °C. For a pressure ratio of 5 and a maximum temperature of 850 °C, determine the back work ratio and the thermal efficiency using the Brayton cycle.

Solution To find the back work ration, we can see that

$$\frac{W_{comp}}{W_{turb}} = \frac{C_p(T_2 - T_1)}{C_p(T_3 - T_4)} = \frac{T_2 - T_1}{T_3 - T_4}$$

The temperatures are $T_1 = 273{+}25 = 298$ K, $T_2 = 273{+}850 = 1123$ K, and

$$T_2 = T_1\left(\frac{P_2}{P_1}\right)^{(\gamma-1)/\gamma} = (298)(5)^{0.2857} = 472.0\,\mathrm{K}$$

and

$$T_4 = T_3\left(\frac{P_4}{P_5}\right)^{(\gamma-1)/\gamma} = (1123)\left(\frac{1}{5}\right)^{0.2857} = 709.1\,\mathrm{K}$$

The back work ratio is then given by

$$\frac{W_{comp}}{W_{turb}} = \frac{472.0 - 298}{1123 - 709} = 0.420 \text{ or } 42.0\%$$

The thermal efficiency is

$$\eta_{th} = 1 - r^{(1-\gamma)/\gamma} = 1 - (5)^{-0.2857} = 0.369 \text{ or } 36.9\%$$

Example 4.3 In an air-standard Brayton cycle with a pressure ratio of 8, the pressure and temperature at the start of the compression are 100 kPa and 300 K. The maximum allowed temperature in the cycle is 1200 K. Determine the energy added per kg air, the work done per kg air, and the thermal efficiency of the cycle. Use Fig. 4.8.

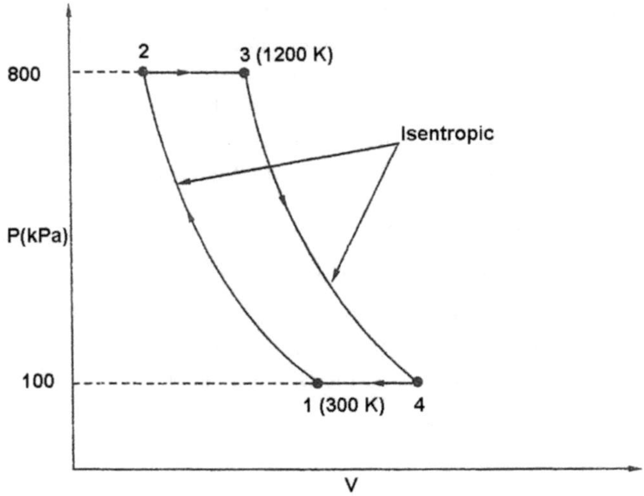

Fig. 4.8 Sketch of Example 4.3

Solution The air-standard Brayton cycle is shown on a *P-V* diagram in figure above. The pressure ratio r_p of the Brayton cycle is given by

$$r_p = \frac{P_2}{P_1} = 8$$

In addition to pressure ratio, the other data given is $P_1 = 100$ kPa, $T_1 = 300$ K, and $T_3 = 1200$ K.

The temperature T_2 of air after isentropic compression process $1 \rightarrow 2$ is given by

$$T_2 = T_1 \left(\frac{P_2}{P_1}\right)^{\frac{\gamma-1}{\gamma}} = 300(8)^{\frac{0.4}{1.4}} = 543.4\,\text{K}$$

The energy

$$q_1 = C_p(T_3 - T_2) = 1.0047(1200 - 543.4) = 659.69\,\text{kJ/kg}$$

The thermal efficiency η_{th} of the Brayton cycle is given by

$$\eta_{th} = 1 - \left(\frac{1}{r_p}\right)^{(1-\gamma)/\gamma} = 1 - \left(\frac{1}{8}\right)^{\frac{0.4}{1.4}} = 0.448$$

The network done W_{net} per kg air is given by

$$W_{net} = q_1 \eta_{th} = (659.69) \times (0.448) = 295.54\,\text{kJ/kg}$$

4.9 The Non-Ideal Brayton Cycle

The ideal air-standard Brayton cycle assumes isentropic compression and expansion processes. So far this has not been achieved in any real device. The isentropic efficiency for these processes is defined as [1]

$$\text{Isentropic efficiency(compression)} = \frac{\Delta h_{\text{isentropic}}}{\Delta h_{\text{actual}}}$$
$$\text{Isentropic efficiency(expansion)} = \frac{\Delta h_{\text{actual}}}{\Delta h_{\text{isentropic}}}$$

Unfortunately, the isentropic efficiency of a compressor or turbine will depend on the pressure ratio for the device. In doing parametric or design studies, it is more useful to define an efficiency that does not depend on the pressure ratio but only on the manufacturing tolerances and efficiencies of individual stages [1]. This small stage, or infinitesimal stage, efficiency is called the polytropic efficiency.

Consider the combined first and second law for an infinitesimal process.

$$dh = vdp + Tds \tag{4.18}$$

The term Tds represents a heat flow for the process. During a compression, the inefficiency of the process represents a heat flow into the system. For an expansion the inefficiency represents a heat flow out of the system. On an infinitesimal basis, we can write [1]

$$\begin{aligned} dh &= vdp + (Tds) = {}^{vdp}/_{e_{\text{c,poly}}} \text{ for a compressor} dh \\ &= vdp + (-Tds) = e_{\text{t,poly}} * vdp \text{ for a turbine} \end{aligned} \tag{4.19}$$

then these two equations can be integrated similar to the way the isentropic relation was integrated. For an isentropic expansion of a calorically perfect ideal gas, we have [1]

$$\begin{aligned} vdp &= \frac{RT}{p} dp \\ dh &= c_p dT = \frac{RT}{p} dp \\ \frac{dT}{T} &= \frac{R}{C_p}\frac{dp}{p} = \frac{\gamma - 1}{\gamma}\frac{dp}{p} \\ \left(\frac{T_2}{T_1}\right) &= \left(\frac{p_2}{p_1}\right)^{\frac{\gamma-1}{\gamma}} \end{aligned}$$

For a polytropic compression we have

$$\begin{aligned}
vdp &= \frac{RT}{e_{c,\text{poly}}p}dp \\
dh = c_p dT &= \frac{RT}{e_{c,\text{poly}}p}dp \\
\frac{dT}{T} = \frac{R\,dp}{C_p\,p} &= \frac{\gamma-1}{e_{c,\text{poly}}\gamma}\frac{dp}{p} \\
\left(\frac{T_2}{T_1}\right) &= \left(\frac{p_2}{p_1}\right)^{\frac{\gamma-1}{e_{c,\text{poly}}\gamma}}
\end{aligned}$$

And for a polytropic expansion we have

$$\begin{aligned}
vdp &= \frac{e_{t,\text{poly}}RT}{p}dp \\
dh = c_p dT &= \frac{e_{t,\text{poly}}RT}{p}dp \\
\frac{dT}{T} = \frac{R\,dp}{C_p\,p} &= \frac{e_{t,\text{poly}}(\gamma-1)}{\gamma}\frac{dp}{p} \\
\left(\frac{T_2}{T_1}\right) &= \left(\frac{p_2}{p_1}\right)^{\frac{e_{t,\text{poly}}(\gamma-1)}{\gamma}}
\end{aligned}$$

Now for a calorically perfect gas, the isentropic efficiency of a compressor is given by

$$\eta_{c,\text{isen}} = \frac{C_p(T_{\text{out,isen}} - T_{\text{in}})}{C_p(T_{\text{out,actual}} - T_{\text{in}})} = \frac{\frac{T_{\text{out,isen}}}{T_{\text{in}}} - 1}{\frac{T_{\text{out,actual}}}{T_{\text{in}}} - 1} = \frac{\left(\frac{p_{\text{out}}}{p_{\text{in}}}\right)^{\frac{\gamma-1}{\gamma}} - 1}{\left(\frac{p_{\text{out}}}{p_{\text{in}}}\right)^{\frac{\gamma-1}{\gamma e_{c,\text{poly}}}}}$$

And the isentropic efficiency of a turbine is given by

$$\eta_{t,\text{isen}} = \frac{C_p(T_{\text{out,actual}} - T_{\text{in}})}{C_p(T_{\text{out,isen}} - T_{\text{in}})} = \frac{\frac{T_{\text{out,actual}}}{T_{\text{in}}} - 1}{\frac{T_{\text{out,isen}}}{T_{\text{in}}} - 1} = \frac{\left(\frac{p_{\text{out}}}{p_{\text{in}}}\right)^{\frac{e_{t,\text{poly}}(\gamma-1)}{\gamma}} - 1}{\left(\frac{p_{\text{out}}}{p_{\text{in}}}\right)^{\frac{\gamma-1}{\gamma e_{c,\text{poly}}}}}$$

Example 4.4 A gas turbine operates in ambient conditions of 14.677 psi, 17 °C, and the maximum cycle temperature is limited to 1000 K. The compressor, which has a polytropic efficiency of 88%, is driven by the gas generator turbine, and a separate power turbine is geared to the power output on a separate shaft; both turbines have polytropic efficiencies of 90%. There is a pressure loss of 2.9 psi bar between the compressor and the gas generator turbine inlet. Neglecting all other losses, and assuming negligible kinetic energy changes, calculates:

(a) The compressor pressure ratio which will give maximum specific power output

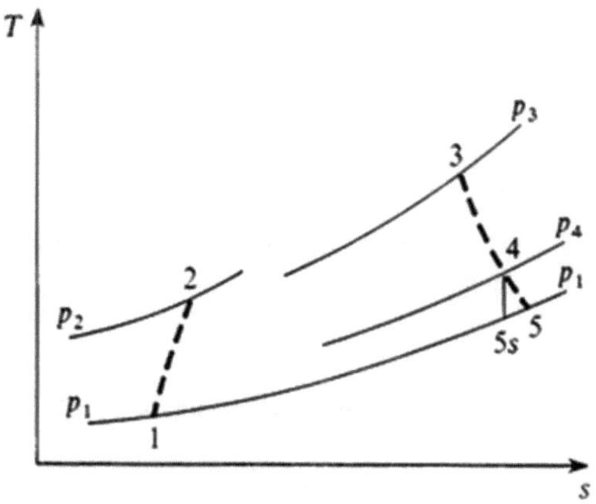

Fig. 4.9 Gas turbine cycle on a *T-s* diagram for Example 4.2

(b) The isentropic efficiency of the power turbine

For the gases in both turbines, take $c_p = 1.15$ kJ/kg K and $\gamma = 1.33$.
For air take $c_p = 1.005$ kJ/kg K and $\gamma = 1.4$.

Solution (a) The cycle is shown on a *T-s* diagram in Fig. 4.9.
Let $\frac{p_2}{p_1} = r$. From 4.14c, we can write

$$T_2 = T_1 r^{(\gamma-1)/\gamma\eta_{\infty c}} = (17 + 273) \times r^{0.4/1.4\times 0.88} = 290r^{0.325}$$

Now

$$p_3 = p_2 - 0.2 = (p_1 \times r) - 0.2 = 1.012 \times r - 0.2$$
$$p_5 = p_1 = 14.677$$
$$\text{or} \quad \frac{p_3}{p_5} = \frac{14.677r - 2.9}{14.677} = r - 0.198$$

Since the polytropic efficiency of both turbines is the same, then using 4.15c provides

$$\frac{T_3}{T_5} = \frac{(p_3)^{(\gamma-1)\eta_{\infty e}/\gamma}}{(p_1)} = (r - 0.198)^{0.9/4}$$

$$T_5 = \frac{1000}{(r - 0.198)^{0.225}}$$

$$\begin{aligned}\text{Turbine specific power output} &= c_p(T_3 - T_5 \\ &= 1.15\left(1000 - \frac{1000}{(r-0.198)^{0.225}}\right) \\ &= 1150\left[1-(r-0.198)^{-0.225}\right]\end{aligned}$$

$$\begin{aligned}\text{Compressor specific power output} &= c_p(T_2 - T_1) \\ &= 1.005\left(290r^{0.325} - 290\right) \\ &= 291.5\left(r^{0.325} - 1\right)\end{aligned}$$

Net specific power output

$$\dot{W} = 1150\left[1-(r-0.198)^{-0.225}\right] - 291.5\left(r^{0.325}-1\right)$$

To find the maximum of the above relation, we set $d\dot{W}/dr = 0$, so we get

$$0.225 \times 1150 \times (r-0.198)^{-1.225} = 0.325 \times 291.5 \times r^{-0.675}$$

Trial and error, or graphical, solution gives $r = 6.65$

The compressor pressure ratio for maximum specific power output = 6.65

(b) $T_2 = 290r^{0.325} = 290(6.65)^{0.325} = 536.8\,\text{K}$

Now we can write the following:

Gas turbine power output = Compressor power input

Therefore

$$\begin{aligned}&1.15(1000 - T_4) = 1.005(536.8 - 290) \\ &\text{i.e.,}\quad T_4 = 784.3\,\text{K}\end{aligned}$$

Then

$$\frac{p_3}{p_4} = \frac{(T_3)^{\gamma/(\gamma-1)\eta_{\infty e}}}{(T_4)} = \frac{(1000)^{4/0.9}}{(7843.3)} = 2.944$$

Also

$$p_3 = 6.65p_1 - 0.290 = (6.65 \times 1.012) - 0.290 = 4.73\,\text{bar}$$

Therefore

$$p_4 = \frac{4.73}{2.944} = 1.607\,\text{bar}$$

Then

$$\frac{T_4}{T_5} = \frac{(p_4)^{(\gamma-1)/\eta_{\infty e}/\gamma}}{(p_5)} = \frac{(1.607)^{0.9/4}}{(14.677)} = 1.110$$

$$\frac{T_4}{T_{5s}} = \frac{(p_4)^{(\gamma-1)/\gamma}}{(p_5)} = \frac{(1.607)^{0.25}}{(14.677)} = 1.123$$

Using the relationship for turbine isentropic efficiency, η_T as below

$$\text{Power turbine isentropic efficiency}\, \eta_{PT} = \frac{c_p(T_4 - T_5)}{c_p(T_4 - T_{5s})} = \frac{(T_4 - T_5)}{(T_4 - T_{5s})}$$

Then we can write

$$\eta_{PT} = \frac{(T_4 - T_5)}{(T_4 - T_{5s})} = \frac{1 - (T_5/T_4)}{1 - (T_{5s}/T_4)} = \frac{1 - (1/1.11)}{1 - (1/1.123)} = 0.905 \text{ or } 90.5\%$$

4.10 The Basic Rankine Cycle

Although the Carnot cycle is the most efficient cycle, its work ratio is low, and there are practical difficulties to follow it [1]. Hence the ideal cycle, which is more suitable as a criterion for actual steam cycles than the Carnot cycle, is called *Rankine cycle*. The basic Rankine (vapor) cycle consists of four processes. The state points are identified in Fig. 4.10.

Process 1 to 2—Isentropic compression of a liquid in a pump
Process 2 to 3—Constant pressure heat addition in a boiler or heat exchanger—creates vapor
Process 3 to 4—Isentropic expansion of the vapor in a turbine
Process 4 to 1—Constant pressure heat extraction in a condenser—returns vapor to liquid

The general layout of the components is pictured in Fig. 4.10. The steam generator, or boiler, consists of two generally separate heat exchangers—the economizer to heat the high pressure water to its boiling temperature and the evaporator to convert the saturated liquid to steam. The heat transfer processes are different for

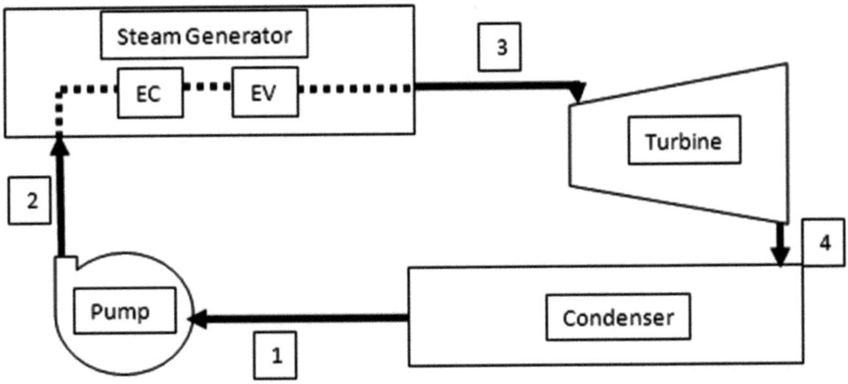

Fig. 4.10 Major components of a basic Rankine cycle

Fig. 4.11 *T-s* diagram for a basic Rankine cycle

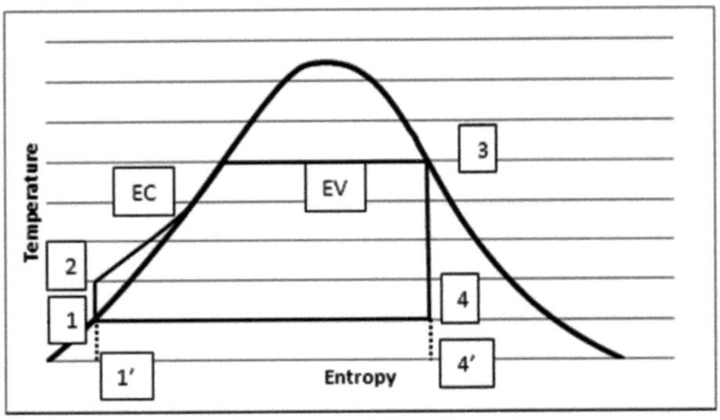

the two different states of the liquid/vapor, so this usually necessitates a different design for each region.

The temperature-entropy diagram for the basic cycle is presented in Fig. 4.11. The work output of the system is represented by the area enclosed in the cycle on the *T-s* diagram by the curve **1–2–3–4–1.** The heat input to the system is represented by the curve **1′–1–2–3–4–4′–1′**. Therefore, the net thermal efficiency of the cycle is represented by the ratio of the two areas.

$$\eta_{\mathrm{th}} = \frac{\text{area } 1-2-3-4-1}{\text{area } 1'-2-3-4'-1'} \tag{4.20}$$

The efficiency of a Rankine cycle must be less than that of a Carnot cycle because the heat addition does not all occur at the peak temperature of the cycle.

One of the major advantages of the Rankine cycle is that the compression process in the pump takes place on a liquid. The energy consumption of this process is almost negligible compared to the work output of the turbine, so a less than isentropic efficiency for the liquid pump is not a major energy loss for the cycle.

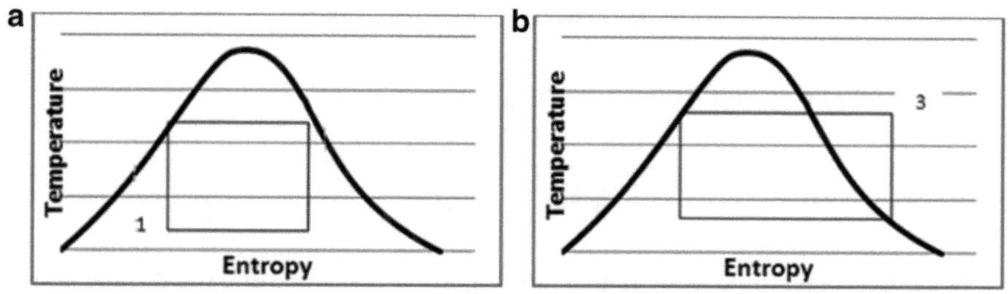

Fig. 4.12 Alternative cycles

Consider the two ways of improving the Rankine cycle efficiency by adding all of the heat at the peak temperature as indicated in Fig. 4.12a, b. In Fig. 4.12a, the cycle will require pumping on a vapor at state point 1, which is much less efficient than pumping on a liquid.

In Fig. 4.12b, to reach state point 3 at a constant temperature from the saturation line, the pressure will have to be continuously lowered, which is very difficult to accomplish. In the vapor region, a constant pressure curve departs the saturation line with a very significant upward slope as will soon be shown. Also, by lowering the pressure to maintain a constant temperature with heat addition, some of the mechanical energy that was used in the pump would be sacrificed. There are several ways of increasing the work output of a Rankine cycle, but they do not improve on the efficiency relative to a Carnot cycle [1].

Now consider how the cycle efficiency is calculated. The heat input is given by

$$q_{\mathrm{in}} = h_3 - h_2 \tag{4.21}$$

The turbine work is given by

$$w_{\mathrm{t}} = h_3 - h_4 \tag{4.22}$$

The enthalpy at state 4 is found by assuming an isentropic expansion process across the turbine so the entropy at state point 4 must be the same as the entropy at state point 3. Once the pressure and entropy at state point 4 are known, the rest of the thermodynamic variables can be found including the enthalpy. Typically, a constant entropy expansion down to the pressure at state point 4 will put the fluid under the vapor dome and in the two-phase region. The entropy is then used to find the quality of the steam in the two-phase region. Once the quality is known, the enthalpy at state 4 can be easily obtained by the standard property expression for variables in the two-phase regime [1].

The pump work is calculated by

$$w_p = h_2 - h_1 = \Delta pv = (p_2 - p_1)v_1 \tag{4.23}$$

Therefore, the cycle thermal efficiency is given by

$$\eta_{th} = \frac{h_3 - h_4 - (p_2 - p_1)v_1}{h_3 - h_2} = \frac{h_3 - h_4 - (p_2 - p_1)v_1}{h_3 - [h_1 + (p_2 - p_1)v_1]} \tag{4.24}$$

Note that for a closed system, the atmospheric pressure exiting the turbine and, in the condenser, will be significantly below atmospheric pressure as the liquid is being condensed to a temperature slightly above the cooling water temperature, which is near the ambient atmospheric temperature.

This can cause atmospheric gases to leak into the condenser and decrease the efficiency of the other components in the system, notably the pump. Therefore, a leak tight condenser is a very desirable component.

Example 4.5 Consider a basic Rankine power cycle with the saturated steam delivered to the turbine at 1000 psi and the condensate water delivered to the pump at 2.5 psi. Assume an isentropic turbine and pump. Calculate the cycle thermal efficiency.

Solution At 1000 psi $T_{sat} = 1004.3$ R, $h_3 = 1195.1$ Btu/lbm, $s_3 = 1.39548$ Btu/lbm/R at 2.5 psi, $T_{sat} = 594.1$ R, $v_f = 0.0163$ m^3/kg, $h_f = 102.6$ Btu/lbm, $h_{fg} = 1019.1$ Btu/lbm, $s_f = 0.18954$ Btu/lbm, $s_{fg} = 1.71547$ Btu/lbm/R

Then

$$\text{Carnot Efficiency } \eta_C = 1 - \frac{594.1}{1004.3} = 0.4084$$

$$x_4 = \frac{1.39548 - 0.18954}{1.71547} = 0.7030 \quad h_4 = 102.6 + 0.7030^{*}1019.1 = 819.0$$

$$\Delta h_{3-4} = 1195.1 - 819.0 = 376.1\,\text{Btu/lbm}$$

$$\Delta h_{pump} = (1000 - 2.5)*144^{*}0.0163/778 = 3.0\,\text{Btu/lbm}$$

$$h_2 = 102.6 + 3.0 = 105.6\,\text{Btu/lbm} \quad \Delta h_{2-3} = 1195.1 - 105.6 = 1089.5\,\text{Btu/lbm}$$

$$\eta_{th} = \frac{376.1 - 3.0}{1089.5} = 0.3425 \quad \eta_{th} = 34.25\% \quad \eta_{II} = \frac{0.3425}{0.4084} = 0.8386$$

In addition to losses required to drive the pump, inefficiencies in the turbine and pump can also affect the thermodynamic cycle efficiency. The change in enthalpy across a non-ideal turbine is given by

$$\Delta h_{actual} = \eta_t \Delta h_{isentropic} \tag{4.25}$$

And for a non-ideal pump, the actual change in enthalpy is given by

$$\Delta h_{actual} = \frac{\Delta h_{isentropic}}{\eta_p} \tag{4.26}$$

Example 4.6 Re-evaluate the thermal efficiency of the Rankine cycle in Example 4.5 assuming a turbine with an adiabatic efficiency of 90% and a pump with an adiabatic efficiency of 80%.

Solution The only changes are the Δh's across the pump and the turbine.

$$\Delta h_{\text{pump}} = \frac{3.0}{0.8} = 3.75\,\text{Btu/lbm} \quad \Delta h_{3-4} = 0.9 * 376.1 = 338.5\,\text{Btu/lbm}$$
$$\eta_{\text{th}} = \frac{338.5 - 3.75}{1089.5} = 0.3073 \quad \eta_{\text{th}} = 30.73\%$$

Now consider possible ways of increasing the efficiency of the basic Rankine cycle. There are really three possibilities: (1) increase the boiler pressure; (2) decrease the condensing temperature, and therefore the pressure, for the condensate entering the pump; and (3) increase the peak temperature of the cycle.

Example 4.7 Change the turbine inlet pressure for Example 4.5 to 1200 psi, and re-evaluate the cycle efficiency.

Solution At 1200 psi, $T_{\text{sat}} = 1026.9$, $h_3 = 1186.8$ Btu/lbm, $s_3 = 1.37253$ Btu/lbm/R

$$\text{Carnot Efficiency } \eta_C = 1 - \frac{594.1}{1026.9} = 0.4215 x_4 = \frac{1.37253 - 0.18954}{1.71547} = 0.6896$$
$$h_4 = 102.6 + 0.6896 * 1019.7 = 805.8\,\text{Btu/lbm} \Delta h_{3-4} = 1186.8$$
$$-805.8 = 381.0\,\text{Btu/lbm} \quad \Delta h_{2-3} = 1186.8$$
$$-105.6 = 1081.2\,\text{Btu/lbm} \eta_{\text{th}} = \frac{381.0 - 3.0}{1081.2} = 0.3496$$
$$\eta_{\text{th}} = 34.96\% \quad \eta_{\text{II}} = \frac{0.3496}{0.4215} = 0.8294$$

Example 4.8 Change the condensate pump input pressure in Example 4.5 to 4.7 psi, and re-evaluate the thermal efficiency of the cycle.

Solution At 1.5 psi, $T_{\text{sat}} = 575.3$ R, $v_f = 0.0162$ m^3/kg, $h_f = 83.8$ Btu/lbm, $h_{fg} = 1029.9$ Btu/lbm, $s_f = 0.15746$ Btu/lbm, $s_{fg} = 1.79023$ Btu/lbm/R

$$\text{Carnot Efficiency } \eta_C = 1 - \frac{575.3}{1004.3} = 0.4272 x_4 = \frac{1.39548 - 0.15746}{1.79023} = 0.6915$$
$$h_4 = 83.8 + 0.6915 * 1029.9 = 796.0\,\text{Btu/lbm} \Delta h_{3-4} = 1195.1$$
$$-796.0 = 399.1\,\text{Btu/lbm} \quad \Delta h_{\text{pump}} = (1000 - 1.5)$$
$$\times * 144 * 0.0162/778 = 2.99 \;\; \text{Btu/lbm} \Delta h_{2-3} = 1195.1$$
$$-(83.8 + 2.99) = 1108.3\,\text{Btu/lbm} \eta_{\text{th}} = \frac{399.1 - 2.99}{1108.3} = 0.3574 \quad \eta_{\text{th}} = 35.74\%$$
$$\eta_{\text{II}} = \frac{0.3574}{0.4272} = 0.8366$$

4.11 The Rankine Cycle with a Superheater

The only way to increase the peak temperature of the Rankine cycle without increasing the boiler pressure is to heat the steam itself to a higher temperature. This requires the addition of another type of heat exchanger called a superheater. This third heat exchanger is added to the steam generator as described in Fig. 4.13. The *T-s* diagram is provided in Fig. 4.14.

The only difference that the superheated Rankine cycle presents when evaluating the cycle thermal efficiency is that the enthalpy at state point 3, the inlet to the turbine, must be obtained from the superheated steam tables.

Example 4.9 Add a superheater to the cycle described in Example 4.5 that takes the turbine cp to 1500 R.

Solution At 1000 psi and 1500 R, $h_3 = 1532.5$ Btu/lbm, $s_3 = 1.67285$ Btu/lbm/R

At 2.5 psi, $T_{\text{sat}} = 594.1$ R, $v_f = 0.0163$ m^3/kg, $h_f = 102.6$ Btu/lbm, $h_{fg} = 1019.1$ Btu/lbm, $s_f = 0.18954$ Btu/lbm, $s_{fg} = 1.71547$ Btu/lbm/R

$$\text{Carnot Efficiency } \eta_c = 1 - \frac{594.1}{1500} = 0.6039 x_4 = \frac{1.67258 - 0.18954}{1.71547} = 0.8645$$
$$h_4 = 102.6 + 0.8645 * 1019.1 = 983.6 \,\text{Btu/lbm} \Delta h_{3-4} = 1532.5$$
$$-983.6 = 548.9 \,\text{Btu/lbm} \quad \Delta h_{2-3} = 1532.5$$
$$-105.6 = 1426.9 \,\text{Btu/lbm} \eta_{th} = \frac{548.9 - 3.0}{1426.9} = 0.3826$$
$$\eta_{th} = 38.26\% \quad \eta_{II} = \frac{0.3826}{0.6039} = 0.6335$$

Note that increasing the superheater outlet temperature further will increase the cycle efficiency further but will decrease the second law efficiency relative to a Carnot cycle.

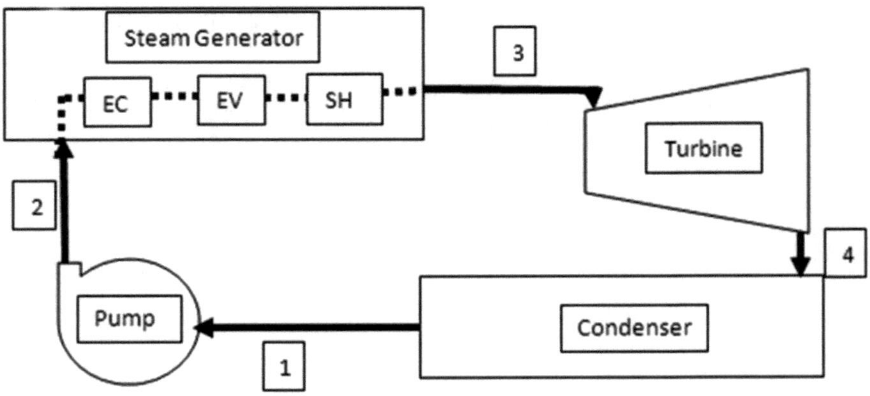

Fig. 4.13 Components for a superheated Rankine cycle

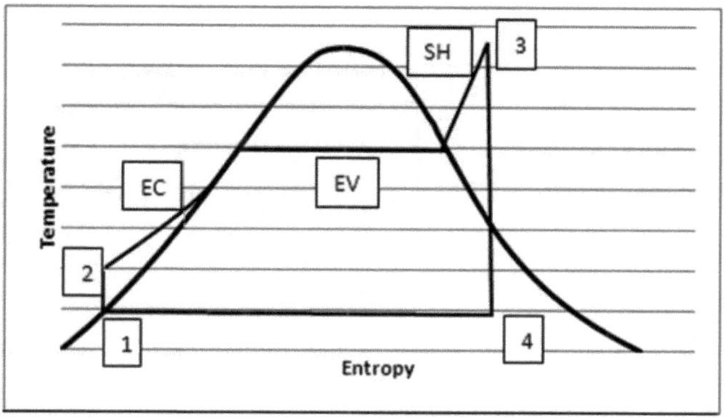

Fig. 4.14 *T-s* diagram for a Rankine cycle with a superheater

Example 4.10 Increase the superheater exit temperature from Example 4.9 to 1600 R, and estimate the increase in cycle efficiency.

Solution At 1000 psi and 1600 R, $h_3 = 1589.3$ Btu/lbm, $s_3 = 1.70929$ Btu/lbm/R

$$\text{Carnot Efficiency } \eta = 1 - \frac{594.1}{1600} = 0.6287 x_4 = \frac{1.70929 - 0.18954}{1.71547} = 0.8859$$
$$h_3 = 102.6 + 0.8859^* 1019.1 = 1005.4\,\text{Btu/lbm}\Delta h_{3-4} = 1589.3$$
$$-1005.4 = 583.9\,\text{Btu/lbm} \quad \Delta h_{2-3} = 1589.3$$
$$-105.6 = 1492.7\,\text{Btu/lbm}\eta_{\text{th}} = \frac{583.9 - 3.0}{1492.7} = 0.3892$$
$$\eta_{\text{th}} = 38.92\% \quad \eta_{\text{II}} = \frac{0.3892}{0.6287} = 0.6190$$

Obviously, the reason the second law efficiency goes down is because the peak temperature of the cycle has been increased, but an even smaller fraction of the total heat input to the cycle has occurred near the peak temperature of the cycle.

4.12 Superheated Rankine Cycle with Reheaters

In order to deliver more of the heat at a temperature close to the peak of the cycle, the steam is usually extracted from the turbine after a less than maximum expansion and reheated to a higher temperature. This involves splitting the turbine into a high-pressure and a low-pressure turbine or into high-, intermediate-, and low-pressure turbines. Three turbines are not uncommon in many plants. They may all be on the same shaft to drive a common generator, but they will have separate cases. With a reheater, the flow is extracted after a partial expansion, run back through the boiler to heat it back up to the peak temperature, and then passed to the lower pressure turbine. A component diagram for a single reheater is provided in Fig. 4.15.

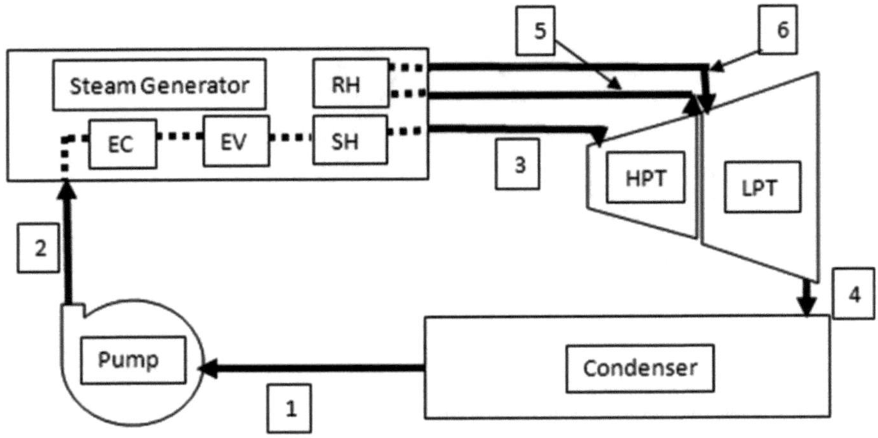

Fig. 4.15 Superheated Rankine cycle with single reheater

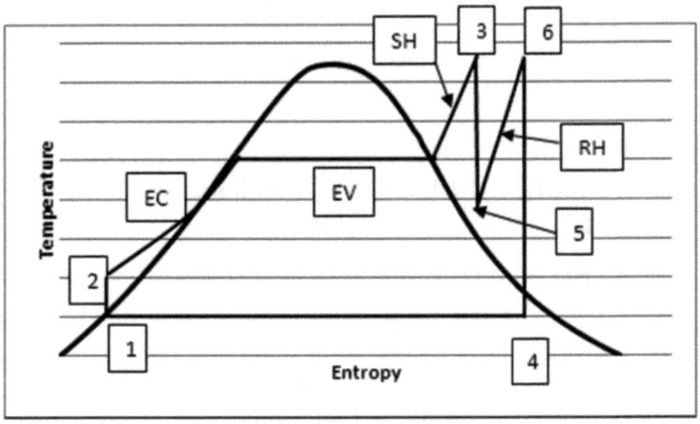

Fig. 4.16 *T-s* diagram for superheated Rankine cycle with single reheater

The fluid is extracted after a partial expansion in the high-pressure turbine at state 5. Typically, the extraction pressure is around one-fourth of the peak pressure. It is rerouted to the boiler and reheated to the peak temperature at state 6. It may even be heated to a higher temperature as the components may be able to stand a higher temperature at the reduced pressure. It is then routed back to the intermediate- or low-pressure turbine to be expanded to condenser pressure [1].

The temperature-entropy diagram is provided in Fig. 4.16. Note that state point 5 may or may not be under the vapor dome. If it is not under the vapor dome, the superheated steam tables must be interpolated based on the known pressure and entropy to find the enthalpy at state point 5.

Most large steam plants today typically use three turbines and two reheaters. The analysis follows that for a single reheater but is repeated second time. In each case, a rule of thumb is that the pressure out of the turbine should be approximately one-fourth of the inlet pressure.

Example 4.11 Add a reheater to the cycle of Example 4.9 after the superheated steam has been expanded to 250 psi.

Solution At 250 psi and $s_3 = 1.67258$ Btu/lbm interpolate to find h_5

$$
\begin{aligned}
&s(200\,\mathrm{psi}, 1000\,\mathrm{R}) = 1.65013,\\
&h(200\,\mathrm{psi}, 1000\,\mathrm{R}) = 1293.6 s(300\,\mathrm{psi}, 1000\,\mathrm{R}) = 1.59813,\\
&h(300\,\mathrm{psi}, 1000\,\mathrm{R}) = 1284.2 s(250\,\mathrm{psi}, 1000\,\mathrm{R}) = 1.64213,\\
&h(250\,\mathrm{psi}, 1000\,\mathrm{R}) = 1288.9 s(200\,\mathrm{psi}, 1000\,\mathrm{R}) = 1.7002,\\
&h(200\,\mathrm{psi}, 1100\,\mathrm{R}) = 1346.2 s(300\,\mathrm{psi}, 1100\,\mathrm{R}) = 1.65091,\\
&h(300\,\mathrm{psi}, 1100\,\mathrm{R}) = 1339.6 s(250\,\mathrm{psi}, 1100\,\mathrm{R}) = 1.67556,\\
&h(250\,\mathrm{psi}, 1100\,\mathrm{R}) = 1342.9 y = \frac{1.67258 - 1.64213}{1.67556 - 1.64213} = 0.911 \quad h_5 = 1288.9\\
&+0.911^*(1342.9 - 1288.9) = 1388.1\,\mathrm{Btu/lbm}\Delta h_{3-5} = 1532.5\\
&-1388.1 = 144.4\,\mathrm{Btu/lbm} \quad h_6(250\,\mathrm{psi}, 1500\,\mathrm{R}) = 1552.8\,\mathrm{Btu/lbm}\\
&\quad s_6 = 1.83603\,\mathrm{Btu/lbm/R}\Delta h_{5-6} = 1552.8\\
&-1388.1 = 164.7\,\mathrm{Btu/lbm} x_4 = \frac{1.83603 - 0.18954}{1.71547} = 0.9598 \quad h_4 = 102.6\\
&+0.9598^*\,1019.1 = 1080.7\,\mathrm{Btu/lbm}\Delta h_{6-4} = 1552.8\\
&-1080.7 = 472.1\,\mathrm{Btu/lbm} \quad \Delta h_{2-3} = 1532.5\\
&-105.6 = 1426.9\,\mathrm{Btu/lbm}\eta_{\mathrm{th}} = \frac{144.4 + 472.1 - 3.0}{1426.9 + 164.7} = 0.3855 \quad \eta_{\mathrm{th}} = 38.55\%\\
&\quad \eta_{\mathrm{II}} = \frac{0.3855}{0.6039} = 0.6384
\end{aligned}
$$

Note that the current generation of LWRs only produces saturated steam and cannot use nuclear-driven superheaters or reheaters. Advanced reactors including liquid metal reactors, gas-cooled reactors, and molten salt reactors may be able to take advantage of superheating and reheating to improve their cycle efficiencies. From a cost of operation perspective, improvements in efficiency are not as important for nuclear plants as they are for fossil plants because the cost of fuel is a much smaller fraction of the total operating cost for a nuclear plant. However, it still pays to improve thermal efficiency in all cases.

4.13 The Carnot Cycle

The Carnot cycle is illustrated in Figs. 4.17 and 4.18. This is the basis for all heat engine cycles and the turbine cycle in particular. The segments in the cycle are

$$
\begin{array}{lll}
a - b & \text{Energy addition,} & \Delta E = Q_2 \\
b - c & \text{Shaft work,} & \Delta E - W_2 \\
c - d & \text{Condensation,} & \Delta E = -Q_1 \\
d - a & \text{Pressurization,} & \Delta E = W_1
\end{array}
$$

The overall efficiency η is

$$\eta = \frac{\text{net work done}}{\text{heat input}} = \frac{W}{Q_2} = \frac{Q_2 - Q_1}{Q_2} \tag{4.27}$$

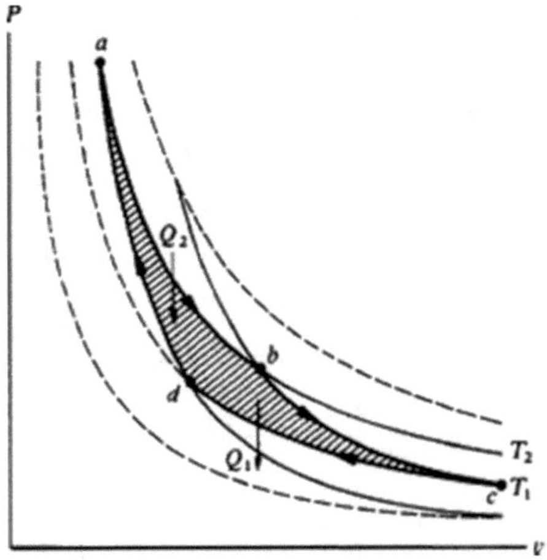

Fig. 4.17 The Carnot cycle

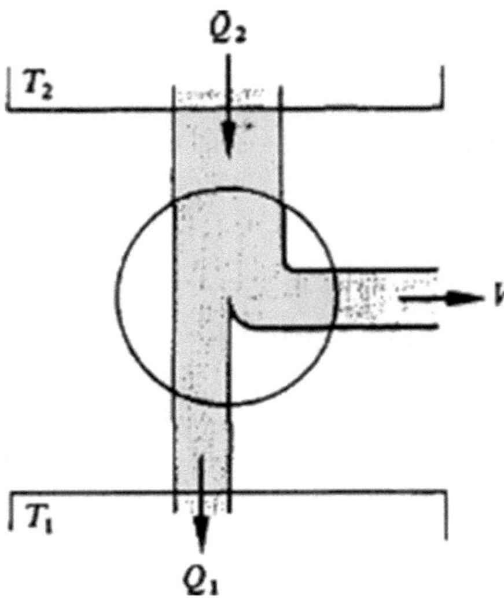

Fig. 4.18 Schematic flow diagram of a heat engine

($W = Q_2 - Q_1$ since there is no net change in energy in a complete cycle, i.e., $\Delta E = 0 = Q - W =$ net heat addition − net work done, i.e., $Q = Q_2 - Q_1 = W$). Q_1 is the rejected heat

It can be shown that [1]

$$\frac{|Q_2|}{|Q_1|} = \frac{T_2}{T_1} \tag{4.28}$$

Thus

$$\eta = \frac{T_2 - T_1}{T_2} = 1 - \frac{T_1}{T_2} \tag{4.29}$$

4.14 Entropy

Since Q_2 is heat flow *in* and Q_1 is heat flow *out*, they are opposite. Thus, from 4.28, we conclude that

$$\frac{T_2}{T_1} = -\frac{Q_2}{Q_1} \quad \rightarrow \quad \frac{Q_1}{T_1} + \frac{Q_2}{T_2} = 0 \tag{4.30}$$

This can be generalized by splitting a general reversible cycle as in Figs. 4.19 and 4.20. We split the cycle up into many small Carnot cycles. The common boundaries cancel. For each small cycle

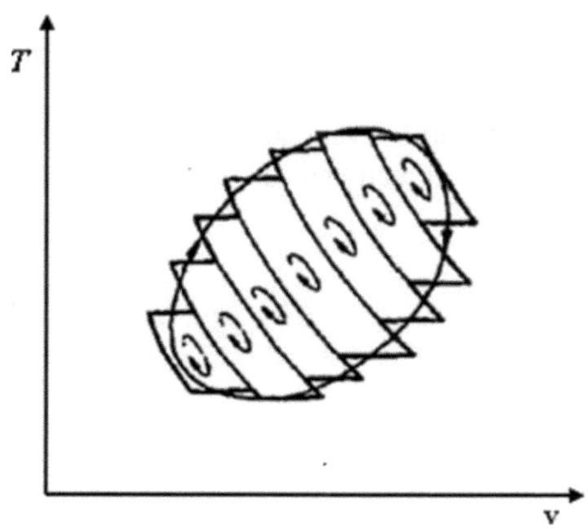

Fig. 4.19 An arbitrary reversible cyclic process can be approximated by a number of small Carnot cycles

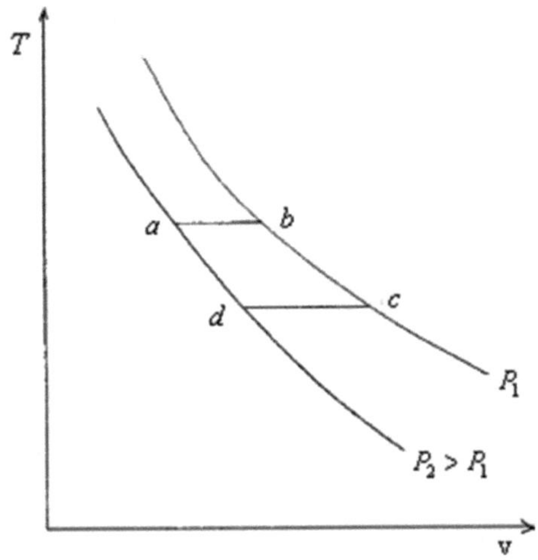

Fig. 4.20 *T-v* diagram for the Carnot cycle

$$\frac{\Delta Q_1}{T_1} + \frac{\Delta Q_2}{T_2} = 0 \tag{4.31}$$

Summing *all* cycles deduces that

$$\begin{aligned} \sum \frac{\Delta Q}{T} &= 0 \\ \oint \frac{dQ}{T} &= 0 \end{aligned} \tag{4.32}$$

Since the closed integral is equal to zero, dQ/T must be an exact differential and must be a *state variable*, i.e., a property of the state of the material, like u, P, T, ρ ,etc. We define this to be the entropy, S. Thus, we can write

$$\oint dS = 0 \tag{4.33}$$

So, since S is a property of a system, we can express any equilibrium state in terms of S plus one other state variable (T, P, or whatever).

The Carnot cycle now becomes as shown in Fig. 4.21. The *T-s* diagram gives the heat flow directly since

$$\int_a^b TdS = \int_a^b dQ = Q_2$$

and

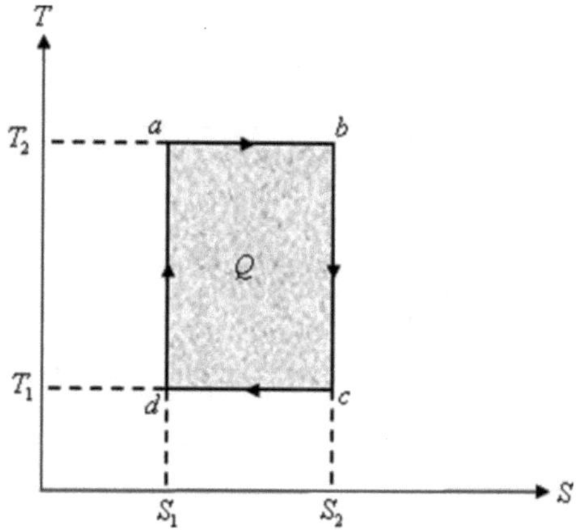

Fig. 4.21 The temperature-entropy diagram for the Carnot cycle

$$\int_{c} TdS = Q_1$$

Therefore

$$\begin{aligned} \oint TdS &= \int_a^b + \int_b^c + \int_c^d + \int_d^c \\ &= Q_2 + 0 - Q_1 + 0 = Q_2 - Q_1 \\ &= \text{net heat flow into the system} \end{aligned} \tag{4.34}$$

Notice this is consistent with the previous definition of efficiency η:

$$\begin{aligned} \eta &= \frac{Q_2 - Q_1}{Q_2} = \frac{T_2(S_2 - S_1) - T_1(S_2 - S_1)}{T_2(S_2 - S_1)} \\ &= \frac{T_2 - T_1}{T_2} = 1 - \frac{T_1}{T_2} \end{aligned} \tag{4.35}$$

4.15 Reactor Power Cycle

The thermodynamic power cycle in reactor systems is similar to the Carnot cycle. As sketched in Fig. 4.22, the steam generator boils the working fluid (water) isothermally (sort of), the turbine expands the fluid automatically and performs shaft work, the condenser extracts the rejected heat and condenses the fluid, and the water pump returns the fluid to the steam generator at pressure. Of course, the cycle is not reversible, but the principles of the cycle are the same. The typical cycle used in power plants is called the Rankine cycle.

The *T*-*s* and *h*-*s* diagrams for an ideal simple Rankine cycle are given in Fig. 4.23.

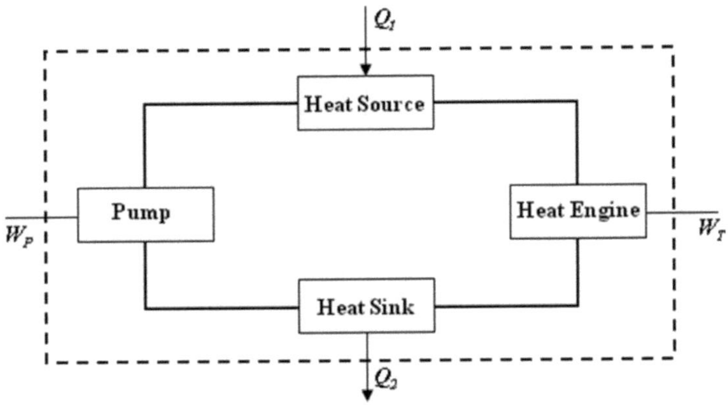

Fig. 4.22 Schematic diagram for a reactor power cycle

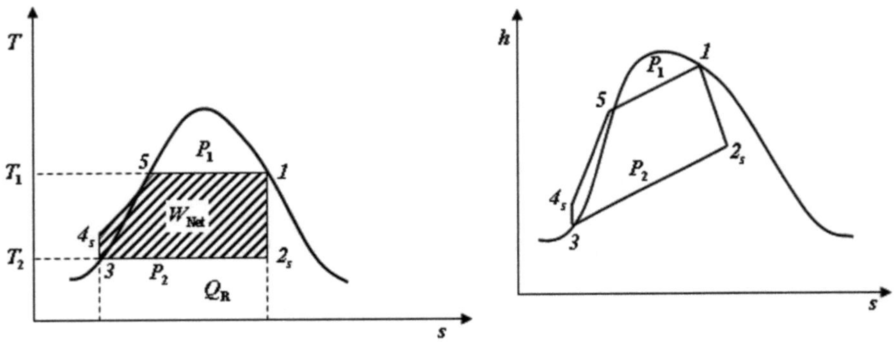

Fig. 4.23 *T-s* and *h-s* diagram representations for the ideal Rankine cycles

Note: In Fig. 4.23, we are assuming fluid velocities are zero; i.e., the diagram illustrates stagnation properties.

The *h-s* diagram is useful for calculation purposes, while the *T-s* diagram is useful for illustration purposes.

In the ideal Rankine cycle, saturated steam (shown as point I) enters the turbine and expands isentropically to position 2_s. At point 2_s the wet steam enters the condenser where heat is removed until the fluid is condensed to a saturated liquid at point 3. After leaving the condenser, the fluid is condensed isentropically from pressure P_2 to the boiler pressure P_1. The high pressure liquid at point 4 enters the boiler, where the fluid is vaporized and emerges as steam at point 1.

The shaded area represents the net work done ($W = Q_2 - Q_1$), the total under the cycle curve represents the heat additional, Q_2, and the unshaded area is the rejected heat, Q_1.

From the *h-s* diagram, it is straightforward to determine (on a unit mass basis)

Shaft work of the turbine $= W_T = h_1 - h_{2s}$
Pumping work $= W_P = h_{4s} - h_3$
Heat input $= Q_2 = h_1 - h_{4s}$

$$\eta = \frac{W_T - W_P}{Q_2} = \frac{W_{NET}}{Q_2} = \frac{(h_1 - h_{4s}) - (h_{4s} - h_3)}{(h_1 - h_{4s})} \tag{4.36}$$

Note: The above expression for η can be arranged to give as expected below:

$$\eta = \frac{(h_1 - h_{4s}) - (h_{4s} - h_3)}{(h_1 - h_{4s})} = \frac{Q_2 - Q_1}{Q_2} \tag{4.37}$$

Turbine performance is frequently given as the turbine heat rate:

$$\text{Turbine heat rate} \equiv \frac{\text{heat supplied to boiler}}{\text{net work out}} = \frac{1}{\eta} \tag{4.38}$$

Power cycle performance is improved in practice by:

(1) Raising the boiler pressure
(2) Lowering the exhaust pressure
(3) Using superheat
(4) Using reheat

(1), (3), and (4) effectively raise the inlet temperature, while (2) effectively lowers the outlet temperature with attendant effect on the cycle efficiency [1].

The condenser pressure is limited by the temperature of available cooling water, size and cost of the condenser, and size of the vacuum pumps required to desecrate the condenser.

Consequently, the practical lower limit in condenser pressure is a few centimeters of H_g abs.

Consequently, (1), (3), and (4) are used in achieving increases in efficiency.

4.16 Raising Boiler Pressure

The effect of increasing the boiler pressure on Rankine cycle efficiency is readily shown on the *T-s* diagram (See Fig. 4.24). Increasing boiler pressure results in an increase in net work (represented by the shaded area) with a corresponding decrease in heat rejected.

However, for the indirect power cycle, the downside of raising the boiler pressure (and temperature since the steam is saturated) is that it forces the primary side temperature up to provide sufficient ΔT to transfer the heat from the primary to secondary side. This higher primary side temperature pushes the fuel closer to its

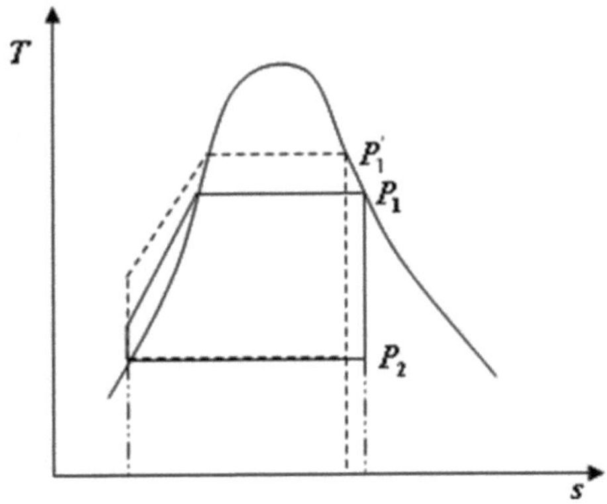

Fig. 4.24 Effects of increasing boiler pressure on the Rankine cycle

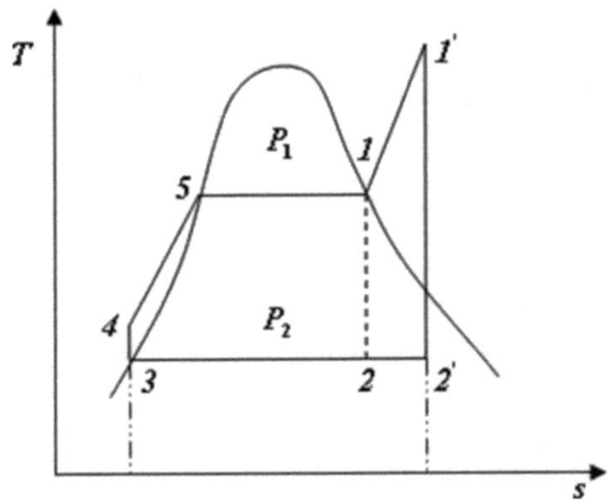

Fig. 4.25 Rankine cycle with superheat

limits and increases the tendency for the fluid to boil. To counter this, if necessary, the primary side pressure would have to be increased, and pressure vessel walls would have to be thicker. In a pressure-vessel-type reactor, this can be costly or lead to reduced reliability. In pressure tube reactors, the main drawback is the increased parasitic neutron absorption and consequent higher burnup.

4.17 Superheat

Figure 4.25 illustrates the Rankine cycle with superheat. Superheat causes a net increase in temperature at which heat is being received with a resulting improvement in cycle efficiency.

Another important factor is that the amount of moisture in the fluid leaving the turbine is reduced which increases turbine efficiency and reduces erosion. However, in order to make use of superheat, one must have a high-temperature heat source or reduce boiler pressure.

4.18 Reheat

The effective temperature of heat addition is increased, and the moisture content further reduced by using reheats in the Rankine cycle. A schematic diagram of the power plant and appropriate temperature-entropy diagram is shown in Fig. 4.26. High-pressure, superheated steam is expanded in a high-pressure turbine to an intermediate pressure P_2', and the fluid then returned to a second-stage boiler and is superheated and reheated to state $1''$. The reheated steam is then expanded in a low-pressure turbine to the final exhaust pressure P_2''. The moisture content of the working fluid is drastically reduced by use of reheat, and this approach is used in all fossil-fueled and many nuclear power plants. The approach used to compute the work and efficiency of reheat cycles is the same as used in the example problem for the simple Rankine cycle. One calculates the work produced in each turbine separately and the required pumping work. Heat is added to the fluid at two different

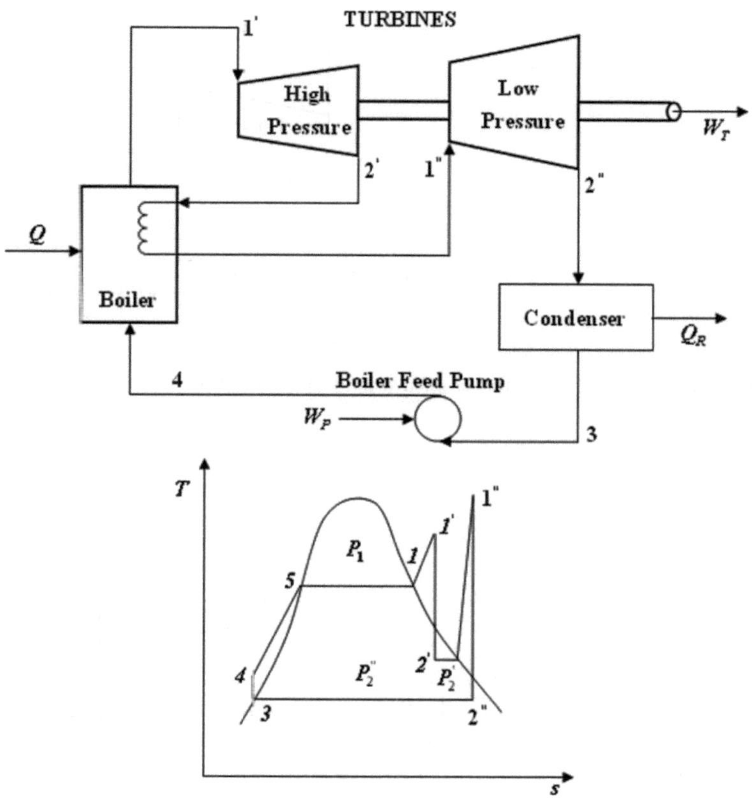

Fig. 4.26 Rankine cycle with reheat

stages of the cycle and is given by the difference in enthalpy between states $1'$ and 4 and states $1''$ and $2''$.

4.19 Regeneration

Modifications to the cycle can also be made to reduce cycle irreversibility. One of the principle sources is the sensible heat addition required to bring the boiler feedwater up to saturation temperature. This is accomplished by using some of the flow through the turbine to heat the feedwater. To achieve reversibility, the setup would be as in Fig. 4.27, but this is impractical. A practical setup is shown in Fig. 4.28. Analysis is beyond the scope of this book.

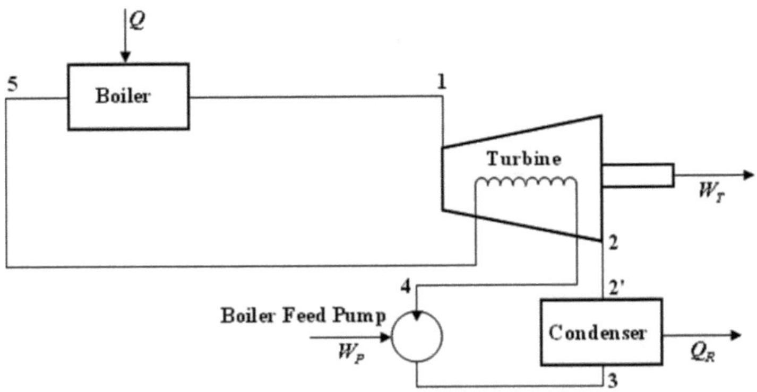

Fig. 4.27 Schematic diagram of a power plant with ideal regeneration

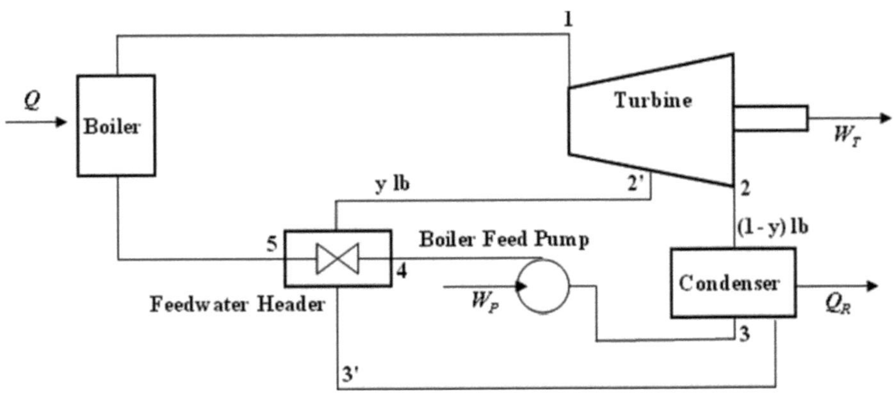

Fig. 4.28 Single heater regenerative cycle

References

1. Zohuri, B., & McDaniel, P. J. (2018). *Thermodynamics in nuclear power plant systems* (2nd ed.). New York: Springer.
2. Zohuri, B. (2017). *Thermal-hydraulic analysis of nuclear reactors* (2nd ed.). New York: Springer.

Chapter 5
Modeling the Nuclear Air Brayton Combined Cycle

Given that the combined cycle (CC) code does a good job of modeling current-generation gas turbine combined cycle (GTCC) plants, it is useful to extrapolate its capabilities to Nuclear Air-Brayton Combined Cycle (NACC) power plants and Nuclear Air-Brayton Recuperated Cycle (NARC) power plants. The combined cycle plants will be dealt with in this chapter and the recuperated plants in the next chapter. In the Nuclear Air-Brayton power plants, the combustion chamber of the gas turbine system is replaced by the nuclear reactor and a heat exchanger. The nuclear reactor will heat a working fluid, and that working fluid will in turn pass through a heat exchanger to heat the air for the turbine. Because the heat transfers process for a nuclear system is in the opposite direction (solid to gas) from that in the gas turbine (gas to solid), the peak temperatures achievable in a Nuclear Air Brayton system will never be as high as those in a gas turbine system. However, the nuclear system can reheat the air multiple times and expand it across multiple turbines to increase the available power.

In comparing nuclear systems, there are multiple advantages for the Nuclear Air-Brayton cycle over conventional light water cycles. The ability to go to higher temperatures and not be trapped under the water vapor dome adds a major flexibility. The higher gas temperatures allow for the implementation of a combined cycle approach to improve system efficiency. There is also a much larger base of gas turbine manufacturers than large steam turbine manufacturers for system designers to draw from. All downstream components after the turbines themselves are currently being used in GTCC plants. A NACC plant will require significantly less water for waste heat removal than a current Light Water Reactor (LWR) system because the efficiency is greater, and a significant amount of the waste heat is directly dumped into the atmosphere without going through a cooling tower. Finally, the NARC system dumps all of its heat into the atmosphere and as a result is not tied to a location near a water source [3].

B. Zohuri, *Heat Pipe Applications in Fission Driven Nuclear Power Plants*,
https://doi.org/10.1007/978-3-030-05882-1_5

5.1 Introduction

In order to demonstrate the possibilities of Nuclear Air-Brayton systems a two-level approach will be pursued. The performance of a Nuclear Air-Brayton system using fairly well-developed sodium technology will be the near-term possibility. The main characteristic for this type of system will be a heat exchanger output temperature of 510 °C, in the range of past deployed systems [1]. A higher temperature system based on the molten salt concepts or the lead bismuth concept as described in Chap. 7 will be considered also. The heat exchanger output temperature for this class of system will be estimated at 660 °C. The High Temperature Gas Reactors will do better, but helium to air heat exchangers have not been considered as they present several different problems [3].

A nominal set of conditions has been chosen as a best estimate for environmental conditions and component performance for each of these systems. Basically, the environment considered is a sea level standard day. For the combined cycle systems, the number of turbines, the turbine exit temperature, the compressor pressure ratio, and the steam pressure in the Rankine bottoming cycle were varied to achieve the maximum thermodynamic efficiency. After the optimum efficiency was determined, the sensitivity of this result to important parameters was estimated. The nominal input parameters for the two systems follow [3].

5.2 Nominal Analysis Parameters

- Number of turbines—Varied
- Turbine inlet temperature—510, 660 °C
- Turbine exit temperature—varied
- Turbine polytropic efficiency—calculated based Wilson-Korakianitis correlation [2]
- Compressor pressure ratio—calculated
- Compressor polytropic efficiency—calculated based on Wilson-Korakianitis correlation [2]
- Main heater pressure ratios—0.99
- Atmospheric pressure—1 atm
- Atmospheric temperature—288 K (15 °C)
- Circulating water input temperature—288 K (15 °C)
- Ratio of exhaust pressure to atmospheric—0.98
- Air pressure ratios across each superheater—0.99
- Air pressure ratio across the evaporator—0.99
- Air pressure ratio across the economizer—0.99
- Pinch point temperature difference—10 K
- Terminal temperature difference at steam exit to superheaters—15 K
- Peak Rankine cycle pressure—varied
- Intermediate Rankine cycle pressure—varied, 1/4 of peak
- Low Rankine cycle pressure—varied, 1/16 of peak

- Condenser pressure—9 kPa
- Steam turbines thermal efficiency—0.90
- Power level—50 MW [3]

The Wilson Korakianitis correlation for turbines is [2]

$$\eta_{\text{poly}} = 0.7127 + 0.03^{*}\ln(d_t) - 0.0093^{*}\ln(1/\text{pr}_t)$$

And for compressors is

$\eta_{\text{poly}} = 0.862{+}0.015^{*}\ \ln(\text{mdot}) - 0.0053^{*}\ \ln(\text{pr}_c)$
d_t = turbine mean diameter in meters
pr_t = turbine pressure ratio
pr_c = compressor pressure ratio
mdot = compressor mass flow rate in kg/s

5.3 Nominal Results for Combined Cycle Model

The turbine exit temperatures and the peak pressure in the steam bottoming cycle were varied for systems using one to five turbines. The best efficiency achievable in each case is plotted in Fig. 5.1. The efficiency is a monotonic function of the number of turbines, with the five-turbine system only very slightly better than the four-turbine system. The four-turbine case was chosen as the baseline representative combined cycle system. Therefore, one reactor supports four heat exchangers, one before each turbine. The peak efficiency for the four-turbine case is 47.78% and for the five-turbine case is 48.4%, for the 660 °C system.

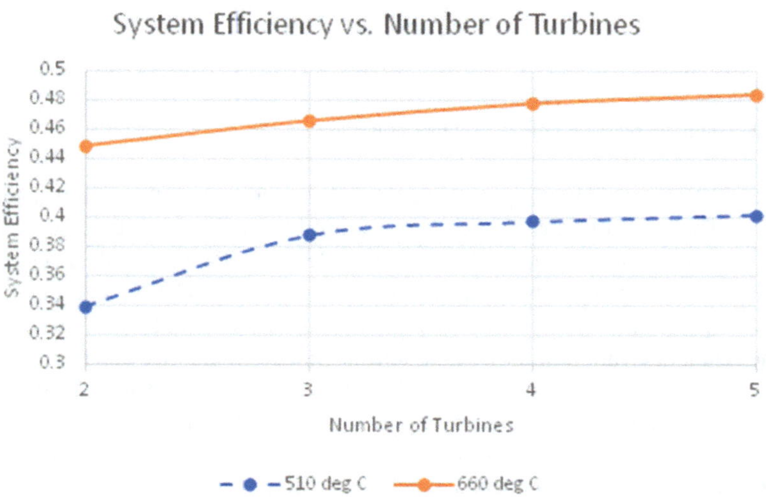

Fig. 5.1 Peak overall thermal efficiency vs. number of turbines

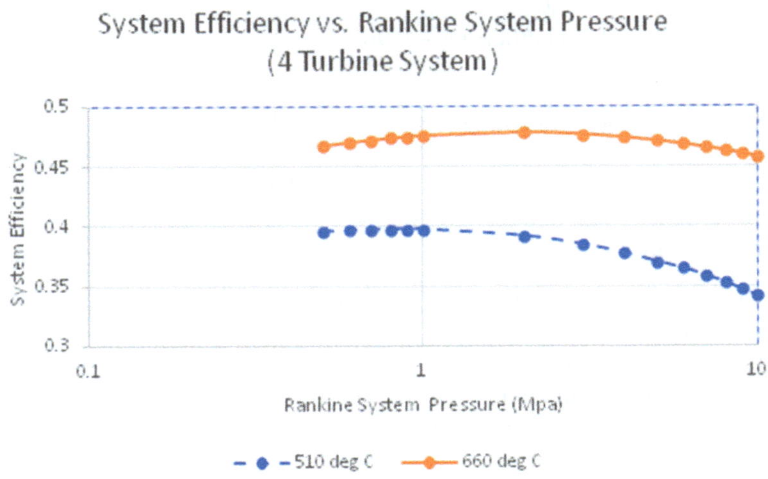

Fig. 5.2 Overall efficiency vs. steam system pressures

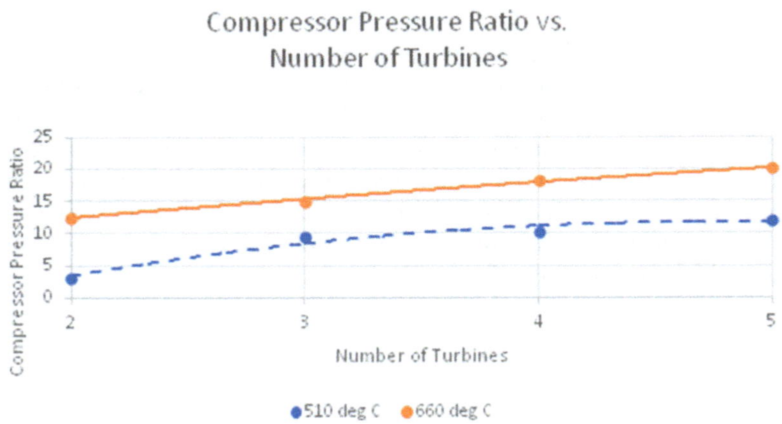

Fig. 5.3 Optimum pressure ratio vs. number of turbines

The 510 °C system is approximately 8% lower across three, four, and five turbines. The optimum turbine exit temperature for the four-turbine case is 515 °C with a best steam pressure in the bottoming cycle of 2 MPa for the advanced, 660 °C systems. The exit temperature for the four-turbine case is 410 °C and a best steam pressure in the bottoming cycle of 0.9 MPa for the near term, 510 °C systems.

The efficiency variation for both systems is plotted in Fig. 5.2 as a function of the bottoming cycle pressure. Obviously, the efficiency is not a strong function of the bottoming steam cycle pressure as long as it is fairly close to the optimum.

The optimum compressor ratio to achieve peak efficiency and keep the exhaust pressure above the environmental pressure is plotted in Fig. 5.3 for each of the

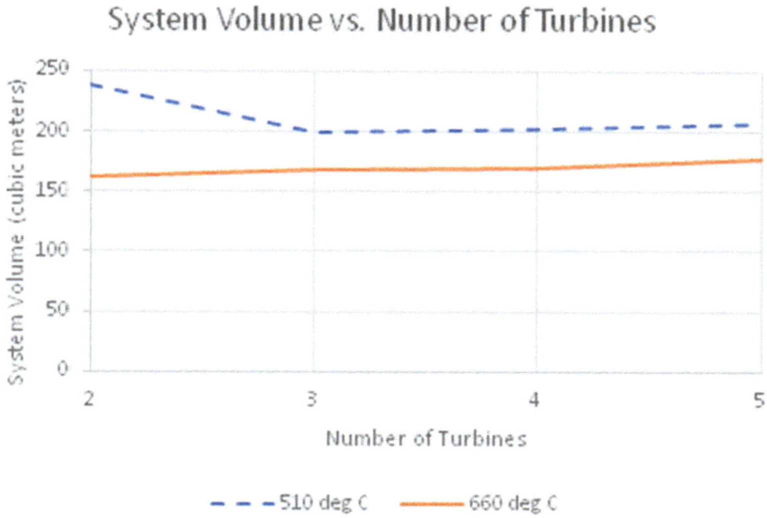

Fig. 5.4 NACC system volume vs. number of turbines

turbine models. Note that there is a definite optimum compressor ratio to obtain peak efficiency for each turbine inlet temperature and the number of turbines. Obviously, the compressor ratio has to increase as the number of turbines increases, as there is only one compressor, and in this model, there are several turbines.

Since the exhaust temperature and the steam pressure in the bottoming cycle are varying in a discrete manner in fairly large steps, the calculated points do not follow a smooth curve, so a power fit to the data has been included.

The baseline system chosen here is a 50 MW(e) Small Modular Reactor. The main purpose of designing the heat exchangers and estimating sizes for pump, compressor, and turbine components was to get an overall estimate of the size for the complete power conversion system. This estimate is provided in Fig. 5.7 below. Note that for the 50 MW(e) system, the air turbines produce 57% of the power, and the steam turbines produce 43% of the power. The estimated system volume as a function of the number of turbines is plotted in Fig. 5.4. For volume estimation, a pool sodium reactor and heat exchanger system were included. Its size was estimated based on historical data reported in Waltar and Reynolds [1]. For comparison the volume of the most advanced Small Modular Reactor, the NuScale design, is approximately 380 m^3. NuScale is baselined at 50 MW(e) also.

The system volume was estimated as a function of electrical power produced for other power levels based on a four-turbine NACC system. This data is presented in Fig. 5.5. The volume varies fairly linearly with power level, and the higher-temperature advanced system will be about 20% smaller.

For the 50 MW(e) system, the reactor fills 52% of the volume, the air-Brayton system fills 13% of the volume, and the Rankine bottoming cycle fills 35% of the volume.

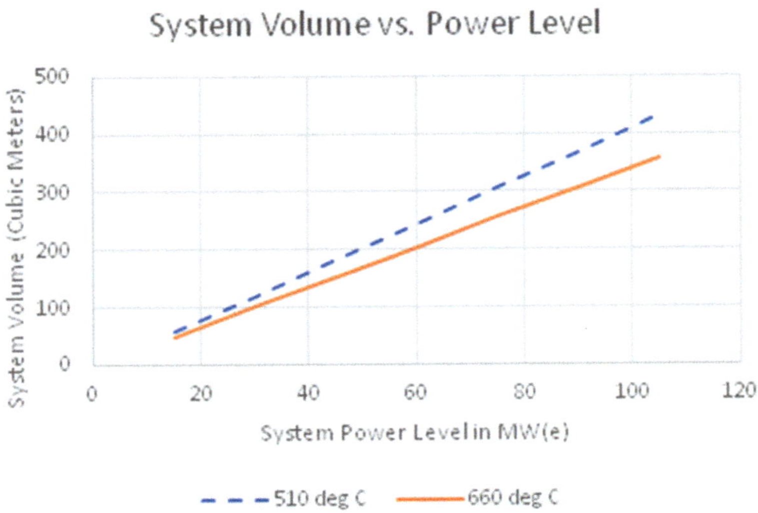

Fig. 5.5 System volume vs. electrical power level

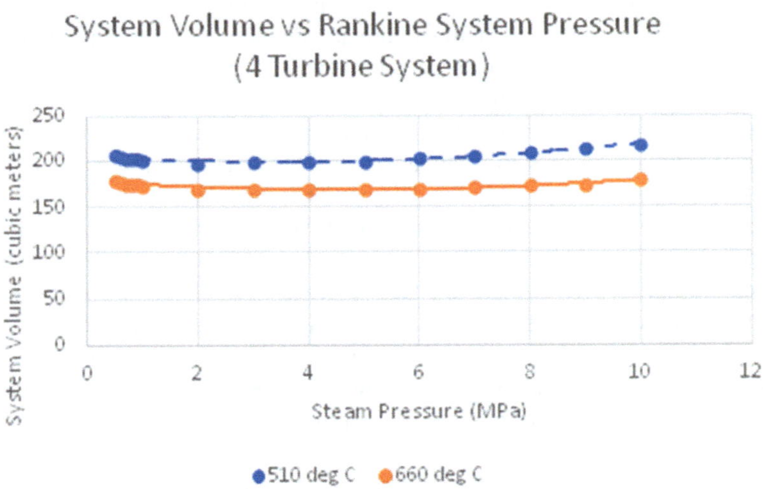

Fig. 5.6 System volume as a function of steam system pressure

Since the Rankine bottoming cycle is the largest component of the power conversion system, the volume filled by this system was investigated as a function of its operating pressure since efficiency does not strongly depend on this pressure. The dependence of the system volume on the Rankine bottoming cycle pressure is presented in Fig. 5.6.

Three major sensitivities were considered—compressor polytropic efficiency, pressure drops (or ratios) in the main heaters, and the environmental temperature. Figure 5.7 presents the dependence of the overall thermal efficiency on the gas compressor polytropic efficiency.

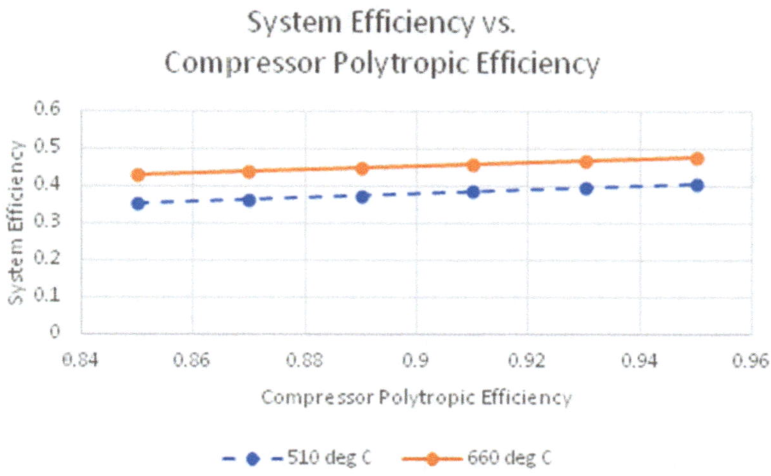

Fig. 5.7 Overall cycle efficiency vs. compressor polytropic efficiency

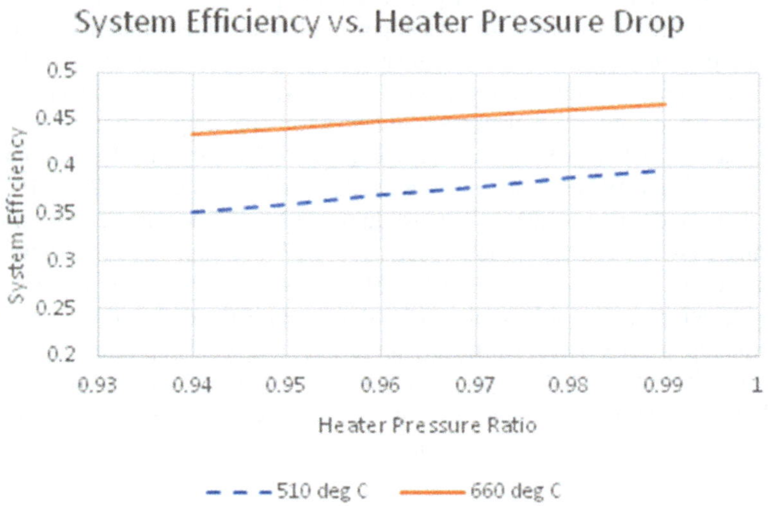

Fig. 5.8 Overall cycle efficiency vs. heater pressure ratio (four turbines)

The overall thermal efficiency increases about 0.5% for every 1% increase in the compressor polytropic efficiency.

The pressure drops in the main heaters are analogous to the pressure drop in a combustion chamber. Nominally the pressure drop in each heater was set to 1%, and a compact heat exchanger was designed to achieve this drop, in order to estimate system volumes. The nominal pressure drop in a gas turbine is on the order of 3–5%, so the heat exchangers were designed to be comparable. The effect of a variation in these pressure drops on the overall system efficiency is plotted in Fig. 5.8.

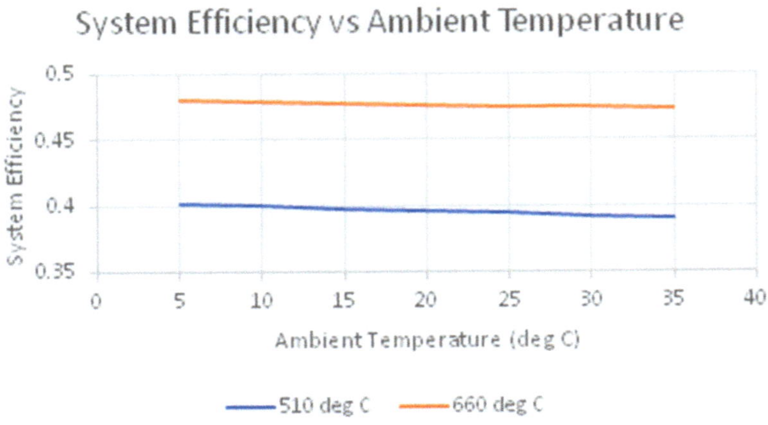

Fig. 5.9 Overall efficiency vs. ambient temperature (four turbines)

The overall efficiency drops approximately 0.9% for every 1% increase in the pressure drop in the main heater for the four-turbine system.

The standard day conditions of 15 °C and 1 atm pressure are not likely to be met very often during the operation of typical power plants, so the variation of efficiency with the ambient temperature was estimated. The results are presented in Fig. 5.9 above. The overall efficiency drops about 1.4% for every 10 K increase in temperature.

It is worth pointing out that a sodium reactor operating with a turbine inlet temperature of 510 °C achieves a thermodynamic efficiency of 39.75% compared to a nominal Light Water Reactor (LWR) at 35% and a NuScale SMR at 31%. This means that a sodium-cooled NACC must dump 43 MW of heat through its circulating water system. The LWR must dump 93 MW through its circulating water system, and the NuScale must dump 111 MW through its circulating water system. The higher temperature molten salt reactor does even better dump only 31.5 MW to its circulating water system. Ultimately all of this heat winds up in the atmosphere either directly with a NACC system or through a cooling tower with any of the closed cycle systems. The main difference is that the NACC systems use far less fresh water to accomplish this heat dump [3].

References

1. Waltar, A. E., & Reynolds, A. B. (1981). *Fast breeder reactors*. New York: Pergamon Press.
2. Wilson, D. G., & Korakianitis, T. (1998). *The design of high-efficiency turbomachinery and gas turbines* (2nd ed., p. 115). New Jersey: Prentice Hall.
3. Zohuri, B., & McDaniel, P. (2018). *Combined cycle driven efficiency for next generation nuclear power plants: An innovative design approach* (2nd ed.). New York: Springer.

Chapter 6
Basic Principles of Heat Pipes and History

The heat pipe is one of the remarkable achievements of thermal physics and heat transfer engineering in this century because of its unique ability to transfer heat over large distances without considerable losses. The main applications of heat pipes deal with the problems of environmental protection and energy and fuel savings. Heat pipes have emerged as an effective and established thermal solution, particularly in high heat flux applications and in situations where there is any combination of nonuniform heat loading, limited airflow over the heat-generating components, and space or weight constraints. This chapter will briefly introduce heat pipe technology and then highlight its basic applications as a passive thermal control device [1].

6.1 Introduction

The original idea of heat pipe was considered in 1944 by Gaugler [2] and in 1962 by Trefethen [3]. Gaugler patented a very lightweight heat transfer device that was essentially a very basic presentation of heat pipe. During that time period, the technology did not require a need for such sophisticated yet constructively simple two-phase and passive heat transfer device, and there was not much attention that was paid to it. It was first suggested by Trefethen [3] in 1962 and then appeared through a patent application of heat pipe again in 1963 by Wyatt [4]. It was not widely considered and publicized until 1964 when George Grove [5] and his co-worker at the Los Alamos National Laboratory independently reinvented the same concept for their existing space program and its application. He is the one who named this most satisfactory and simplistic heat transmission device "heat pipe" and developed its applications.

Heat pipes are two-phase flow heat transfer devices where a process of liquid to vapor and vice versa circulates between evaporator and condenser with high effective thermal conductivity. Due to the high heat transport capacity, heat exchanger

B. Zohuri, *Heat Pipe Applications in Fission Driven Nuclear Power Plants*,
https://doi.org/10.1007/978-3-030-05882-1_6

with heat pipes has become much smaller than traditional heat exchangers in handling high heat fluxes. With the working fluid in a heat pipe, heat can be absorbed on the evaporator region and transported to the condenser region where the vapor condenses releasing the heat to the cooling media. Heat pipe technology has found increasing applications in enhancing the thermal performance of heat exchangers in microelectronics; energy saving in classical heating, ventilating, and air conditioning (HVAC) systems for operating rooms, surgery centers, hotels, clean rooms, etc.; temperature regulation systems for the human body; and other industrial sectors including spacecraft and various types of nuclear reactor technologies as a fully inherent cooling apparatus. The heat pipe is a self-contained structure which achieves very high thermal energy conductance by means of two-phase fluid flow with capillary circulation. A heat pipe operates within a two-phase flow regime as an evaporation-condensation device for transferring heat in which the latent heat of vaporization is exploited to transport heat over long distances with a corresponding small temperature difference.

Heat added to the evaporator is transferred to the working fluid by conduction and causes vaporization of the working fluid at the surface of the capillary structure. Vaporization causes the local vapor pressure in the evaporator to increase and vapor to flow toward the condenser, thereby transporting the latent heat of vaporization. Since energy is extracted at the condenser, the vapor transported through the vapor space is condensed at the surface of the capillary structure, releasing the latent heat. Closed circulation of the working fluid is maintained by capillary action and/or bulk forces. An advantage of a heat pipe over other conventional methods to transfer heat such as a finned heat sink is that a heat pipe can have an extremely high thermal conductance in steady-state operation. Hence, a heat pipe can transfer a high amount of heat over a relatively long length with a comparatively small temperature differential. Heat pipe with liquid metal working fluids can have a thermal conductance of a thousand or even tens of thousands folds better than the best solid metallic conductors, silver or copper. In a heat pipe, energy is transported by utilizing phase change of the working substance instead of a large temperature gradient and without external power. Also, the amount of energy transferred through a small cross section is much larger than that by conduction or convection. Heat pipes may be operated over a broad range of temperatures by choosing an appropriate working fluid (Fig. 6.1).

However, this useful device has some operating limitations such as the sonic, the capillary, the entrainment, and finally the boiling limit, which will be discussed throughout the book. When any of these limitations is encountered, the capillary structure may dry out leading to failure of the heat pipe. In addition to these limitations, when liquid metal is used as the working fluid, start-up difficulty may take place due to possible solid state of the working fluid and extremely low vapor density.

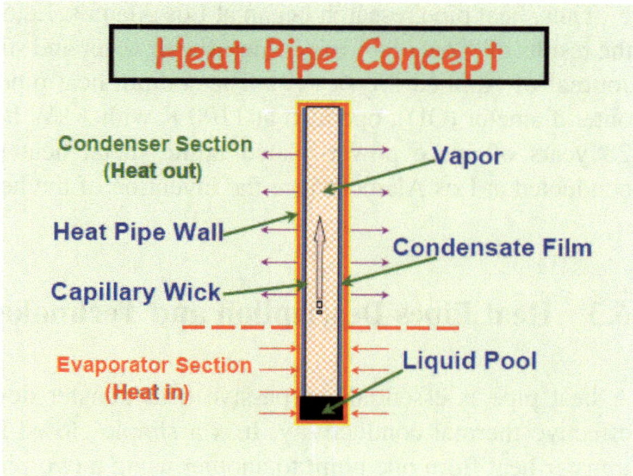

Fig. 6.1 Heat pipe concept

6.2 Heat Pipe History

Early research in heat pipes conducted at Los Alamos was directed to applications in space-based thermionic energy conversion systems operating in excess of 1500 K. Heat pipes were considered for heating thermionic emitters, for cooling thermionic collectors, and for the ultimate radiation of heat to space fluids, and materials were tailored to this temperature regime. Experiments with a Nb-1%Zr heat pipe, with lithium operating at 1573 K, 207 W/cm^2 evaporator radial heat flux; a 1.95 kW/cm^2 axial heat flux and an Ag-Ta operating at 2273 K, 410 W/cm^2 evaporator radial heat flux; and a 4 kW/cm^2 axial heat flux are reported in Deverall and Kemme [6]. The results of early thermionic-related heat pipe fluid-wall compatibility and life test studies with systems of In–W at 2173 K for 75 h, Ag–Ta at 2173 K for 100 h, Cs–Ti at 673 K in excess of 2000 h, Na-stainless steel at 1073 K for 500 h, and Li-Nb-1% Zr at 1373 K for 4300 h are summarized in Grover et al. [7], Deverall and Kemme [6], Grover et al. [8], Cotter et al. [9], and Ranken and Kemme [10]. A study characterizing both potassium and sodium heat pipes with various wick structures and a treatment of the limitations to heat pipe start-up and operation is contained in Kemme [11].

On July 24, 1963, George Grover made the following entry into his laboratory notebook: "Heat transfer via capillary movement of fluids. The 'pumping' action of surface tension forces may be sufficient to move liquids from a cold temperature zone to a high-temperature zone (with subsequent return in vapor form using as the driving force the difference in vapor pressure at the two temperatures) to be of interest in transferring heat from the hot to the cold zone. Such a closed system, requiring no external pumps, may be of particular interest in space reactors in moving heat from the reactor core to a radiating system. In the absence of gravity, the forces must only be such as to overcome the capillary and the drag of the returning vapor through its channels."

Thus, heat pipe research began at Los Alamos. Later that year, Grover submitted the results of "heat pipe" experiments with water and sodium as working fluids to the Journal of Applied Physics [7]. The sodium heat pipe, 90 cm long with a 1.9 cm outer diameter (OD), operated at 1100 K with 1 kW heat input. This paper reviews 28 years of space power-related liquid metal heat pipe research that has been conducted at Los Alamos since the invention of the heat pipe.

6.3 Heat Pipes Description and Technology

A heat pipe is essentially a passive heat transfer device with an extremely high effective thermal conductivity. It is a simple closed-loop device that can quickly transfer heat from one point to another using a two-phase flow schema. It is a high thermal conductance device as well which transfers heat by two-phase fluid circulation. The operating temperature range of a heat pipe is determined by the type of working fluid used and its optimum design envelope. They are often referred to as the "superconductors" of heat as they possess an extraordinary heat transfer capacity and rate with almost no heat loss. Of the various means of transmitting heat, the heat pipes are known as one of the most satisfactory devices to carry on such a task. In a simple form of its structure, this device is transporting heat from one point to another via evaporation and condensation, and the heat transport fluid is recirculated by capillary forces which automatically develop as induction of the heat transport process.

This closed loop of heat pipe is consisting of a sealed hollow tube with two zones, namely, evaporator and condenser, in a very simple case of such device that inside walls are lined with a capillary structure known as wick. A thermodynamic working fluid having a substantial vapor pressure at the desired operating temperature saturates the pores of the wick. When heat is applied to any portion of the heat pipe evaporator, this fluid is heated, and it evaporates, readily filling the hollow center of the pipe. The vapor then diffuses throughout the heat pipe. Condensation of the vapor occurs on the pipe wall whenever the temperature is even slightly below that of the evaporation area. As it condenses, the liquid gives up the heat it acquired and returns to the evaporator section or heat source by means of capillary action within the wick. This tends to produce isothermal operation and a high effective thermal conductance. When a heat sink is attached to a portion of the heat pipe, condensation takes place preferentially at this point of heat loss, and a vapor flow pattern is then established.

The system, proven in aerospace application, transmits thermal energy at rates hundreds of times greater than the most efficient solid conductor does and at a far superior energy-to-weight ratio.

In terms of thermal conduction, a heat pipe is designed to have very high thermal conductance. Heat is transported from the heat source (evaporator section of the heat pipe) to the heat sink (condenser section of the heat pipe) by means of a condensable fluid contained in a sealed chamber.

Liquid is vaporized, absorbing heat in the evaporator section. Then the vapor flows to the condenser section, where it condenses and releases its latent heat. The liquid is drawn back to the evaporator section by capillary action, where it is re-vaporized to continue the cycle. The temperature gradient along the length of pipe is minimized by designing for a very small vapor pressure drop as the vapor flows from the evaporator section to the condenser section. Thus, the saturation temperatures (temperatures at which evaporation and condensation take place) are very nearly the same in both sections.

The idea of heat pipes was first suggested by Gaugler in 1942 [2]. However, it was not until 1962, when Grover et al. [5] invented it, that its remarkable properties were appreciated and serious development began. It consists of a sealed aluminum or copper container whose inner surfaces have a capillary wicking material. A heat pipe is similar to a thermosyphon. It differs from a thermosyphon by virtue of its ability to transport heat against gravity by an evaporation-condensation cycle with the help of porous capillaries that form the wick. The wick provides the capillary driving force to return the condensate to the evaporator. The quality and type of wick usually determine the performance of the heat pipe, for this is the heart of the product. Different types of wicks are used depending on the application for which the heat pipe is being used.

The spectrum of heat pipe working fluids extends from cryogens to liquid metals, the choice of fluid being such that its saturation temperature, at the heat pipe operating pressure, be compatible with the heat pipe's application. Also, the fluid is chosen to be chemically inert when wetting the pipe and capillary wick. Ideally, the fluid would have a high thermal conductivity and latent heat. It should have a high surface tension and low viscosity.

Heat transfer in a heat pipe is limited by the rate at which liquid can flow through the wick; "choking" (the inability to increase vapor flow with increasing pressure differential, also called "sonic limit"); entrainment of liquid in the vapor stream, such that liquid flow to the evaporator is reduced; and the rate at which evaporation can take place without excessive temperature differentials in the evaporator section.

Isothermalizer heat pipes will transport heat in either direction, and for a given configuration, the heat flow will depend entirely on the temperature difference between the heat source and heat sink. The isothermalizer heat pipe is therefore a basically passive device with a fixed conductance, provided that none of its limiting conditions are exceeded (sonic, entrainment, capillary, and boiling limits) (see Figs. 6.2 and 6.3). The isothermalization function of a heat pipe can be modified to produce active devices in two ways: diode heat pipes, where the pipe operates as an isothermalizer in the forward mode and shuts off in the reverse mode, or variable conductance heat pipes, where the conductance in the forward mode can be actively controlled and again shuts down in the reverse mode.

As illustrated in Fig. 6.2, a traditional heat pipe is a hollow cylinder filled with a vaporizable liquid. (a) Heat is absorbed in the evaporating section. (b) Fluid boils to vapor phase. (c) Heat is released from the upper part of cylinder to the environment; vapor condenses to liquid phase. (d) Liquid returns by gravity to the lower part of cylinder (evaporating section).

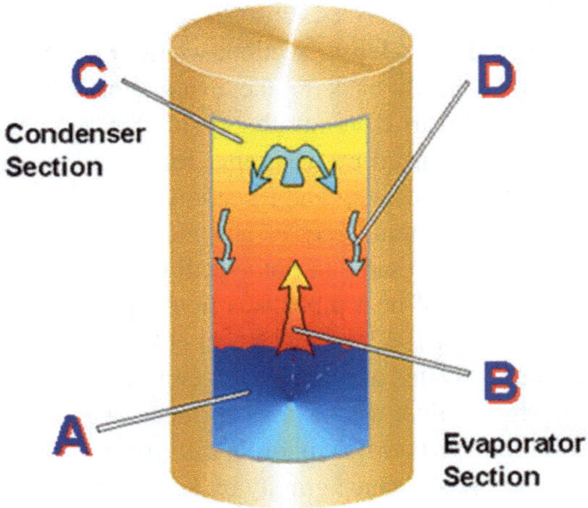

Fig. 6.2 A traditional heat pipe schematic

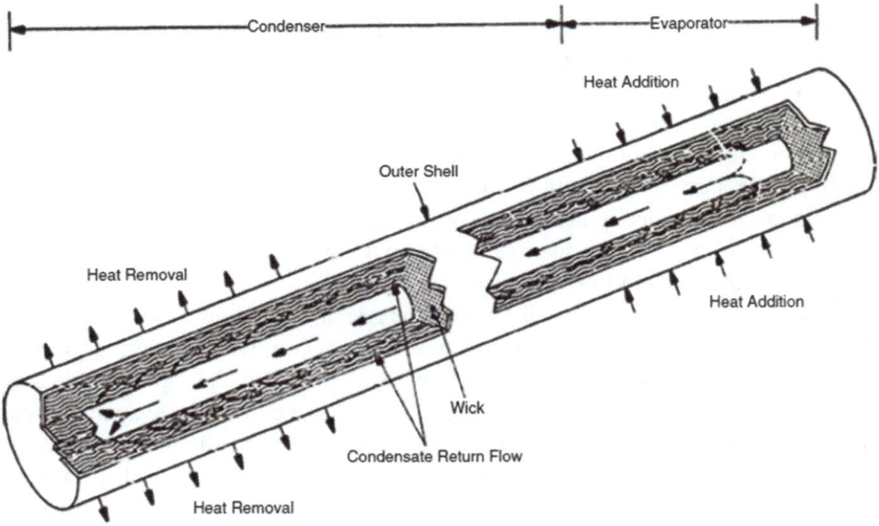

Fig. 6.3 Parts and functions of basic heat pipe

But today's heat pipes can work both vertically and horizontally as well as arbitrary angle of installation and operation of its application. Recent applications of them in zero gravity in particular in satellite are enhanced and proven by NASA and Air Force in collaboration with Los Alamos National Laboratory and other contractors such as TRW and Honeywell and others. Figures 6.2, 6.3, and 6.4 are some examples of such applications.

Figure 6.5 shows application of loop heat pipe containing two parallel evaporators and two parallel condensers with passive and self-regulating as well as heat load sharing between evaporators. This configuration was implemented to NASA's New

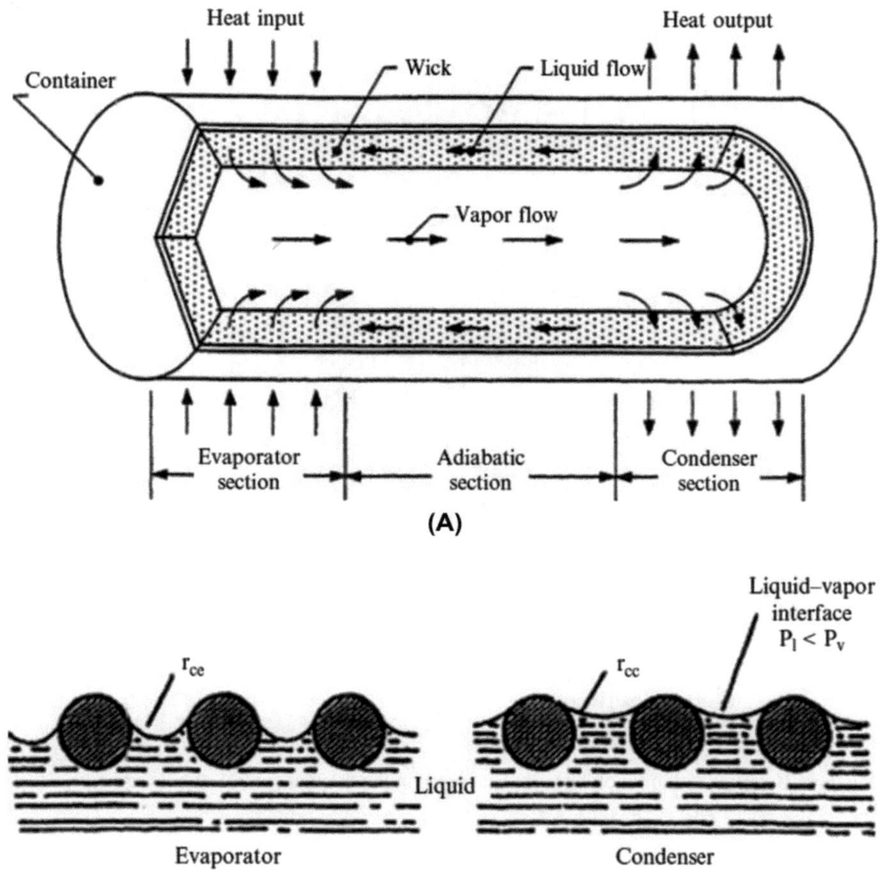

Fig. 6.4 (**a**) Components and principle of operation of a conventional heat pipe; (**b**) radii of curvature of the liquid–vapor interface in the condenser and evaporator [6]

Millennium Program, the Space Technology 8 [12] (ST8) mission. Part of the new spacecraft has been illustrated in Fig. 6.6 where the loop heat pipe was used.

Depending on its applications in particular in the nuclear reactor industry where these reactors are a source of electric power generation, heat pipes are used as a cooling part of secondary loop of inherent shutdown system, which typically you see them as liquid heat pipe (i.e., mercury or sodium as cooling environment within heat pipe); there might be some consideration for a section of heat pipe that is known as adiabatic zone where heat pipe is cast into structure of cooling assembly. A typical example of such approach could be seen in the early study of companies such as Westinghouse Electric on their core design of liquid metal fast breeder reactor (LMFBR); the mercury heat pipe was considered as part of a fully inherent shutdown system of this particular reactor design (where the author was involved with such design and Westinghouse was awarded few patents). This sort of approach was giving better safety factor for any incidental meltdown of reactor and was providing

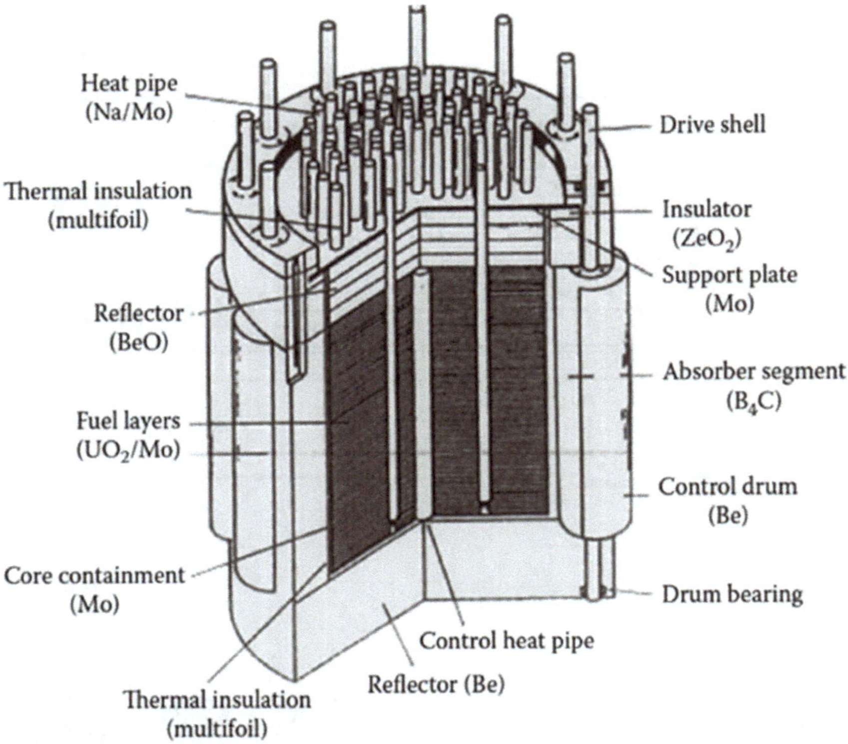

Fig. 6.5 Usage of sodium/molybdenum heat pipes in the thermal control of nuclear power reactor [7]

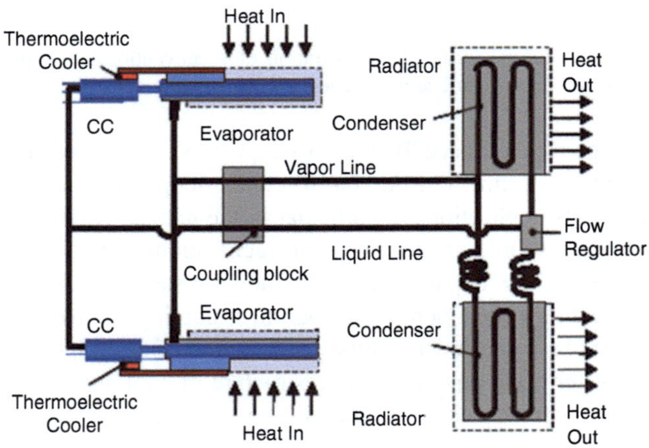

Fig. 6.6 Loop heat pipe (LHP) containing two parallel evaporators and two parallel condensers

better tool to release excessive heat and help to reduce the core temperature below critical point without any operator within the loop.

Modern design and new generation of LMFBR such as French-built Phoenix are utilizing such heat pipes. A typical conventional heat pipe with its adiabatic section is depicted in Fig. 6.4. These types of heat pipes with their adiabatic zone are designed for use in thermal control of nuclear reactor cores that are used in the form of convection, conduction, and radiation heat transfer device, and they are shown in Fig. 6.5. In case of rapid reduction of core temperature, additional radiation surface area in the form of fin is built on top of evaporation section of heat pipe, or variable heat pipes are utilized which are described in the later part of this section.

This type of loop heat pipe has been utilized as part of new NASA series of experiments for space-worthy value of heat pipe studies.

Loop heat pipe operation involves complex physical processes such as:

- Fluid dynamics, heat transfer, and thermodynamics
- Gravitational, inertial, viscous, and capillary forces

The first orbital test demonstrating heat pipe operation under zero-gravity conditions took place in 1967. The launch vehicle for the ATS-A satellite carried a heat pipe with thermocouples to determine its temperature uniformity and performance under varying heat loads throughout different portions of its Earth orbit. This successful demonstration was followed 1 year later by the launching of GEOS-2, using heat pipes designed by Johns Hopkins. GEOS-2 was the first satellite designed to use heat pipes as an integral part of its overall thertil control system (see Fig. 6.7).

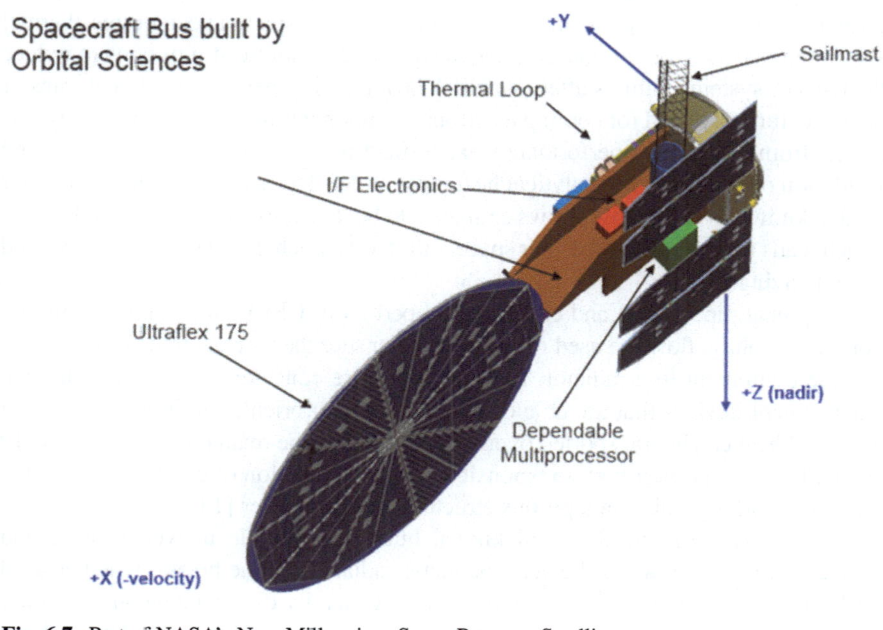

Fig. 6.7 Part of NASA's New Millennium Space Program Satellite

The large-scale satellite and the International Space Station have made great progress in the past several decades. One of the problems urgent to be solved is the heat dissipation. There exists a large amount of heat that should be transferred and radiated into the outer space. Single-phase liquid loops were the major method used in large-scale spacecraft for heat transfer and dissipation in the past decade. Since the 1980s researches worldwide have focused their efforts on two-phase liquid loop technology to be used in the spacecraft thermal control systems on the International Space Station, telecommunication, and technological satellites.

Space nuclear systems require large area radiators to reject the unconverted heat to space. A conceptual design of a waste heat radiator has been developed for a thermoelectric space nuclear power system [12]. The basic shape of the heat pipe radiator was a frustum of a right circular cone. The design included stringer heat pipes to carry rejected heat from the thermoelectric modules to the radiator skin that was composed of small-diameter, thin-walled cross heat pipes. The stringer heat pipes were armored to resist puncture by a meteoroid. The cross heat pipes were designed to provide the necessary unpunctured radiating area at the mission end with a minimum initial system mass. Several design cases were developed in which the individual stringer survival probabilities were varied and the radiator system mass was calculated. Results are presented for system mass as a function of individual stringer survival probability for six candidate container materials, three candidate heat pipe fluids, two radiator operating temperatures, two meteoroid shield types, and two radiating surface cases. Results are also presented for radiator rejected heat as a function of system mass, area, and length for three system sizes. More details can be found in book by Zohuri [1].

Heat pipe operation on Earth is dominated by the force of gravity, which makes it difficult to predict the performance of a heat pipe in space and requires thermal system engineers to adopt conservative designs and ground-test programs to reduce the risks of systems failures after launch. A heat pipe is a very efficient heat transfer device commonly used for cooling electronic components and sensors. Recently, the results from a heat pipe performance experiment have led to the development and validation of an improved analytical heat pipe model. The accuracy of this computer model, known as GAP, now allows engineers to be less conservative in their designs, which leads to fewer heat pipes per spacecraft, thereby achieving significant cost and weight savings.

Loop heat pipe (LHP) and capillary pumped loop (CPL) with "natural" circulation of two-phase flow are used on satellites to ensure the thermal transfer from core module equipment to a radiator. LHP and CPL are considered as reliable thermal management devices that are able of operating at any orientation in a gravitational field, and heat can be transported over long distances. The main components of LHP and CPL are evaporator that is responsible for the generation of capillary forces that drive the working fluid via a porous structure and condenser [13].

Electric blankets are also well known but often provide uneven heating and subject the user to low-level electromagnetic radiation. Nonetheless, whether used for heating or cooling, such suits and blankets generally disadvantageously require external connections and/or require external power sources. Numerous devices have

Fig. 6.8 A method for temperature regulation in the hand

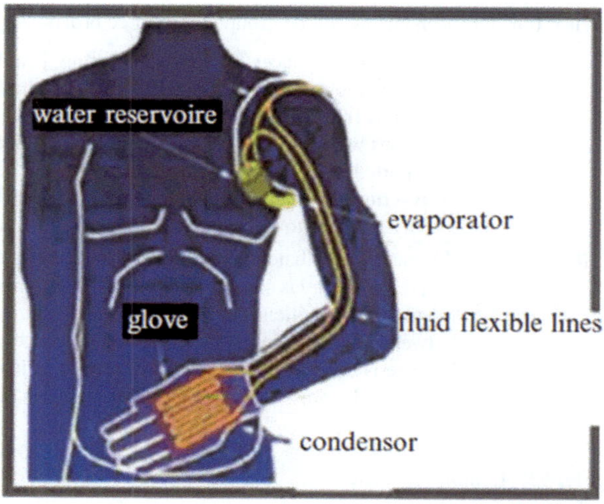

been developed for regional, therapeutic heat transfer. Faghri's invention [14] meets the need for lightweight, comfortable suits and blankets for body temperature regulation by using heat pipes to redistribute body heat and to provide supplemental heat from external sources. A temperature regulation system for the human body will result, taking the form of garments, blankets, and pads.

This invention further provides for an improved pad incorporating heat pipes for use in regional, therapeutic heat transfer. The heat pipes are positioned to provide heat transfer between one or more separate portions of the body. A garment for use in cold environments, such as a body suit, pants, or jacket, may, thus, include heat pipes which extend from the torso of the body, which is typically warmer, to an extreme level. For example, in a garment such as a body suit, this fourth embodiment provides a means to overcome problems with damage to heat-sensitive organs when whole body hyperthermia is induced for medical treatment. Heat may be applied to major portions of the body to induce hyperthermia with one heat exchanger having means for heating, while portions of the body facing heat-sensitive organs may be cooled with another heat exchanger having means for cooling (Fig. 6.8). Temperature control is of particular interest where the present invention is used for deliberate inducement of hyperthermia for medical treatment or to provide controlled heating or cooling for hypothermia or hyperthermia patients.

Note that:

Heat transfer mechanisms can be grouped into three broad categories:	
Conduction	Regions with greater molecular kinetic energy will pass their thermal energy to regions with less molecular energy through direct molecular collisions, a process known as conduction. In metals, a significant portion of the transported thermal energy is also carried by conduction-band electrons.
Convection	When heat conducts into a static fluid, it leads to a local volumetric expansion. As a result of gravity-induced pressure gradients, the expanded fluid parcel becomes

(continued)

Heat transfer mechanisms can be grouped into three broad categories:	
	buoyant and displaces, thereby transporting heat by fluid motion (i.e., convection) in addition to conduction. Such heat-induced fluid motion in initially static fluids is known as **free convection**. For cases where the fluid is already in motion, heat conducted into the fluid will be transported away chiefly by fluid convection. These cases, known as **forced convection**, require a pressure gradient to drive the fluid motion, as opposed to a gravity gradient to induce motion through buoyancy.
Radiation	All materials radiate thermal energy in amounts determined by their temperature, where the energy is carried by photons of light in the infrared and visible portions of the electromagnetic spectrum. When temperatures are uniform, the radiative flux between objects is in equilibrium, and no net thermal energy is exchanged. The balance is upset when temperatures are not uniform, and thermal energy is transported from surfaces of higher to surfaces of lower temperature.

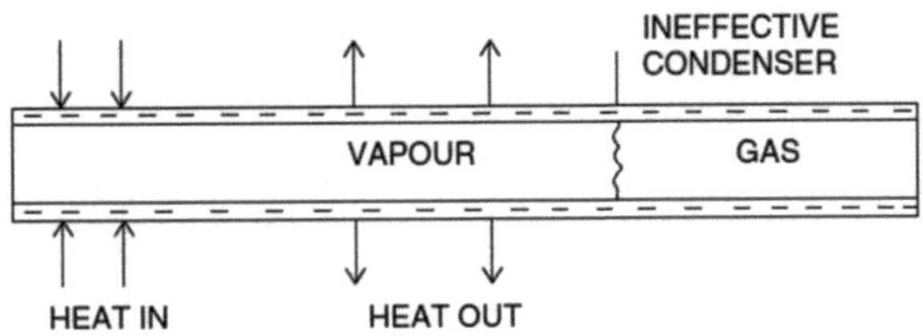

Fig. 6.9 Equilibrium state of gas-loaded heat pipe [6]

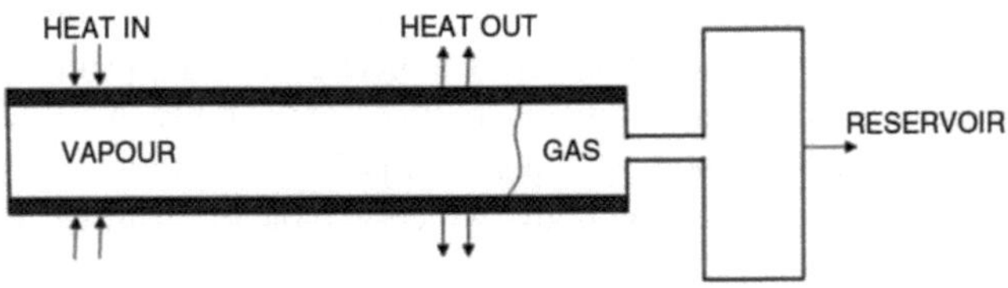

Fig. 6.10 Cold-reservoir variable conductance heat pipe [6]

Generally speaking, typically, there are two classes of heat pipes, variable conductance heat pipe (VCHP) or "conventional" heat pipe also known as constant conductance heat pipe (CCHP) or fixed conductance heat pipe (FCHP). A typical conventional heat pipe is illustrated in Fig. 6.9, while a variable conductance heat pipe is depicted in Fig. 6.10. The distinctive feature of these types of heat pipes from conventional ones are their abilities and functionality to operate in a specific desired temperature range along certain portions of the pipe, in spite of variations in the source and sink conditions. When such conditions are desired based on the application of heat pipe, it is important to actively or passively control the heat pipe so that the desired temperature range can be maintained.

Figure 6.10 is the simple form of variable heat pipe structure with gas-buffered section on the condenser part of the pipe. Later on, there was an add-on reservoir structured to give a downstream form of condenser (Fig. 6.10) which was allowing the heat pipe to have adequate effective length on condenser side of the pipe to operate at maximum of its capability and provide more accurate control of the vapor temperature [15].

In the early days of designing the cold-reservoir variable conductance heat pipe (VCHP), there was a problem associated with diffusion of vapor into the reservoir, followed by condensation. These all are discussed in more details about variable heat pipe in book by Zohuri [1] and some approaches taken by the designer of such heat pipes to improve their structure and applications in the industry. Computer codes were generated to have better feeling for these types of heat pipes, and of the most known computer code, "GASPIPE" was developed by Marcus [16] and associates at TRW where they studied and developed VCHP for NASA in the early days of the 1970s.

A new type of variable conductance heat pipe, the liquid-controlled heat pipe (LCHP), has been developed. While the gas-controlled heat pipe is able to stabilize the temperature of the heating zone, the LCHP limits the temperature of the cooling zone to a certain adjustable value. The physical principle is to regulate the heat transfer capability by regulating the amount of liquid inside the heat pipe. The liquid is partly stored in a reservoir with a variable volume, as, for example, bellows. The temperature of the cooling zone, corresponding to the vapor pressure inside the heat pipe, can be adjusted by the outer pressure (gas or spring) as it is demonstrated in Fig. 6.11. The LCHP is applicable where heat is needed at a constant temperature or where the vapor pressure inside a heat pipe has to be limited.

In Fig. 6.11a, a gradual increase in working temperature is accompanied by a rapid increase in saturated vapor pressure of the working fluid. In contrast there is only a slight increase in the temperature on the non-condensable gas, and, since the relationship between pressure and cubic capacity is constant, the boundary surface is

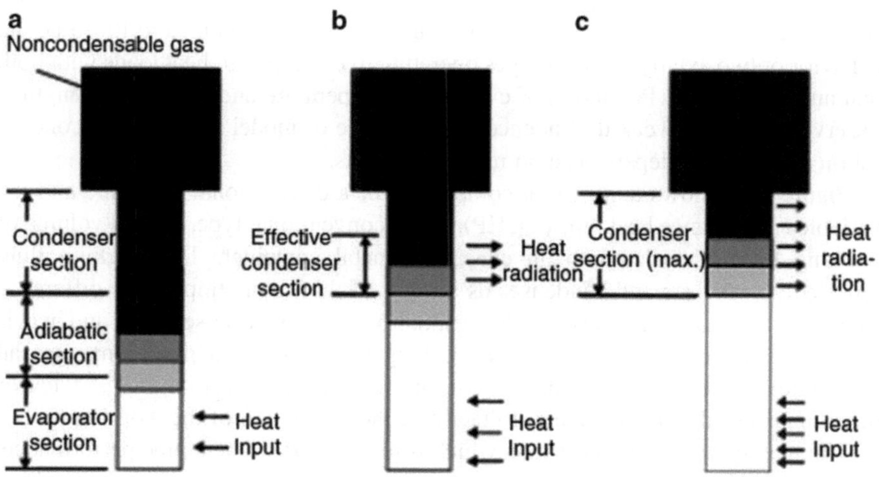

Fig. 6.11 Heat transfer during VCHP operation

pushed out by the working fluid so that the "condenser section that is effective for radiation with admixture of working fluid" migrates toward the gas reservoir.

In Fig. 6.11b, the region occupied by the working fluid is reduced, and the internal heat transfer rate is still small. If the temperature is further increased, the condenser section that is effective for radiation becomes larger, and the heat radiation rate increases. At the point of maximum heat radiation, the non-condensable gas is totally contained within the gas reservoir.

In Fig. 6.11c, if the heat radiation rate is greater than the maximum heat input to the evaporator section by the heat source, sufficient heat radiation can be obtained, and the temperature of the evaporator section will not increase further [14].

As we mentioned before, constant conductance heat pipe (CCHP) is also known as fixed conductance heat pipe (FCHP), and these types of heat pipes transport heat from a heat source to a heat sink with a very small temperature difference. Axial groove capillary wick structures are utilized because of the relative ease of manufacturing (aluminum extrusions) and their demonstrated heritage in spacecraft and instrument thermal control applications. CCHPs can transport heat in either direction and are typically used to transfer heat from specific thermal loads to a radiator panel or as part of an integrated heat pipe radiator panel. Common working fluids include ammonia, propylene, ethane, and water.

Fixed conductance heat pipes (FCHPs) filled with working fluid at low or moderate temperatures develop a volume of excess liquid when operated at high temperatures. The excess liquid forms as either a puddle or a slug at the coldest end of the condenser and creates a temperature differential between the evaporator and the condenser end cap. Simple algebraic expressions are presented for predicting the thermal performance of an FCHP operating with a liquid slug formed by the combined influence of liquid density temperature dependence and meniscus depression.

Both differential and two-node models are developed to account for condensation modeled either as a constant flux processor based on an isothermal vapor with a constant internal film coefficient. Numerical examples are included to illustrate the behavior of two axially grooved pipes operating over a range of heat loads with both real and ideal fluids. Prediction of evaporator temperature and liquid slug length is observed to have a weak dependence on the choice of model and mode of condensation and a strong dependence on real fluid effects.

Figure 6.12 shows a structural comparison of a conventional heat pipe and the variable conductance heat pipe (VCHP). In the conventional type, a small volume of working fluid is sealed into an evacuated metal container. The working fluid repeatedly vaporizes and condenses as a result of the small temperature difference (or temperature gradient) between the evaporator and condenser sections, and heat is transferred by the latent heat of the working fluid. The heat pipe comprises an evaporator section, an adiabatic section, and a condenser section, and a wick or mesh is provided in the container to facilitate the circulation of the working fluid.

The maximum heat transfer rate, which is the measure of heat pipe performance, is determined by the wick, the volume of working fluid, and so on, while the working

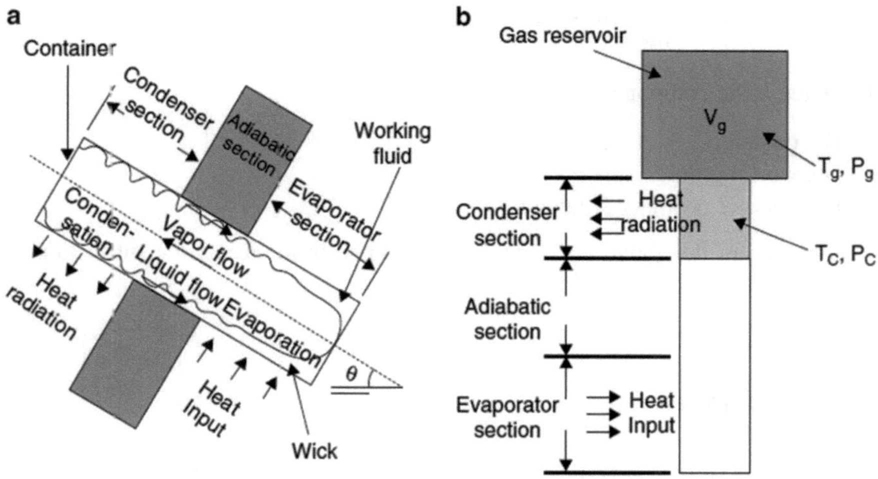

Fig. 6.12 Structural comparison of conventional and variable conductance heat pipes [10]

Fig. 6.13 Heat radiation characteristics of conventional and variable conductance heat pipes [10]

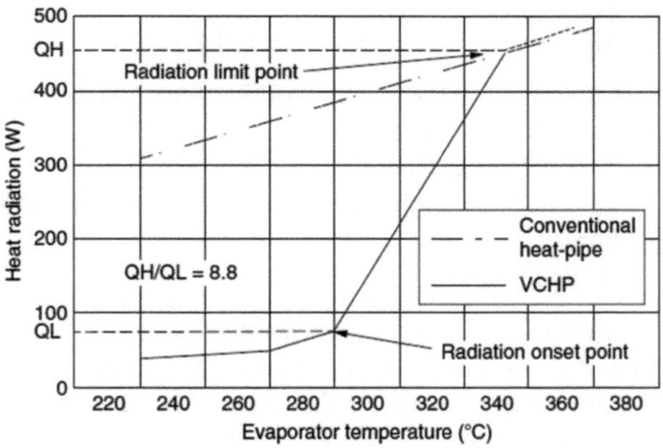

temperature is determined passively by the external heat source, the temperature of the heat radiator (condenser) section, and so on [14].

Figure 6.13 compares the heat radiation characteristics of conventional and variable conductance heat pipes. Unlike the conventional heat pipe, in which the heat radiation rate has a constant gradient relative to temperature, the radiation rate in the VCHP increases rapidly from a given temperature (the radiation onset temperature) until the radiation limit is reached. When the radiation limit is exceeded, it traces a constant gradient, like a conventional heat pipe. The term "variable conductance" has its origin in this characteristic, and the slope of the line between the radiation onset point and the radiation limit point (hereinafter referred to as the radiation gradient) is an important characteristic of the VCHP [14].

6.4 Heat Pipes Principle Operations

The three basic components of a heat pipe are:

1. The container
2. The working fluid
3. The wick or capillary structure

Each one of the above three issues is explained in the following subsections as below.

6.4.1 Container

The function of the container is to isolate the working fluid from the outside environment. It has to therefore be leakproof, maintain the pressure differential across its walls, and enable transfer of heat to take place from and into the working fluid.

Selection of the container material depends on many factors. These are as follows:

- Compatibility (both with working fluid and external environment)
- Strength-to-weight ratio
- Thermal conductivity
- Ease of fabrication, including welding, machineability, and ductility
- Porosity
- Wettability

Most of the above are self-explanatory. A high strength-to-weight ratio is more important in spacecraft applications. The material should be nonporous to prevent the diffusion of vapor. A high thermal conductivity ensures minimum temperature drop between the heat source and the wick.

6.4.2 Working Fluid

A first consideration in the identification of a suitable working fluid is the operating vapor temperature range. Within the approximate temperature band, several possible working fluids may exist, and a variety of characteristics must be examined in order to determine the most acceptable of these fluids for the application considered [1].

The prime requirements are:

- Compatibility with wick and wall materials
- Good thermal stability
- Wettability of wick and wall materials
- Vapor pressure not too high or low over the operating temperature range

- High latent heat
- High thermal conductivity
- Low liquid and vapor viscosities
- High surface tension
- Acceptable freezing or pour point

The selection of the working fluid must also be based on thermodynamic considerations which are concerned with the various limitations to heat flow occurring within the heat pipe such as viscous, sonic, capillary, entrainment, and nucleate boiling levels.

In heat pipe design, a high value of surface tension is desirable in order to enable the heat pipe to operate against gravity and to generate a high capillary driving force. In addition to high surface tension, it is necessary for the working fluid to wet the wick and the container material, i.e., contact angle should be zero or very small. The vapor pressure over the operating temperature range must be sufficiently great to avoid high vapor velocities, which tend to set up large temperature gradient and cause flow instabilities [1].

A high latent heat of vaporization is desirable in order to transfer large amounts of heat with minimum fluid flow and hence to maintain low-pressure drops within the heat pipe. The thermal conductivity of the working fluid should preferably be high in order to minimize the radial temperature gradient and to reduce the possibility of nucleate boiling at the wick or wall surface. The resistance to fluid flow will be minimized by choosing fluids with low values of vapor and liquid viscosities. Tabulated below are a few mediums with their useful ranges of temperature (Fig. 6.14) [1].

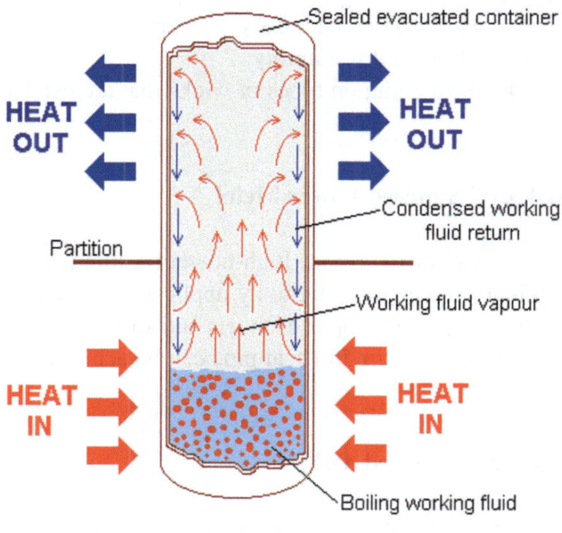

Fig. 6.14 Schematic representation of the heat pipe

6.4.3 *Wicker or Capillary Structure*

It is a porous structure made of materials like steel, aluminum, nickel, or copper in various ranges of pore sizes. They are fabricated using metal foams and more particularly felts, the latter being more frequently used. By varying the pressure on the felt during assembly, various pore sizes can be produced. By incorporating removable metal mandrels, an arterial structure can also be molded in the felt.

Fibrous materials, like ceramics, have also been used widely. They generally have smaller pores. The main disadvantage of ceramic fibers is that they have little stiffness and usually require a continuous support by a metal mesh. Thus, while the fiber itself may be chemically compatible with the working fluids, the supporting materials may cause problems. More recently, interest has turned to carbon fibers as a wick material. Carbon fiber filaments have many fine longitudinal grooves on their surface, have high capillary pressures, and are chemically stable. A number of heat pipes that have been successfully constructed using carbon fiber wicks seem to show a greater heat transport capability.

The prime purpose of the wick is to generate capillary pressure to transport the working fluid from the condenser to the evaporator. It must also be able to distribute the liquid around the evaporator section to any area where heat is likely to be received by the heat pipe. Often these two functions require wicks of different forms. The selection of the wick for a heat pipe depends on many factors, several of which are closely linked to the properties of the working fluid.

The maximum capillary head generated by a wick increases with decrease in pore size. The wick permeability increases with increasing pore size. Another feature of the wick, which must be optimized, is its thickness. The heat transport capability of the heat pipe is raised by increasing the wick thickness. The overall thermal resistance at the evaporator also depends on the conductivity of the working fluid in the wick. Other necessary properties of the wick are compatibility with the working fluid and wettability.

The most common types of wicks that are used are as follows.

6.4.3.1 Sintered Power Metal

This process will provide high-power handling, low-temperature gradients, and high capillary forces for antigravity applications. Figure 6.15 shows a complex sintered wick with several vapor channels and small arteries to increase the liquid flow rate. Very tight bends in the heat pipe can be achieved with this type of structure.

6.4.3.2 Grooved Tube

The small capillary driving force generated by the axial grooves is adequate for low-power heat pipes when operated horizontally or with gravity assistance. The

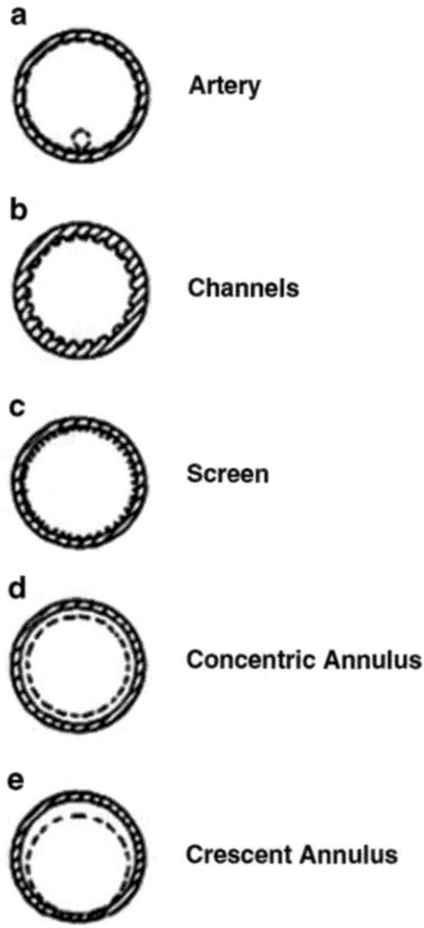

Fig. 6.15 Cross section of various wick structures [15]

tube can be readily bent. When used in conjunction with screen mesh, the performance can be considerably enhanced.

6.4.3.3 Wire Screen Mesh

This type of wick is used in the majority of the products and provides readily variable characteristics in terms of power transport and orientation sensitivity, according to the number of layers and mesh counts used (see Fig. 6.16).

Figure 6.17 demonstrates several common wicking structures presently in use, along with more advanced concepts under development [18].

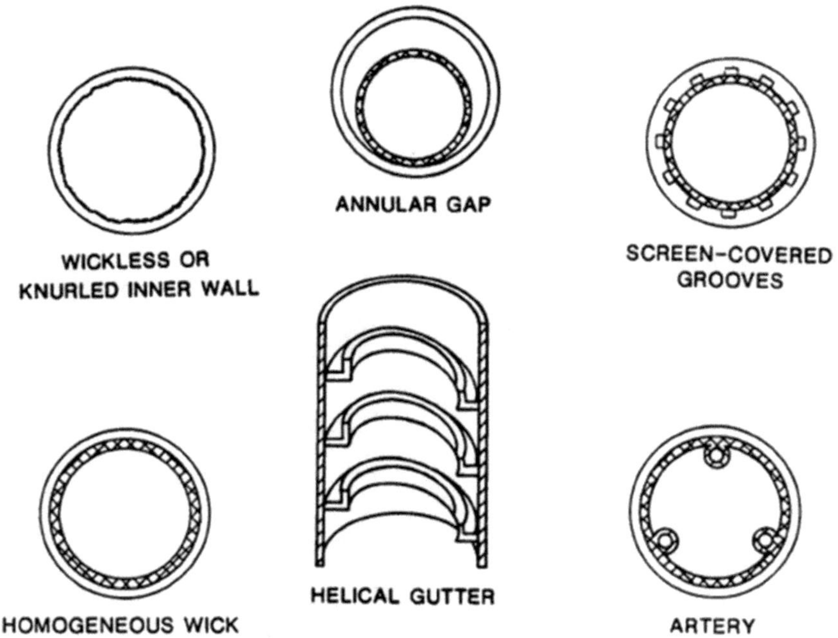

Fig. 6.16 Heat pipe liquid return geometries [17]

6.5 How the Heat Pipe Is Working

Inside the container is a liquid under its own pressure that enters the pores of the capillary material, wetting all internal surfaces. Applying heat at any point along the surface of the heat pipe causes the liquid at that point to boil and enter a vapor state. When that happens, the liquid picks up the latent heat of vaporization. The gas that then has a higher pressure moves inside the sealed container to a colder location where it condenses. Thus, the gas gives up the latent heat of vaporization and moves heat from the input to the output end of the heat pipe (see Fig. 6.18) [1].

Heat pipes have an effective thermal conductivity many thousands of times that of copper. Its "axial power rating (APR)" specifies the heat transfer or transport capacity of a heat pipe. It is the energy moving axially along the pipe. The larger the heat pipe diameter, the greater is the APR. Similarly, the longer the heat pipe, the lesser is the APR. Heat pipes can be built in almost any size and shape.

A simple Heat Pipe Assemblies Design Guidelines can be found in so many manufactures of heat pipe. Figure 6.19 is the best that the author has found.

Web site of Aavid Engineering, which was founded in 1964 as subsidiary of Aavid Thermal Technologies, Inc., is recommending certain and simple criteria as rule of thumb for any heat pipe.

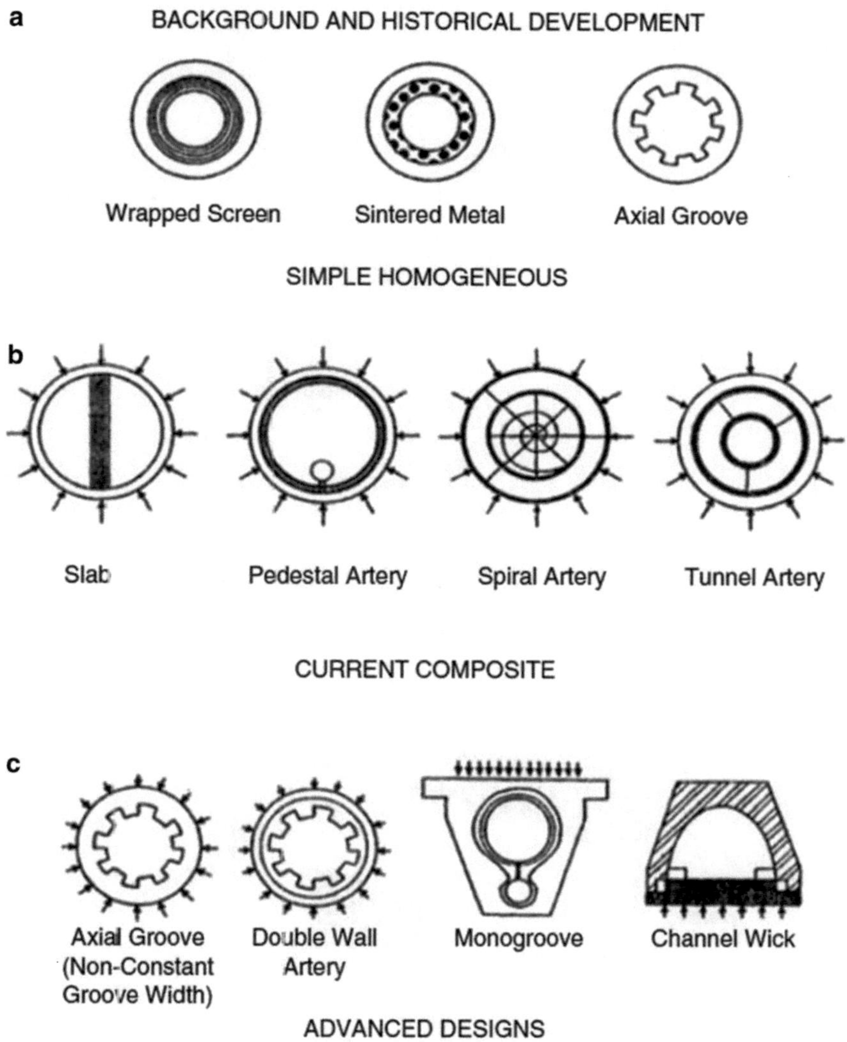

Fig. 6.17 Typical heat pipe wicking configurations and structures [12]

6.6 Heat Pipe Assemblies Design Guidelines

The following approach is sort of a quick and back-of-the-envelope-type analysis of choosing your design before you model your heat pipe for best optimum point of your design that falls within the constraint and limit of operating range of heat pipe such as sonic, entrainment, wicking, and boiling limits.

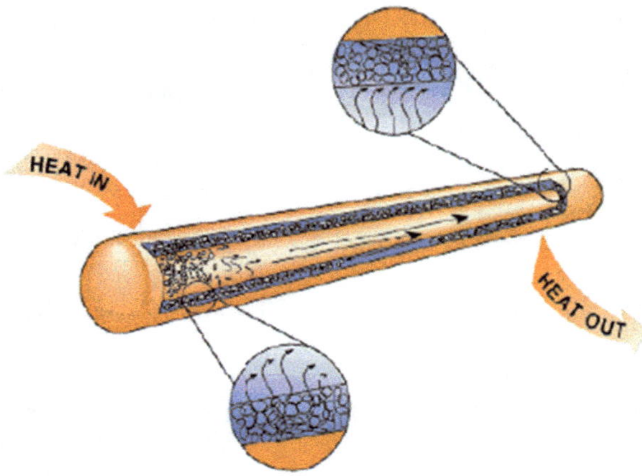

Fig. 6.18 Principle of conserved energy and heat transfer in heat pipe

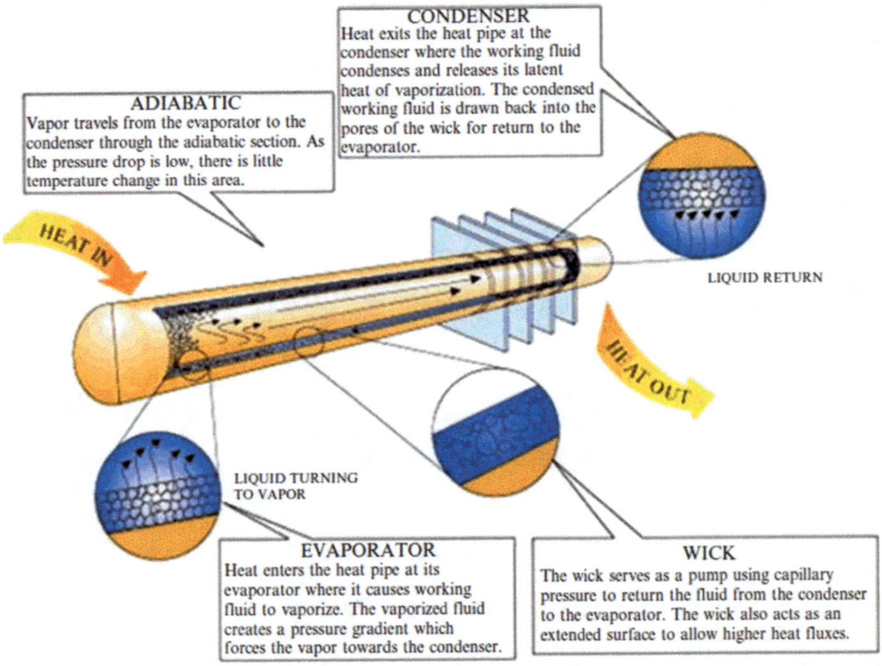

Fig. 6.19 Heat pipe operation [13]

6.6.1 Orientation with Respect to Gravity

For the best performance, the application should have gravity working with the system; that is, the evaporator section (heated) should be lower, with respect to gravity, than the condenser (cooling) section. In other orientations where gravity is not aiding the condensed liquid return, the overall performance will be degraded.

Performance degradation depends on a number of factors including wick structure, length, and working fluid of the heat pipe along with heat flux of the application. Careful design can minimize the performance loss and allow an accurate prediction performance.

6.6.2 Temperature Limits

Most pipes use water and methanol/alcohol as the working fluids. Depending on the wick structure, pipes will operate in environments with as low as −40 °C. Upper temperature limits depend on the fluid, but 60–80 °C is the average limit.

6.6.3 Heat Removal

Heat can be removed from the condenser using air cooling in combination with extrusion, bonded-fin heat sinks, or flat-fin stock. Enclosing the condenser in a cooling jacket allows liquid cooling.

6.6.4 Reliability

Heat pipes have no moving parts and have demonstrated life of over 20 years. The largest contributor to heat pipe reliability comes from control of the manufacturing process. The seal of the pipe, purity of the materials used in the wick structure, and cleanliness of the internal chamber have measurable effect on the long-term performance of a heat pipe. Any leakage will eventually render the pipe inoperable.

Contamination of the internal chamber and wick structure will contribute to the formation of non-condensable gas (NCG) that will degrade performance over time. Well-developed processes and rigorous testing are required to ensure reliable heat pipes.

6.6.5 Forming or Shading

Heat pipes are easily bent or flattened to accommodate the needs of the heat sink design. Forming heat pipes may affect the power handling capability as the bends and flattening will cause a change in fluid movement inside the pipe. Therefore, design rules that take heat pipe configurations into consideration and the effect on thermal performance ensure the desired solution performance.

6.6.6 Effects of Length and Pipe Diameter

The vapor pressure differential between the condenser end and the evaporator end controls the rate at which the vapor travels from one end to the other. Diameter and length of the heat pipe also affect the speed at which the vapor moves and must be considered when designing with heat pipes. The larger the diameter, the more cross-sectional area available to allow vapor to move from the evaporator to the condenser.

This allows for greater power carrying capacity. Conversely, length when in opposition to gravity has a negative effect on heat transport as the rate at which the working fluid returns from the condenser end to the evaporator end is controlled by the capillary limit of the wick, which is an inverse function of the length of the pipe. Therefore, shorter heat pipes carry more power than longer pipes when used in application not assisted by gravity.

6.6.7 Wick Structure

Heat pipe inner walls can be lined with a variety of wick structures. The four most common wicks are:

(a) Groove tube
(b) Wire screen mesh
(c) Sintered powder metal
(d) Fiber/spring

Among the above common ones, the three most common ones that are used in the industry to fabricate heat pipes are grooved tube and wire screen mesh that were described and presented in Sects. 4.4.3.2 and 4.4.3.3 or (a) to (c) here.

The wick structure provides a path for liquid to travel from condenser to the evaporator using capillary action. Wick structures have performance advantages and disadvantages depending on the desired characteristics of the heat sink design.

Some structures have low capillary limits making them unsuitable for applications where they must work without gravity assist. The plots of Fig. 6.20a, b are demonstration standard operating range of simple heat pipe.

6.7 Heat Pipe Working Process

No matter what type of application you may consider the heat pipe for, there are certain limitations that one has to look upon it for the heat pipe to operate properly and be able to perform the requirements that are imposed on the heat pipe and its application in that environment.

These limitations are briefly described below:

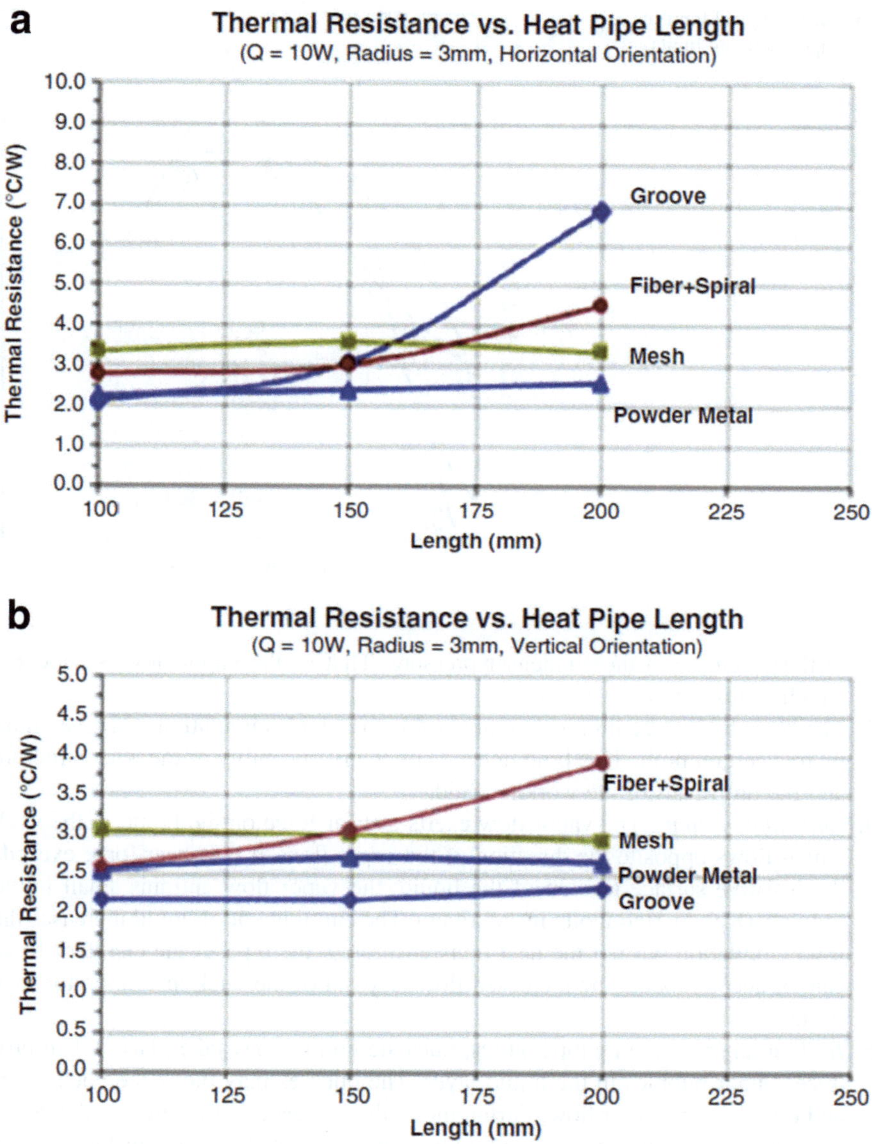

Fig. 6.20 Standard operating range of simple heat pipe

1. *Viscous limit*: In long pipes and at low temperatures, the vapor pressure is low, and the effect of viscous friction on the vapor flow may dominate over the inertial forces. In this situation, the circulation of the working fluid is limited, which, consequently, limits the heat transfer through the pipe.
2. *Sonic limit*: At low vapor pressures, the velocity of the vapor at the exit of the evaporator may reach the speed of sound. Then the evaporator cannot respond to

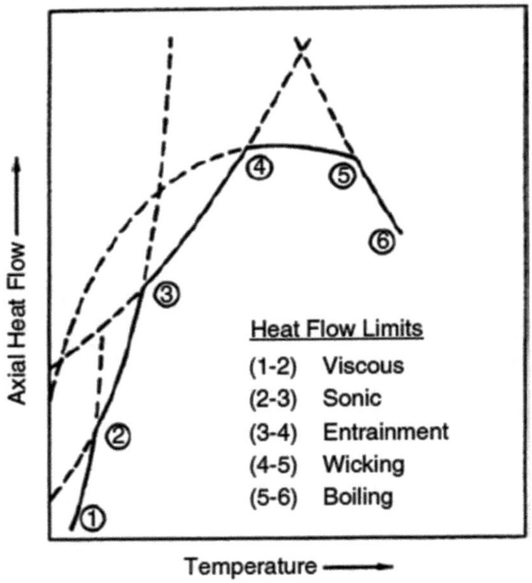

Fig. 6.21 Typical limit of heat pipe operating limits

further decrease in the condenser pressure. That is, the vapor flow is chocked, which limits the vapor flow rate.

3. *Capillary limit*: A capillary structure is able to provide circulation of a given fluid up to a certain limit. This limit depends on the permeability of the wick structure and the properties of the working fluid.
4. *Entrainment limit*: The vapor flow exerts a shear force on the liquid in the wick which flows opposite the direction of the vapor flow. If the sheer force exceeds the resistive surface tension of the liquid, the vapor flow entrains small liquid droplets (Kelvin-Helmholtz instabilities). The entrainment of liquid increases the fluid circulation but not the heat transfer through the pipe. If the capillary force cannot accommodate the increased flow, dry out of the wick in evaporator may occur.
5. *Boiling limit*: At high temperatures, nucleate boiling may take place which produces vapor bubbles in the liquid layer. The bubbles may block the wick pores and decrease the vapor flow. Furthermore, the presence of the bubbles decreases the conduction of heat through liquid layer which limits the heat transfer from the heat pipe shell to the liquid which is by conduction only (see Fig. 6.21).

Chapters of the book by Zohuri [1] are showing how to calculate each of these limits and its mathematical modeling. These analyses will allow the designer to plot the above curves, and the best design and fabrication of a heat pipe is where it operates within these envelope limits which is the area below all the curves. This area is known as best optimum design of heat pipe for the particular application of designer.

6.8 Constraints

There are many parameters which affect the performance of a heat pipe like compatibility of materials, operating temperature range, heat transport limitation, thermal resistance, operating orientation, dimension and geometric constraints, etc. For example, in miniature heat pipes, the maximum heat transport capacity was found to be primarily governed by the capillary pressure [18].

All heat pipes have three physical elements in common. These include an outer container, a small amount of working fluid, and a capillary wick structure. In addition to these basic components, heat pipes may also include gas reservoirs (variable conductance/diode heat pipes) and liquid or gas traps (diodes). Functionally, the heat pipe consists of three sections: evaporator, condenser section, and adiabatic regions. The evaporator section is mounted to the heat-producing components, while the condenser is thermally coupled to a heat sink or radiator. The adiabatic section allows heat to be transferred from the evaporator to the condenser with very small heat losses and temperature drops.

Figure 6.22 depicts the basic heat pipe.

Heat pipes can operate in the fixed conductance, variable conductance, or diode mode. The fixed conductance heat pipe can transfer heat in either direction and operates over broad temperature ranges but has no inherent temperature control capability. Constant conduction heat pipes allow isothermalization of shelves, radiators, and structures; spread heat from high heat dissipating components; and conduct heat away from heat-producing devices embedded within instruments and satellites.

In the variable conductance heat pipe (VCHP), a small quantity of non-condensable gas (NCG) is loaded into the heat pipe. The VCHP can be used to control the temperature of equipment within very narrow limits; control is possible to less than 1 K by using careful design techniques. This is accomplished by

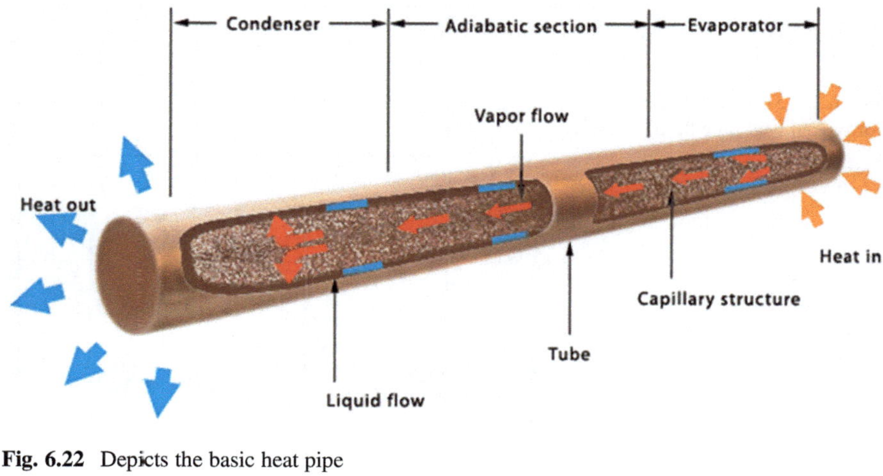

Fig. 6.22 Depicts the basic heat pipe

controlling the location of the NCG/vapor interface within the condenser end of the heat pipe, thereby varying the active length of the condenser and causing a modulation in the condenser heat rejection capability. Temperature control of the attached device is achieved by an active feedback system consisting of a temperature sensor at the heat source and a controller for a heater at the NCG reservoir. The heater causes the gas in the reservoir to expand, thus moving the gas/vapor interface. Diode heat pipes permit heat to flow in one direction and inhibit heat flow in the opposite direction.

Specific benefits of heat pipes are:

1. Heat pipes have enormously more heat transfer capability than other methods on a weight and size basis.
2. Heat pipes permit configuration flexibility in contact areas with heat sources and heat sinks.
3. Heat can be transported over considerable distances with insignificant temperature drop.
4. Capillary pumping in the wick is generated by the heat transfer process and requires no other power or moving parts to pump the condensate.
5. Heat pipes operate satisfactorily in a zero-gravity environment.

The choice of working fluid is dictated by several considerations, including operating temperature, latent heat of vaporization, liquid viscosity, toxicity, chemical compatibility with container material, wicking system design, and performance requirements.

Figures 6.23 and 6.24 and Table 6.1 depict some of the above characteristics for several fluids.

The highest performance from a heat pipe is obtained by utilizing a working fluid that has a high surface tension (s), a high latent heat (l), and a low liquid viscosity (N_1). These fluid properties are contained in the parameter N_1, the liquid transport factor. Figure 6.25 is a plot of N_1 for five typical heat pipe working fluids. These data are used as selection criteria for heat pipe working fluids.

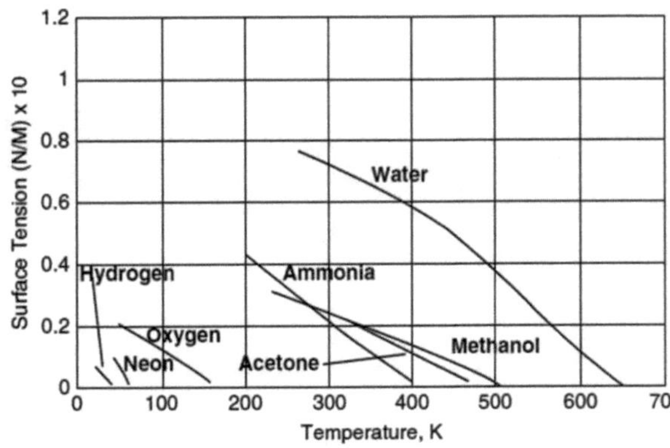

Fig. 6.23 Surface tension for typical heat pipe fluid

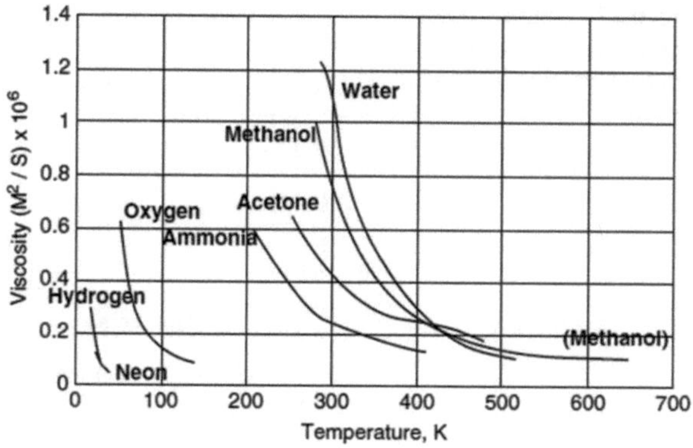

Fig. 6.24 Viscosity for typical heat pipe fluid

Table 6.1 Comparison of latent heat to specific heat for typical heat pipe fluids

Fluid properties				
Fluid	Boiling point K	Latent heat (kJ/kg) h_{fg}	Specific heat (kJ/kg K) c_p	Ratio (K) h_{fg}/c_p
Helium	4	23	4.60	5
Hydrogen	20	446	9.79	46
Neon	27	87	1.84	47
Oxygen	90	213	1.90	112
Nitrogen	77	198	2.04	97
Argon	87	162	1.14	142
Propane	231	425	2.20	193
Ethane	184	488	2.51	194
Methane	111	509	3.45	147
Toluene	384	363	1.72	211
Acetone	329	518	2.15	241
Heptane	372	318	2.24	142
Ammonia	240	1180	4.80	246
Mercury	630	295	0.14	2107
Water	373	2260	4.18	541
Benzene	353	390	1.73	225
Cesium	943	49	0.24	204
Potassium	1032	1920	0.81	2370
Sodium	1152	3600	1.38	2608
Lithium	1615	19,330	4.27	4526
Silver	2450	2350	0.28	8393

Once an application is defined, the heat pipe designer reviews the requirements and selects the best working fluid. Below the freezing point of water and above about 200 K, ammonia is an excellent working fluid. Regardless of the fluid chosen, minimum purity must be at least 99.999%. A careful analysis of the purity of the ammonia should be obtained from an independent laboratory prior to use.

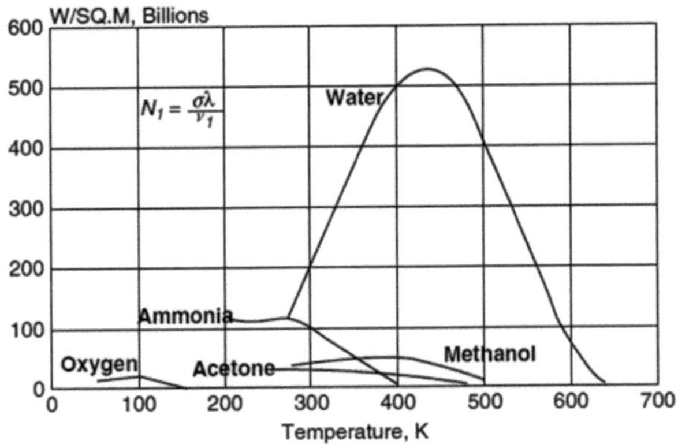

Fig. 6.25 Comparison of liquid transport factor for typical heat pipe working fluids

Table 6.2 Material composite for heat pipe/fluid combinations

	Aluminum	Stainless steel	Copper	Nickel	Titanium
Water	I	C*	C	C	
Ammonia	C	C		C	
Methanol	I	C	C	C	
Acetone	C	C	C		
Sodium		C		C	I
Potassium				C	I

The outer container usually consists of a metal tube to provide mechanical support and pressure containment. The chosen design and processing of the container are extremely important in selecting the metal, because they can affect the useful life of the heat pipe. In addition, a compatibility must exist between the pipe material and the working fluid. For heat pipes, working fluid/container compatibility issues encompass any chemical reactions or diffusion processes occurring between the fluid and wall/wick materials that can lead to gas formation and/or corrosion.

Table 6.2 lists the compatibilities of several metals and working fluids. Along with the metal/fluid compatibility, other considerations in the metal selection are ease of working the material, extrusion capability of the material, and its weldability. Proper container cleaning and heat pipe processing procedures are of extreme importance, since residual contamination within the heat pipe may also lead to gas generation. Steps must also be taken to ensure the purity of the fluid charge; trace amounts of water in ammonia can lead to a reaction with the aluminum container and the formation of hydrogen gas. Chi [6] and B&K Engineering [13] list standard cleaning and filling methods for a variety of working fluid-wall material combinations. Special consideration must be given to the processing of heat pipes to be used at temperatures below 250 K. As the temperature drops, the vapor pressure of the fluid falls off. This allows any non-condensable gas created by contamination to expand, thus creating an even larger problem.

The fluid names are separated by horizontal lines that extend to the right.

In descending order, the fluid names are *water*, *ammonia*, *methanol*, *acetone*, *sodium*, and *potassium*.

Five columns are on the right of the table. Each column is headed by a name of a metal. The names of the metals are separated by vertical lines that extend downward. From left to right, the names of the metals are *aluminum*, *stainless steel*, *copper*, *nickel, and titanium*.

The extended horizontal lines and the extended vertical lines intersect to form boxes. There are five boxes in a horizontal line adjacent to each fluid name. Each box contains one to two symbols or is left blank.

In Table 6.2, the symbols are identified as follows:

C = Compatible
I = Incompatible
* = Sensitivity to cleaning

As we stated in above, the symbols of C, compatible; I, incompatible; and *, sensitivity to cleaning a column of six fluid names are on the left of the table.

The fluid names and the contents of the horizontal row of boxes adjacent to them are as follows:

"*Water*" *aluminum,* I*; stainless steel,* C**; copper,* C*; nickel,* C*; itanium, blank*

"*Ammonia*" *aluminum,* C; *stainless steel,* C; copper, blank; *nickel,* C; *titanium,* blank

"*Methanol*" *aluminum,* I; *stainless steel,* C; *copper,* C; *nickel,* C; *titanium,* blank

"*Acetone*" *aluminum,* C; *stainless steel,* C; *copper*, C; *nickel,* blank; *titanium,* blank

"*Sodium*" *aluminum,* blank; *stainless steel,* C; copper, blank; nickel, C; *titanium,* I

"*Potassium*" *aluminum,* blank; *stainless steel,* blank; *copper,* blank; *nickel,* C; titanium

The heat pipe wick structure provides a porous medium for the formation of liquid menisci (which cause the capillary pumping action) and a vehicle for returning the working fluid from the condenser to the evaporator. To accomplish these wick functions effectively, the designer must provide pores, cavities, or channels of the right size, shape, quantity, and location. An optimization technique is used in wick design to find the desired combination of ultimate heat transfer capacity, pumping capability, and temperature drop. The designer must also consider ease of wick fabrication, compatibility with the working fluid, wetting angle, and permeability of the selected wick material.

Figure 6.26 depicts a cross-sectional view of an axial groove wick; this design probably is the most commonly used for space application.

In addition, X-ray certification of all welds at the end caps and fill tube is required to ensure good weld penetration and the absence of voids. The heat pipe container must be pressure tested to at least twice its maximum expected operating pressures (MEOP) prior to filling [19]. Other qualification procedures include performance tests at adverse tilt angles to demonstrate proper wick function and gas pocket tests performed with the heat pipe in the reflux mode. Heat pipes should be handled with

Fig. 6.26 Axially grooved wick

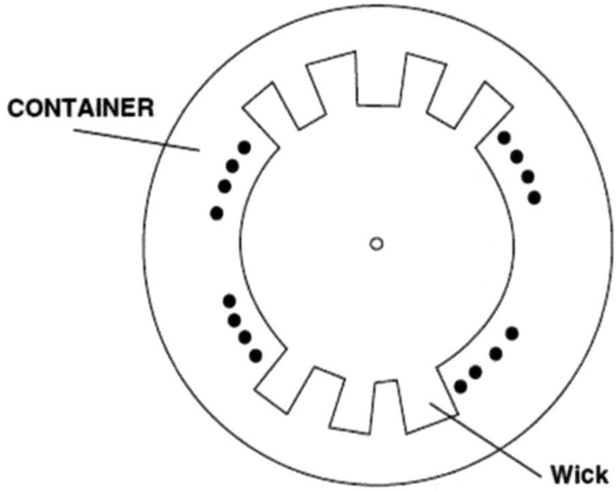

care, especially those that contain ammonia or other high vapor pressure fluids. They should be treated as any other pressure vessel, and appropriate safety precautions must be exercised. Exposure to ammonia vapor can cause severe irritation to eyes and other mucous membranes. Exposure to ammonia liquid can cause severe burns to the skin. Whenever possible, heat pipes should be stored in a cold, dry environment.

This will inhibit any internal chemical reactions which produce non-condensable gas.

6.9 Lesson(s) Learned

Heat transfer by means other than the heat pipe can have the following impacts:

1. A price paid with respect to weight and size of the heat transfer equipment
2. Significant heat lost in transfer over considerable distances
3. Electromotive devices, such as liquid pumps, required to move the heat
4. Possible problems presented by operation in zero gravity

Nonadherence to the implementation methods presented above could result in the following possible impacts: improper cleaning and processing of the aluminum container could result in contaminants reacting with the ammonia to form NCG, which will interfere with the flow of vapor and reduce the heat transfer effectiveness [20].

Contaminants reacting with ammonia normally produce hydrogen, and the gas collects in the condenser region. As more and more of the condenser is blocked, the surface area available for heat rejection decreases, reducing the heat transfer effectiveness; ultimately, the heat pipe may cease to function. Failure to certify welds at the end caps and the fill tube could result in improper or defective welds permitting leaks or catastrophic failure of the pressure vessel. For long-term space missions, working fluids in the appropriate temperature range, such as methanol and water, exhibit an incompatibility with aluminum and should not be used.

6.10 Applications

There are many applications for heat pipes, which are well proven and may now be regarded as routine. In conventional use, heat pipes are integrated into a total thermal subsystem to transport heat from the heat source to remote areas. The heat pipes' ability to act as a primary heat conductive path allows engineers to solve thermal problems in applications with space constraints or other limitations. Thus, you can use heat pipes to carry heat away from the heat-sensitive components to the finned array or a heat sink located in an area where more space for heat dissipation is allowed—leaving room for electronic layout flexibility.

A high-capacity power electronic cooler is an example of a thermal solution where no sufficient space is available to mount directly a finned heat sink to the heat source.

In addition to acting as a heat conductive path and aiding in remote heat transfer, heat pipes can improve thermal solution efficiency. You can accomplish this by embedding heat pipes into the heat sink base or passing the heat pipes through the fins. In most cases, embedding heat pipes into the conventional thermal solution results in size or weight reduction.

The most appropriate application for heat pipe integration into the heat sink base is when the base is large compared with the heat source. In such applications, the heat source location produces the highest temperature. The smaller the heat source, the more spreading has to occur over the heat sink base, resulting in a greater temperature rise in the center of the base. Integrating heat pipes into the base of the heat sink decreases the temperature gradient across the base, yielding a more efficient solution.

You can also improve heat sink fin efficiency with heat pipe integration. Fin efficiency is related to the rate at which the fin can dissipate heat energy. The maximum rate at which the fin can dissipate energy is the rate that would exist if the fin were at base temperature. Therefore, the efficiency of the fin can be improved by passing a heat pipe through the fin. Compared with the traditional finned heat sink, the use of a heat pipe configuration with implementation of fin as part of condenser reduces footprint area of the power heat sink and improves heat dissipation capability.

While external factors such as shock, vibration, force impact, thermal shock, and corrosive environment can affect heat pipe life, its integration into a thermal solution also delivers many benefits. If manufactured and designed properly, heat pipes are highly reliable and have no moving parts. In addition, heat pipes are economical, having little effect on the overall cost of the total thermal [21].

The heat pipe itself is not a heating or cooling device. Heat pipe assemblies are used for moving heat away from the input area (cooling—the most common application) or for moving heat into the output area (heating). Heat pipe assemblies typically have three parts:

1. The heat input assembly
2. The heat transport assembly (the heat pipe)
3. The heat output assembly

Heat pipe assemblies provide thermal management solutions in all mediums: liquid, solid, and gas. Compared to heat pipes, the traditional methods of cooling (extruded metal heat sinks, fans, water, air conditioning, etc.) all have inherent limitations in size, weight, and efficiency. More and more, the limiting factor in systems of all kinds is the inability to dissipate heat. The desire for more power in smaller packages with less weight often cannot be achieved because of excess heat.

Use of heat pipes in high-power (>150 W) cooling applications has been limited to custom applications requiring either low thermal resistance or having a severely restricted enclosure area. The cost of these larger diameter heat pipes was high due to a limited number of manufacturers and handmade assembly times.

A new and valuable addition to the heat transfer community, a heat transport device known as a loop heat pipe (LHP), is discussed in this work. This body of research is very important as the LHP is becoming increasingly prevalent in heat transfer applications. US commercial use of the loop heat pipe will begin on the next generation of communications satellites being developed and built by Hughes Space and Communications Company. These satellites take advantage of the passive nature of the LHP, requiring no external means of pumping, along with its ability to transport large quantities of heat over significant distances. This device comes to the heat transfer community at an ideal time, as the aerospace industry is demanding higher and higher power payloads and this increasing power must be handled by the most efficient means possible. The LHP is also being investigated for uses in ground-based applications such as solar collectors and computer cooling.

This dissertation focuses on experimentation conducted with a space-based satellite application in mind; however, results are applicable to other implementations as well. The LHP is a descendant of the conventional heat pipe. The LHP utilizes the advantages of the conventional heat pipe while overcoming some of the conventional heat pipe's inherent disadvantages. This dissertation serves as a complete body of work on this new device, from background and literature review on the development and history of the LHP to important computer simulation and experimental work, both ground-based and space-based, performed on the LHP in an effort to gain a thorough understanding of the workings of the loop heat pipe and to investigate novel new applications for the LHP such as the ability to control the temperature of an entire spacecraft payload with a minute fraction of the heater power once required. The LHP introduces important new opportunities to the heat transfer community, and the research presented here furthers the knowledge and understanding of this breakthrough device.

The application of heat pipes can be as diverse as their structure and shapes.

We can claim these unique heat transfer devices are used in many fields in the industry, and they have played very important roles from simple heat exchanger to electronic, space application, nuclear reactor, oil lines, and even for constructing ice pontoons through marshes and the foundations of drilling towers, as well as roads in permafrost regions. Reference [22] has a variety of flavors of heat pipe applications in the industry and the companies and manufactures who are involved with their unique design and application of such devices.

For example, within the United States, there are applications and development in progress in a drill for ultra-deep drilling of a bore in the form of a miniature fast-neutron reactor cooled by means of heat pipes.

Other application of heat pipe can be seen in centrifugal heat pipe shapes that are used for cooling asynchronous motors with short-circuited cast rotors. Such motors are used mainly in mechanical engineering. With the use of centrifugal heat pipes in a rotor, it has become possible to control the motor speed electrically, eliminating the need for complex transmissions and gear trains [19]. Investigations are currently in progress in the USSR to explore the possible use of heat pipes to cool transformers, both air-filled and oil-filled, miniature, and high power, and for cooling of electrical busbars.

The West German firm Brown Boveri Corporation has developed a system of electronic devices with heat pipes.

1. Thyristor systems of power greater than 1 kW; the thermal resistance R of the heat pipe is 0.035 K/W and the cooling air velocity is V¼ 6 m/s.
2. A device for a portable current rectifier system (700 W, thermal resistance; 0.055, cooling air; velocity, $V = 6$ m/s).

Heat pipes have proved adaptable to the incorporation of electronic equipment, thereby increasing the cooling effect by factors of 10.

Products of the British SRDE laboratory (Signal Research and Development Establishment) include the following: heat pipes in the form of planar electrical insulators, a heat pipe of very small diameter, and various combinations of heat pipes and thermally insulating modules.

Very interesting possibilities have opened up for producing static batteries and thermal energy converters based on heat pipes, thermal diodes, vapor chambers, etc. and materials which vary their aggregate state (fused salts, metals, sulfur with halogens, etc.); operating temperature is 500–800, the material of the heat pipes is stainless steel, and the heat transfer agent is sodium. The thermal power stored can be up to 10–100 kW/h. High-temperature heat pipes using alkali metals can be employed successfully as electrodes in plasma generators.

In the energy industry, there is a trend to build electric stations using solar energy and hot springs. At present, an electric station of power of at least 100 kW is under construction in the southern United States; it takes the form of a battery of high-temperature heat pipes, heated by the sun, and working into water vapor generators or thermoelectric converters. Such batteries of heat pipes, linked to heat-storage units, will make it possible to develop electrical energy around the clock. There are plans to use heat pipes as electric cables and distribution lines.

On October 4, 1974, a sounding rocket was launched into space (the Black Brant sounding rocket), which carried heat pipes made by the NASA/Goddard Space Flight Center, ESRO, GFW, Hughes.

Aircraft Company and NASA/Ames:

1. ESRO constructed two aluminum heat pipes of length 885 mm and diameter 5 mm. The wick was a single layer of stainless steel mesh with an artery diameter of 0.5 mm. One pipe was filled with ammonia and the other with acetone. The acetone heat pipe transmitted 8.4 W of power, and the ammonia pipe transmitted 21 W. The heat sink was an aluminum block.
2. GFW (Geselfsehaft fiir Weltraumforschung) of the West German Ministry of Technology constructed a flat aluminum heat pipe in the form of a disk of diameter 150 mm and a titanium heat pipe of length 600 mm, charged with methanol, with its end face joined to the disk by an aluminum tube. The flat heat pipe was filled with acetone, and the other end was joined to a heat-storage device (a canister with a molten substance)—"eicosane"—with a fusion temperature of 35 °C. This system transmitted 26 W of power.
3. The Hughes Aircraft Company constructed two flexible heat pipes made of stainless steel (6.4 mm in diameter; 270 cm in length). The working liquid is methanol, and the wick is a metal mesh.
4. NASA/Ames constructed two stainless steel heat pipes of length 910 mm and diameter 12.7 mm. The liquid is methanol, and the inert gas is nitrogen. The wick is a screw thread on the body, and the artery is a wafer of metallic felt. This kind of artery is insensitive to the presence of non-condensable gas.
5. NASA constructed a cryogenic heat pipe made of aluminum with longitudinal channels of length 910 mm and diameter 16 mm, charged with methanol.

Thus, in the international experiment on October 4, 1974, the organizations NASA/GSFC (Grumman and TRW), NASA/Ames (Hughes), Hughes (Hughes), ESRO (the IKE Institute in Stuttgart), and GFW (Dornier) took part in the testing of heat pipes in space. Of these, Grumman constructed five different groups of heat pipes, and TRW constructed three.

In addition to the sounding rockets, NASA has used a number of satellites for testing heat pipes, to evaluate the effect of long-term weightless conditions on heat pipe parameters (the spacecraft Skylab, OAO-III, ATS-6, CTS, etc.).

The French National Center for Space Research, CNS, independently of the American and European Space Center (USA), developed and operated a program of space experiments with heat pipes, constructed by the Aerospatiale and SABCA companies. In November 1974, the French sounding rocket ERIDAN 214 was launched, carrying a radiator of heat pipes. The aim of the experiment was to verify the operational capability of heat pipes under weightless conditions, to verify that the heat pipes would be ready to operate at the start of a rocket flight, and to select various heat pipe structures for spacecraft equipment.

Three types of heat pipe were investigated:

1. A curved heat pipe made by SABCA, of length 560 mm and diameter 3.2 mm, made of steel stairs, the filter being a stainless steel mesh, with ammonia as the heat transfer agent. The transmitted power was 4 W. The pipe was flexible.
2. A heat pipe made by the CENG organization (the atomic center in Grenoble) of length 270 mm and diameter 5 mm, made of copper, with a wick made of sintered bronze powder. The heat transfer agent was water. The transmitted power was 20 W.
3. A SABCA heat pipe, similar to No. 1, but straight. The transmitted power was 5 W.

The heat sink was a box with a variable-phase fusible substance:

$T_f = 28.5\_C$ (*n*-octadecane). The energy source was an electric battery with $U = 27$ V. The total weight of the experimental equipment was 2.3 kg.

These investigations point very clearly to positive gains at present, and we can confidently assert that heat pipes will find wide applications in space in the near future. For example, the United States plans to use heat pipes for thermal control and thermal protection of the reusable shuttle and also for the Spacelab space laboratory.

For these, the heat-sensitive equipment will be located in boxes or canisters within which the temperature will be held constant by means of heat pipes located in the walls of the enclosure. Reference [22] provides a vast variety application of heat pipes in the present industry and future trend of it.

6.11 Summary

Heat Pipes

General

A heat pipe is a passive energy recovery heat exchanger that has the appearance of a common plate-finned water coil except the tubes are not interconnected. Additionally, it is divided into two sections by a sealed partition. Hot air passes through one side (evaporator) and is cooled, while cooler air passes through the other side (condenser). While heat pipes are sensible heat transfer exchangers, if the air conditions are such that condensation forms on the fins, there can be some latent heat transfer and improved efficiency (Fig. 6.27).

(continued)

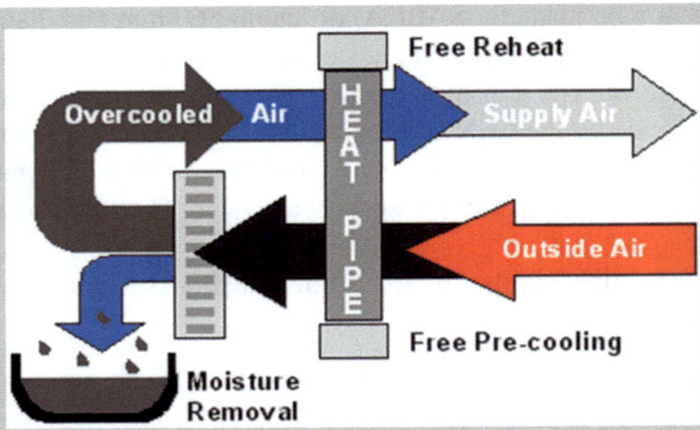

Fig. 6.27 Heat pipe application concept

Heat pipes are tubes that have a capillary wick inside running the length of the tube, are evacuated and then filled with a refrigerant as the working fluid, and are permanently sealed. The working fluid is selected to meet the desired temperature conditions and is usually a Class I refrigerant. Fins are similar to conventional coils—corrugated plate, plain plate, and spiral design. Tube and fin spacing are selected for appropriate pressure drop at design face velocity. HVAC systems typically use copper heat pipes with aluminum fins; other materials are available.

Advantages

- Passive heat exchange with no moving parts.
- Relatively space efficient.
- The cooling or heating equipment size can be reduced in some cases.
- The moisture removal capacity of existing cooling equipment can be improved.
- No cross-contamination between air streams.

Disadvantages

The use of the heat pipe:

- Adds to the first cost and to the fan power to overcome its resistance
- Requires that the two air streams be adjacent to each other
- Requires that the air streams must be relatively clean and may require filtration

Applications

Heat pipe heat exchanger enhancement can improve system latent capacity. For example, a 1 °F dry bulb drop in air entering a cooling coil can increase the latent capacity by about 3%. The heat pipe's transfer of heat directly from the entering air to the low-temperature air leaving the cooling coil saves both cooling and reheating energy. It can also be used to precool or preheat incoming outdoor air with exhaust air from the conditioned spaces.

(continued)

Best Applications

- Where lower relative humidity is an advantage for comfort or process reasons, the use of a heat pipe can help. A heat pipe used between the warm air entering the cooling coil and the cool air leaving the coil transfers sensible heat to the cold exiting air, thereby reducing or even eliminating the reheat needs. Also, the heat pipe precools the air before it reaches the cooling coil, increasing the latent capacity and possibly lowering the system cooling energy use.
- Projects that require a large percentage of outdoor air and have the exhaust air duct in close proximity to the intake can increase system efficiency by transferring heat in the exhaust to either precool or preheat the incoming air.

Possible Applications

- Use of a dry heat pipe coupled with a heat pump in humid climate areas
- Heat pipe heat exchanger enhancement used with a single-path or dual-path system in a supermarket application
- Existing buildings where codes require it or they have "sick building" syndrome and the amount of outdoor air intake must be increased
- New buildings where the required amount of ventilation air causes excess loads or where the desired equipment does not have sufficient latent capacity

Applications to Avoid

- Where the intake or exhaust air ducts must be rerouted extensively, the benefits are likely not to offset the higher fan energy and first cost.
- Use of heat pipe sprays without careful water treatment. Corrosion, scale, and fouling of the heat pipe where a wetted condition can occur need to be addressed carefully.

Technology Types (Resource)

Hot air is the heat source, flows over the evaporator side, is cooled, and evaporates the working fluid. Cooler air is the heat sink, flows over the condenser side, is heated, and condenses the working fluid. Vapor pressure difference drives the evaporated vapor to the condenser end, and the condensed liquid is wicked back to the evaporator by capillary action. Performance is affected by the orientation from horizontal. Operating the heat pipe on a slope with the hot (evaporator) end below horizontal improves the liquid flow back to the evaporator. Heat pipes can be applied in parallel or series.

Efficiency

Heat pipes are typically applied with air face velocities in the 450–550 ft per minute range, with 4–8 rows deep and 14 fins per inch, and have an effectiveness of 45–65%. For example, if entering air at 77 °F is cooled by the

(continued)

heat pipe evaporator to 70 °F and the air off the cooling coil is reheated from 55 to 65 °F by the condenser section, the effectiveness is 45 % [=(65 − 55)/(77 − 55) = 45%]. As the number of rows increases, effectiveness increases but at a declining rate. For example, doubling the rows of a 48% effective heat pipe increases the effectiveness to 65%.

Tilt control can be used to:

- Change operation for seasonal changeover.
- Modulate capacity to prevent overheating or overcooling of supply air.
- Decrease effectiveness to prevent frost formation at low outdoor air temperatures.

Tilt control (6° maximum) involves pivoting the exchanger about its base at the center with a temperature-actuated tilt controller at one end. Face and bypass dampers can also be used.

Manufacturers

Heat Pipes

1. American Heat Pipes, Inc.
   6914 E. Fowler Ave.
   Suite E
   Tampa, FL 33617
   1-800-727-6511
2. Dectron Inc
   4300 Blvd. Poirier
   Montreal, PQ H4R 2C5
   Canada
   (514) 336-9609
   mail@dectron.com
3. Des Champs Laboratories Inc
   P.O.Box 220
   Douglas Way
   Natural Bridges Station, VA 24579
   (703) 291-1111
4. EcoTech Consultants, Inc.
   3466 Holcombe Bridge Road
   Suite 1000
   Norcross, GA 30092
   (404) 723-6564
5. Heat Pipe Technology Inc
   P.O. Box 999
   Alachua, FL 32615-0999

(continued)

1-800-393-2041

6. Munters Dry Cool
   16900 Jordan Rd.
   Selma, TX 78154-1272
   1-800-229-8557
   moreinfo-dc@americas.munters.com
7. Nautica Dehumidifiers, Inc.
   9 East Carver St.
   Huntington, NY 11743
   (516) 351-8249
   dehumidify@aol.com
8. Octagon Air Systems
   1724 Koppers Road
   Conley, GA 30288
   (404) 609-8881
9. Power-Save International
   P.O. Box 880
   Cottage Grove, OR 97424
   1-800-432-5560
10. Seasons 4 Inc.
    4500 Industrial Access Road
    Douglasville, GA 30134
    (770) 489-0716
11. Temprite Industries
    1555 Hawthorne Lane
    West Chicago, IL 60185
    1-800-552-9300
12. Venmar CES
    2525 Wentz Ave.
    Saskatoon, SK S7K 2K9
    Canada
    1-800-667-3717
    customerservice@venmarvent.com

References

1. Zohuri, B. (2016). *Heat pipe design and technology: Modern applications for practical thermal management* (2nd ed.). New York: Springer.
2. Gaugler, R. S. (1944, June 6). *Heat transfer device*. U.S. Patent 2, 350, 348.
3. Trefethen, L. (1962, February). *On the surface tension pumping of liquids or a possible role of the candlewick in space exploration* (G. E. Tech. Info., Ser. No. 615 D114).
4. Wyatt, T. Wyatt (Johns Hopkins/Applied Physics Lab.). *Satellite Temperature Stabilization System. Early development of spacecraft heat pipes for temperature stabilization*. U.S. Patent No. 3,152,774 (October 13, 1964), application was files June 11, 1963.

5. Grove, G. M., Cotter, T. P., & Erikson, G. F. (1964). Structures of very high thermal conductivity. *Journal of Applied Physics, 35*, 1990.
6. Chi, S. W. (1976). *Heat pipe theory and practice*. New York: McGraw-Hill.
7. Dunn, P. D., & Reay, D. A. (1982). *Heat pipes* (3rd ed.). New York: Pergamon.
8. Marcus, B. D. (1971, July). *Theory and design of variable conductance heat pipes: Control techniques* (Research Report No. 2). NASA 13111-6027-R0-00.
9. Bennett, G. A. (1977, September 1). *Conceptual design of a heat pipe radiator*. LA-6939-MS Technical Report. Los Alamos Scientific Lab., NM, USA.
10. Gerasimov, Y. F., Maidanik, Y. F., & Schegolev, G. T. (1975). Low-temperature heat pipes with separated channels for vapor and liquid. *Engineering Physics Journal, 28*(6), 957–960. (in Russian).
11. Watanabe, K., Kimura, A., Kawabata, K., Yanagida, T., & Yamauchi, M. (2001). *Development of a variable-conductance heat-pipe for a sodium-sulfur (NAS) battery*. Furukawa Review, No. 20.
12. Peterson, G. P. (1994). *An introduction to heat pipes: Modeling, testing, and applications* (pp. 175–210). New York: Wiley.
13. Garner, S. D., P. E., Thermacore Inc.
14. Brennan, P. J., & Kroliczek, E. J. (1979). *Heat pipe design handbook*. Towson, MD: B & K Engineering.
15. Kemme, J. E. (1969, August 1). *Heat pipe design considerations*. Los Alamos Scientific Laboratory report LA-4221-MS.
16. MIL-STD-1522A (USAF). (1984, May). *Military standard general requirements for safe design and operation of pressurized missile and space systems*.
17. Woloshun, K. A., Merrigan, M. A., & Best, E. D. *HTPIPE: A steady-state heat pipe analysis program: A user's manual*.
18. Faghri, A. *Temperature regulation system for the human body using heat pipes*. US patent 5269369.
19. Grover, G. M., Cotter, T. P., & Erickson, G. F. (1964). Structures of very high thermal conductance. *Journal of Applied Physics, 35*(6), 1990–1991.
20. Ranken, W. A., & Kemme, J. E. (1965). Survey of Los Alamos and EURATOM heat pipe investigations. In *Proceedings of the IEEE Thermionic Conversion Specialist Conference, San Diego, California, October 1965*. Los Alamos Scientific Laboratory, report LA-DC-7555.
21. Kernme, J. E. (1966). Heat pipe capability experiments. In *Proceedings of Joint AEC Sandia Laboratories report SC-M-66-623, 1, October 1966*. Expanded version of this paper, Los Alamos Scientific Laboratory report LA-3585-MS (August 1966), also as LA-DC-7938. Revised version of LA-3583-MS, Proc. EEE Thermionic Conversion Specialist Conference, Houston, Texas, (November 1966).
22. Grover, G. M., Bohdansky, J., & Busse, C. A. (1965). *The use of a new heat removal system in space thermionic power supplies*. European Atomic Energy Community—EURATOM report EUR 2229.e.

Chapter 7
Direct Reactor Auxiliary Cooling System

Historically, the idea of implementing a loop known as Direct Reactor Auxiliary System (DRACS) as a passive heat removal system in a nuclear power plant for safety purpose is nothing new. The DRAC system originally was derived from the Experimental Breeder Reactor-II (EBR-II), and then it was improved in later fast reactor designs such as Clinch River Breeder Reactor Project (CRBRP) by Westinghouse around 1970s time frame on their Liquid Metal Fast Breeder Reactor (LMFBR) and then later on was manufacture for French reactor known as Phoenix-II, which went into operation in France during 1978 time frame. The DRACS has been proposed for Advanced High-Temperature Reactor (AHTR) as the passive decay heat removal system. The DRACS features three coupled natural circulation/convection loops relying completely on buoyancy as the driving force. In the DRACS, two heat exchangers, namely, the DRACS Heat Exchanger (DHX) and the Natural Draft Heat Exchanger (NDHX), are used to couple these natural circulation/convection loops. In addition, a fluidic diode is employed to restrict parasitic. In addition, a fluidic diode is employed to restrict parasitic flow during normal operation of the reactor and to activate the DRACS in accidents.

7.1 Introduction

The Direct Reactor Auxiliary Cooling System (DRACS) has been proposed for Advanced High-Temperature Reactor (AHTR) as the passive decay heat removal system. The DRACS features three coupled natural circulation/convection loops relying completely on buoyancy as the driving force. In the DRACS (see Fig. 7.1), two heat exchangers, namely, the DRACS Heat Exchanger (DHX) and the Natural Draft Heat Exchanger (NDHX), are used to couple these natural circulation/convection loops. In addition, a fluidic diode is employed to restrict parasitic. In addition, a fluidic diode is employed to restrict parasitic flow during normal operation of the reactor and to activate the DRACS in accidents.

B. Zohuri, *Heat Pipe Applications in Fission Driven Nuclear Power Plants*,
https://doi.org/10.1007/978-3-030-05882-1_7

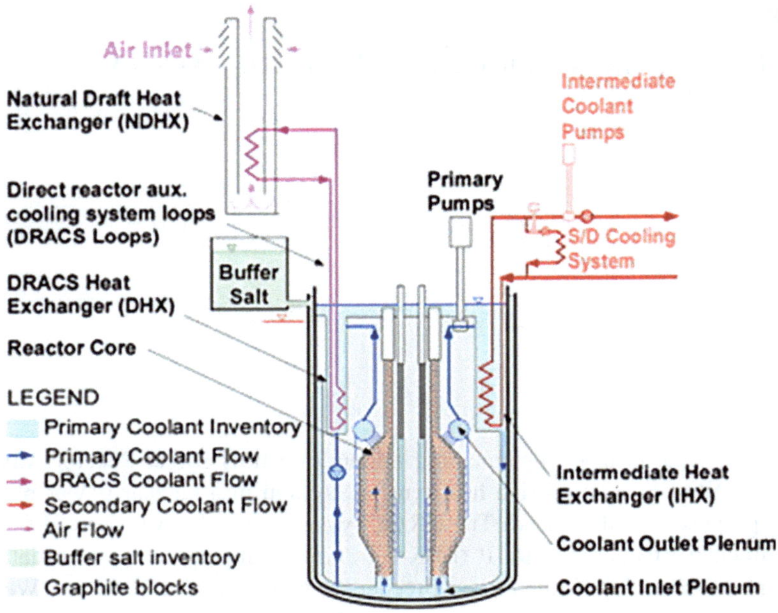

Fig. 7.1 Typical Direct Reactor Auxiliary Cooling System

While the DRACS concept has been proposed, there are no actual prototypic DRACS systems for AHTRs built and tested in the literature. In this report, a detailed modular design of the DRACS for a 20-MW(th) fluoride-salt-cooled high-temperature reactor (FHR) is first developed. As a starting point, the DRACS is designed to remove 1% of the nominal power, i.e., the decay power being 200 kW.

The design process for the prototypic DRACS involves selection of the salts, identification of the reactor core, design of the DHX and NDHX, design of the fluidic diode, design of the air chimney, selection of the loop pipes, and finally determination of the loop height based on pressure drop analysis. FLiBe with high enrichment in Li-7 and FLiNaK have been selected as the primary and secondary salts, respectively. A 16-MW(th) pebble bed core proposed by University of California at Berkeley (UCB) is adopted in the design. Shell-and-tube heat exchangers have been designed based on Delaware Method for the DHX and NDHX.

A vortex diode that has been tested with water in the literature is adopted in the present design. Finally, pipes with inner diameter of 15 cm are selected for both the primary and secondary loops of the DRACS. The final DRACS design features a total height less than 13 m. The design presented here has the potential to be used in the planned small-scale FHR test reactor and will also benefit and guide the DRACS design for a commercial AHTR.

Following the prototypic DRACS design is the detailed scaling analysis for the DRACS, which will provide guidance for the design of scaled-down DRACS test facilities. Based on the Boussinesq approximation used in the field of buoyancy-driven flow (i.e. also known as natural convection) in one-dimensional formulation, the governing equations, i.e., the continuity, integral momentum, and energy equations, are nondimensionalized by introducing appropriate dimensionless parameters, including the dimensionless length, temperature, velocity, etc. The key dimensionless numbers, i.e., the Richardson, friction, Stanton, time ratio, Biot, and heat source numbers that characterize the DRACS system, are obtained from the nondimensional governing equations. Based on the dimensionless numbers and nondimensional governing equations, similarity laws are proposed. In addition, a scaling methodology has also been developed, which consists of the core scaling and loop scaling [1].

7.2 Decay Heat Removal System in Various Reactor Designs

As Clinch River Breeder Reactor Project (CRBRP) design by Westinghouse around the 1970s time frame on their liquid metal fast breeder reactor (LMFBR) (see Fig. 7.2), there are three backup systems to remove decay heat in case of non-availability of the normal heat sink.

1. The first one is the Protected Air-Cooled Condenser (PACC) system that cools the steam drum directly.
2. A second heat sink can be made available by opening the safety relief valve in steam line, thereby venting steam to atmosphere.
3. The third system is a completely separate Overflow Heat Removal System (OHRS) provided to extract heat directly from the in-vessel primary loop. The heat sink for this system is provided by an air-cooled heat exchanger. However, after episode of There Mile Island (TMI) around 1978 the first author of this book (Zohuri), while was working at Westinghouse Advanced Reactor in Waltz Mill, suggested installation of Mercury heat pipe as an alternative for the heat exchanger and designed series of these heap pipes if form of variable conductance (VCHP) one as an alternative (see Fig. 7.3) [2].

The third system above provides a decay heat removal (DHR) system in situations where the steam generator (SG) is not available. The DHR system of a loop type such as Japan sodium-cooled fast reactor (JSFR) consists of a combination of one loop of a Direct Reactor Auxiliary Cooling System (DRACS) and two loops of a Primary Reactor Cooling System (PRACS) as shown in Fig. 7.4.

The heat exchanger (HX) of DRACS is located in the reactor vessel's upper plenum. Each heat exchanger of PRACS is located in intermediate heat exchanger (IHX) upper plenum. These systems operate fully by natural convection and are activated by opening of dc-power-operated dampers, and the cold pool are thermally coupled by the PRACS, which is composed of heat exchangers, fluidic diodes, and connecting pipes as shown in Fig. 7.5. The fluidic diode reduces leakage flows under primary loop forced circulation.

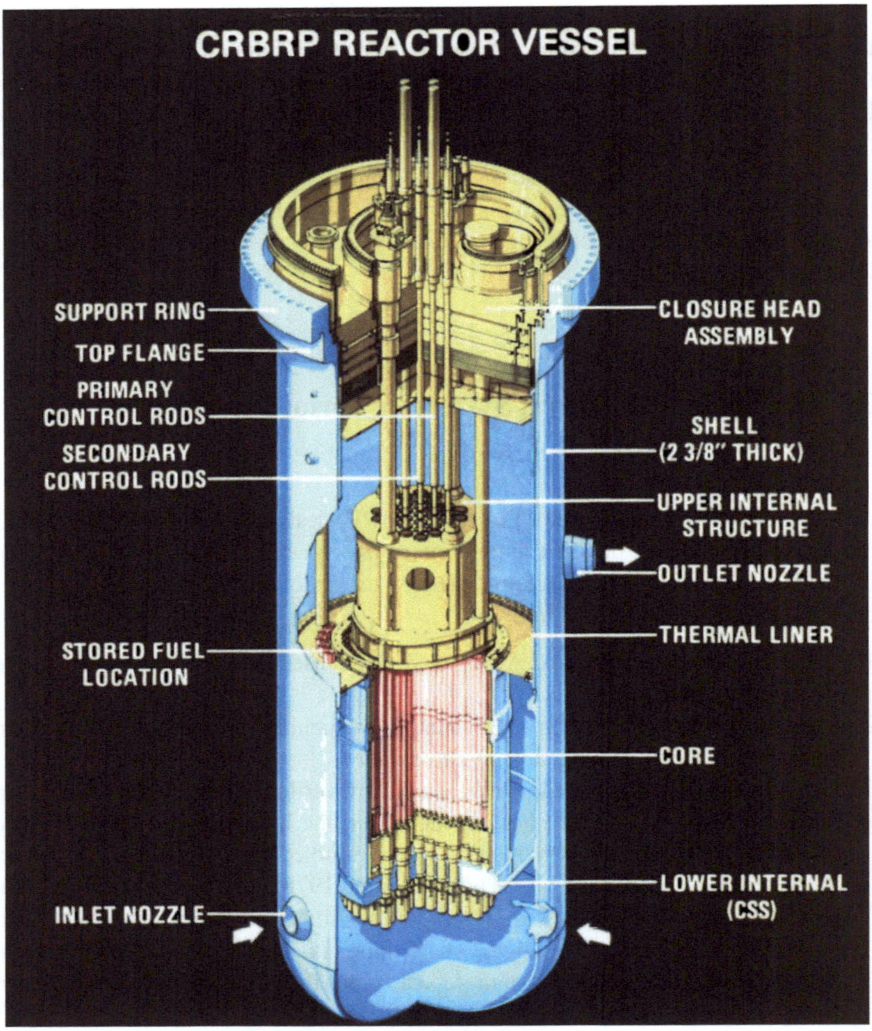

Fig. 7.2 Clinch River Breeder Reactor Project Vessel

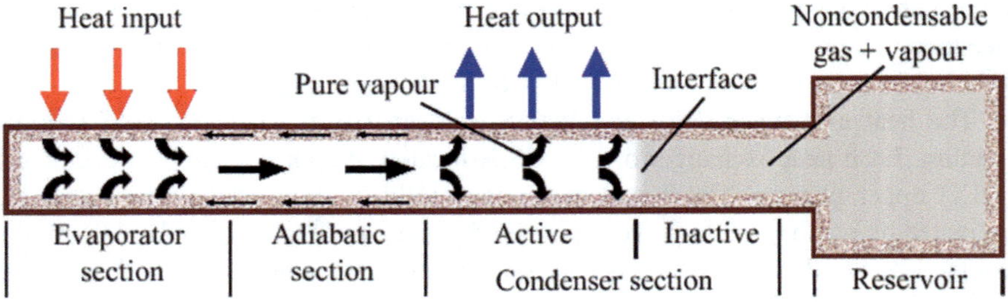

Fig. 7.3 Variable conductance heat pipe infrastructure depiction

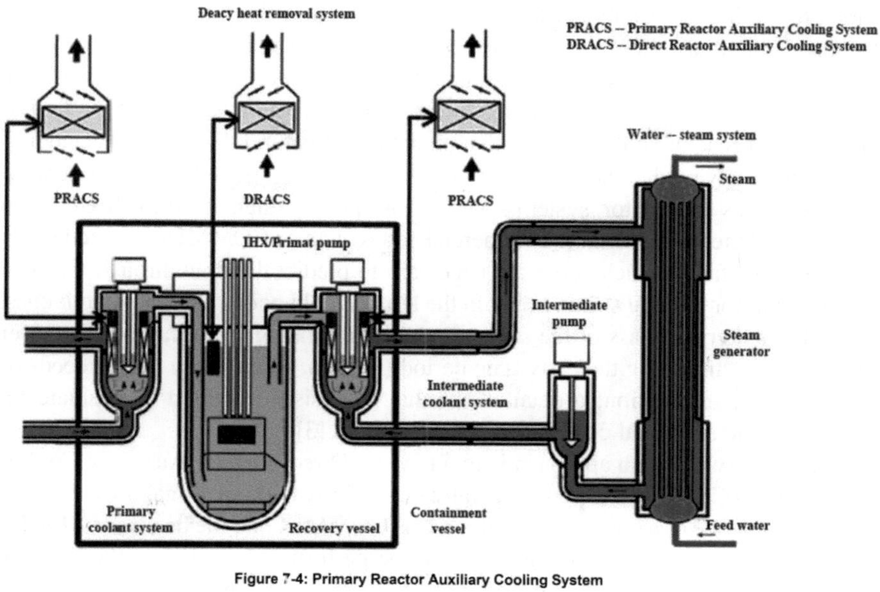

Fig. 7.4 Primary reactor auxiliary cooling system

Fig. 7.5 Depiction of the innovative design of the DHR system

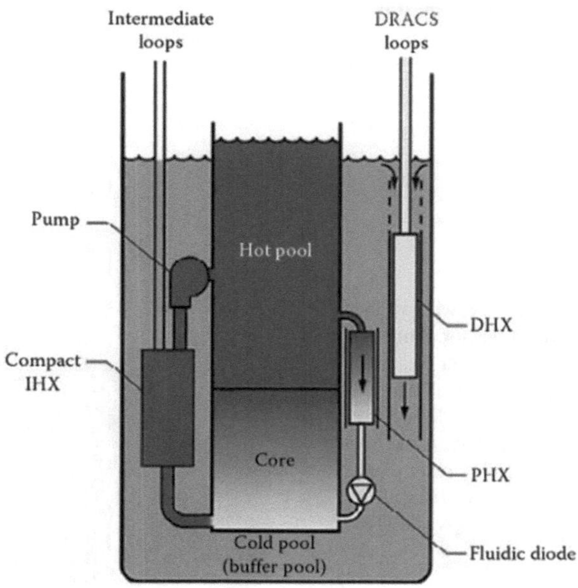

As part of Generation IV (GEN-IV), the two main features of the Advanced Liquid Metal Reactor (ALMR) designs are a large heat capacity sodium pool and the utilization of decay heat removal system (DHRS) that operate on natural convection.

The use of highly reliable passive decay heat removal system such as Direct Removal Auxiliary Cooling System (DRACS), combined with Heat Pipe (HP), either fixed or variable type can significantly reduce the risk profile of an ALMR as a fully inherent shutdown system, when we do not require an actor in the loop to take steps of reactor shutdown due to a manmade of natural disaster and accident.

To accurately predict the behavior of the reactor sodium pool during the operation of these types of reactor systems—and consequently their performance—three-dimensional thermal-hydraulic commercial codes such as COMMIX are needed [3].

More specifically, such codes are necessary to predict the flow through the core, the potential for thermal stratification in the hot and cold pools, and the effectiveness of the large thermal mass of the sodium pool in mitigating the decay power-driven thermal transient during the early time period ($\approx$20 h), when decay heat exceeds on the heat loss. In addition, a detailed thermal analysis is required to evaluate the stresses on the structural components of the system [3].

The objective of such approach is to analyze a Direct Reactor Auxiliary Cooling System (DRACS) test to support the validation of any computational code either in house written one or a commercial one such as COMMIX and the design of ALMRs. This is one of a series of tests that one can be performed to generate the thermal-hydraulic data for the validation of these type of codes and to demonstrate the performance of the Reactor Vessel Auxiliary Cooling System (RVACS) and DRACS of ALMR designs like Sodium Advanced Fast Reactor (SAFR) and Power Reactor Inherently Safe Module (PRISM) as part of safety and commercial licensing of such production power plants. See Fig. 7.6.

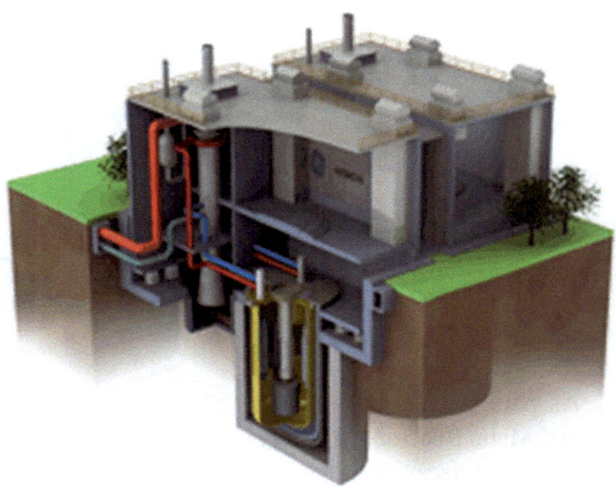

Fig. 7.6 Drawing of the PRISM reactor. **Note**: PRISM (Power Reactor Innovative Small Module, sometimes S-PRISM from Super PRISM) is the name of a nuclear power plant design by GE Hitachi Nuclear Energy (GEH)

7.3 Experimental Validation of Passive Decay Heat Removal Technology for FHR

Like other reactors, fluoride-salt-cooled high-temperature reactors (FHRs) require decay heat cooling systems. A Direct Reactor Auxiliary Cooling System (DRACS) can be used to transfer heat from the reactor coolant to the atmosphere. In a sodium-cooled reactor, it is typically a sodium loop with primary sodium providing the decay heat through a heat exchanger to the natural circulation DRACS sodium loop that dumps its heat to a sodium-to-air heat exchanger.

In an FHR there are added challenges. First, the reactor peak temperature is typically near 700 °C. Second, the mainline FHR coolant is a molten salt with a freezing point of about 460 °C. It is important not to freeze the primary coolant because that could stop circulation through the reactor core that then could result in high temperatures in the reactor core with fuel failure. Third, neutron interaction with the salt coolant generates tritium that can diffuse through metal heat exchangers—escaping to the environment via DRACS. Last, whatever the fluid in the DRACS, one must consider chemical compatibility with the molten salt in the even there is a heat exchanger failure.

One solution that we as authors of this book can propose is utilization of multiple heat pipes either fixed or variable type depending on heat duty transport that these pipes need to manage.

The mainline heat-pipe coolant options for DRACS would be sodium and potassium—but a survey of fluoride salts would also be conducted (initial analysis has not found likely candidates, but a more in-depth search would be made). A vertical heat pipe has a fluid that boils, the vapors travel upward, the vapors are condensed, and the liquid flows back down the pipe walls. Heat pipes have the characteristic that they can be designed to increase heat transfer by more than an order of magnitude over a small temperature range—what we want for a fluoride-salt-cooled high-temperature reactors direct reactor auxiliary cooling system, where there is very low heat losses at low temperatures and high heat removal rates above a preset temperature. NASA has developed sodium, potassium, and cesium heat pipes for proposed space reactors with one of the goals to "turn on" above a preset temperature. Such a system would require many parallel heat pipes to remove the desired quantities of heat providing redundancy. That is a major advantage for this application. It implies that the inventory of coolant in any heat pipe is small, and thus potential quantities of sodium or potassium that could go into the primary system if heat pipe failure would be small. The addition of sodium or potassium metal into the primary coolant would change the chemical redox (primary concern) as well as some neutronic impacts.

Separate from the DRACS heat transfer is the requirement to prevent tritium diffusion through metal pipes to the environment. The base-case tritium barrier will be a double-wall heat exchanger purged with an inert gas containing small quantities of oxygen. The oxygen would convert any tritium to $3H_2O$ that does not diffuse through metal walls. An oxide tritium barrier may be included where the low oxygen level helps preserve the tritium barrier. The backup is a tritium getter between the walls—an option that has been investigated for some systems. Trade-off studies will be done before selection of what combination of methods will be used for tritium control.

It is worth to mention that nuclear reactor power systems could revolutionize space exploration and support human outpost on the Moon and Mars. This paper reviews current static and dynamic energy conversion technologies for use in space reactor power systems and provides estimates of the system's net efficiency and specific power and the specific area of the radiator. The suitable combinations of the energy conversion technologies and the nuclear reactors classified based on the coolant type and cooling method, for best system performance and highest specific power, are also discussed. In addition, four space reactor power system concepts with both static and dynamic energy conversion are presented. These systems concepts are for nominal electrical powers up to 110 kW(e) and have no single point failures in reactor cooling, energy conversion, and heat rejection. Two power systems employ liquid metal heat pipe cooled reactors, Thermoelectric (TE) and Alkali-Metal Thermal-to-Electric Conversion (AMTEC) units for converting the reactor power to electricity and potassium heat pipe radiators. The third power system employs SiGe TE converters and a liquid metal cooled reactor, with a core divided into six identical sectors. Each sector has a separate energy conversion loop, a heat rejection loop, and a rubidium heat pipe radiator panel. The fourth power system has a gas cooled reactor, with a sectored core. Each of the three sectors in the core is coupled to a separate Closed Brayton Cycle (CBC) loop with He-Xe (40 g/mol) working fluid and a NaK-78 secondary loop and two separate water heat pipe radiator panels [4].

7.4 The Mainline Heat Pipe Coolant Options

As we stated in Chap. 4 of this book, the principle of the heat pipe is very simple. Phenomena involved are surface tension and the latent heat of the working fluid. There are no moving parts. Its effective thermal conductivity is hundreds of times more than conventional solid heat conductors. Substantial research in this field did not begin until the 1960s. Already there have been efforts made to employ heat pipes in various areas. Government agencies such as the NAVY can benefit from the useful characteristics of heat pipes. Several potential naval applications are discussed in the paper.

More and more opportunities for the employment of the heat pipes could be realized as time goes on. Heat pipe technology may play a decisive role in improving the overall economics and public perception on nuclear desalination, specifically on seawater desalination. When coupled to the Low-Temperature Multi-Effect Distillation process, heat pipes could effectively harness most of the waste heat generated in various types of nuclear power reactors. Indeed, the potential application of heat pipes could be seen as a viable option to nuclear seawater desalination where the

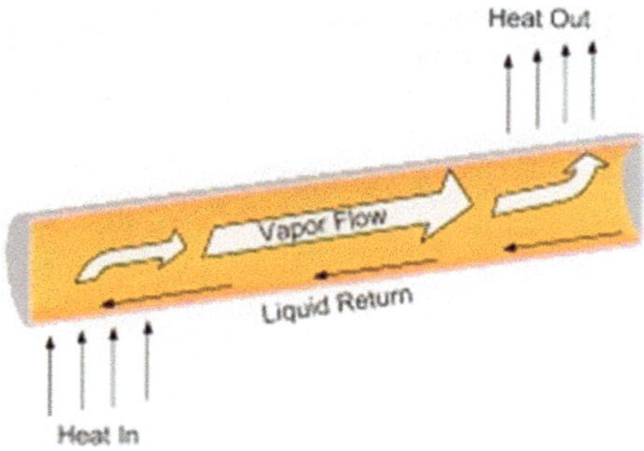

Fig. 7.7 A simple physical configuration of heat pipe

efficiency to harness waste heat might not only be enhanced to produce larger quantities of potable water but also to reduce the environmental impact of nuclear desalination process. Furthermore, the use of heat pipe-based heat recovery systems in desalination plant may improve the overall thermodynamics of the desalination process, as well as help to ensure that the product water is free from any contamination which occur under normal process, thus preventing operational failure occurrences as this would add an extra loop preventing direct contact between radiation and the produced water. In this paper, a new concept for nuclear desalination system based on heat pipe technology is introduced, and the anticipated reduction in the tritium level resulting from the use of heat pipe systems is discussed [2].

In summary, a heat pipe is a two-phase heat transfer device with a very high effective thermal conductivity. It is a vacuum-tight device consisting of an envelope, a working fluid, and a wick structure. As shown in Fig. 7.7, the heat input vaporizes the liquid working fluid inside the wick in the evaporator section. The saturated vapor, carrying the latent heat of vaporization, flows toward the colder condenser section. In the condenser, the vapor condenses and gives up its latent heat. The condensed liquid returns to the evaporator through the wick structure by capillary action. The phase change processes and two-phase flow circulation continue as long as the temperature gradient between the evaporator and condenser is maintained.

Heat pipes function by absorbing heat at the evaporator end of the cylinder and boiling and converting the fluid to vapor. The vapor travels to the condenser end, rejects the heat, and condenses to liquid. The condensed liquid flows back to the evaporator, aided by gravity. This phase change cycle continues as long as there is heat (i.e., warm outside air) at the evaporator end of the heat pipe. This process occurs passively, and there is no external electrical energy required.

At the hot interface of a heat pipe a liquid in contact with a thermally conductive solid surface turns into a vapor by absorbing heat from that surface. The vapor then travels along the heat pipe to the cold interface and condenses back into a liquid—releasing the latent heat. The liquid then returns to the hot interface through either capillary action, centrifugal force, or gravity, and the cycle repeats. Due to the very high heat transfer coefficients for boiling and condensation, heat pipes are highly effective thermal conductors. The effective thermal conductivity varies with heat

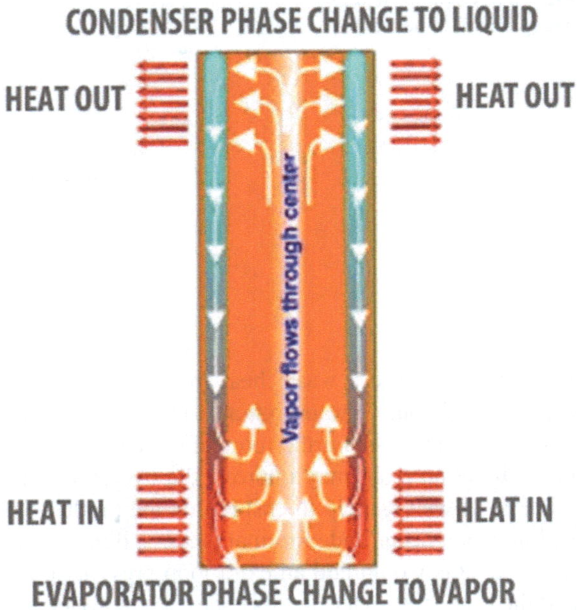

Fig. 7.8 Internal schematic heat pipe structure

pipe length and can approach 100 kW/(m K) for long heat pipes, in comparison with approximately 0.4 kW/(m K) for copper.

Heat pipes employ evaporative cooling to transfer thermal energy from one point to another by the evaporation and condensation of a working fluid or coolant. Heat pipes rely on a temperature difference between the ends of the pipe and cannot lower temperatures at either end below the ambient temperature (hence they tend to equalize the temperature within the pipe) (Fig. 7.8).

Heat pipes have an envelope, a wick, and a working fluid. Heat pipes are designed for very long-term operation with no maintenance, so the heat pipe wall and wick must be compatible with the working fluid. Some material/working fluid pairs that appear to be compatible are not. For example, water in an aluminum envelope will develop large amounts of non-condensable gas over a few hours or days, preventing normal operation of the heat pipe.

7.4.1 Heat Pipe Materials and Working Fluids

Since heat pipes were rediscovered by George Grover in 1963, extensive life tests have been conducted to determine compatible envelope/fluid pairs, some going on for decades. In a heat pipe life test, heat pipes are operated for long periods of time, and monitored for problems such as non-condensable gas generation, material transport, and corrosion.

7.4.2 *Different Types of Heat Pipes*

In addition to the standard Constant Conductance Heat Pipes (CCHPs), there are a number of other types of heat pipes:

- Vapor chambers (planar heat pipes), which are used for heat flux transformation and isothermalization of surfaces
- Variable conductance heat pipes (VCHPs), which use a non-condensable gas (NCG) to change the heat pipe's effective thermal conductivity as power or the heat sink conditions change
- Pressure-controlled heat pipes (PCHPs), which are a VCHP where the volume of the reservoir or the NCG mass can be changed, to give more precise temperature control
- Diode heat pipes, which have a high thermal conductivity in the forward direction and a low thermal conductivity in the reverse direction
- Thermosyphons, which are heat pipes where the liquid is returned to the evaporator by gravitational/accelerational forces
- Rotating heat pipes, where the liquid is returned to the evaporator by centrifugal forces

7.4.3 *Nuclear Power Conversion*

Grover and his colleagues were working on cooling systems for nuclear power cells for space craft, where extreme thermal conditions are encountered. These alkali metal heat pipes transferred heat from the heat source to a thermionic or thermoelectric converter to generate electricity.

Since the early 1990s, numerous nuclear reactor power systems have been proposed using heat pipes for transporting heat between the reactor core and the power conversion system. The first nuclear reactor to produce electricity using heat pipes was first operated on September 13, 2012, in a demonstration using flattop fission.

In Nuclear power plant application, heat pipes can be used as a passive heat transfer system for performing as overall thermal hydraulic and natural circulation sub-system in an Inherent Shutdown, Heat Removal System (ISHRS) in the core (i.e. installed on top of the core doom) of nuclear reactor such as Molten Salt or Liquid Metal Fast Breeder type reactors, as a secondary fully inherent shutdown system loop acting like heat exchanger from safety point of view so the reactor never reaches to its melting point in case of accidental events.

7.4.4 Benefits of the Heat Pipes

The following are overall benefits of heat pipes:

- High thermal conductivity (10,000–100,000 W/m K)
- Isothermal
- Passive
- Low cost
- Shock/vibration tolerant
- Freeze/thaw tolerant

7.4.5 Limitations of Heat Pipes

The followings are limitation of heat pipes in general:

- Heat pipes must be tuned to particular cooling conditions. The choice of pipe material, size, and coolant all have an effect on the optimal temperatures at which heat pipes work.
- When used outside of its design heat range, the heat pipe's thermal conductivity is effectively reduced to the heat conduction properties of its solid metal casing alone—in the case of a copper casing, around 1/80 of the original flux. This is because below the intended temperature range, the working fluid will not undergo phase change; and above it, all of the working fluid in the heat pipe vaporizes and the condensation process ceases.
- Most manufacturers cannot make a traditional heat pipe smaller than 3 mm in diameter due to material limitations.

7.4.6 Conclusion

Overall, a heat pipe is a heat transfer device that combines the principles of both thermal conductivity and phase transition to effectively transfer heat between two solid interfaces (Fig. 7.9)

Phase-change processes and the two-phase flow circulation in the heat pipe (HP) will continue as long as there is a large enough temperature difference between the evaporator and condenser sections. The fluid stops moving if the overall temperature is uniform but starts back up again as soon as a temperature difference exists. No power source (other than heat) is needed.

In some cases, when the heated section is below the cooled section, gravity is used to return the liquid to the evaporator. However, a wick is required when the evaporator is above the condenser on earth. A wick is also used for liquid return if there is no gravity, such as in NASA's micro-gravity applications.

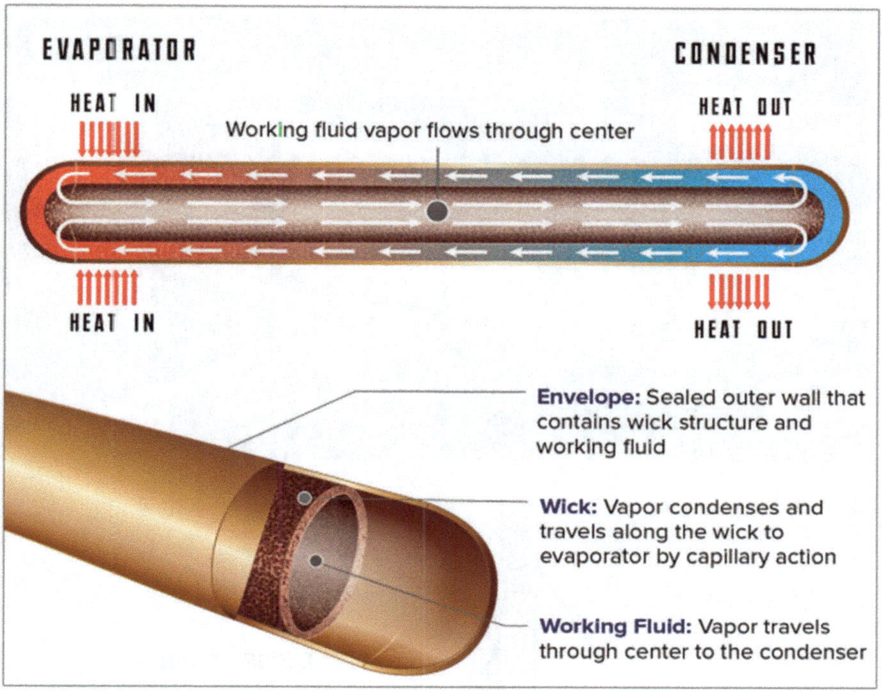

Fig. 7.9 Tope view depiction of heat pipe

It is fruitful to state that at part of application of nuclear power system in space, would be efficient to look at and study, the Brayton cycle (i.e. See Chap. 4 for this cycle) vapor chamber in form of heat pipe radiator. It is recognized that the heat rejection systems required for the relatively large space nuclear power plants application in the future comprise a significant portion of the total spacecraft weight and area. Thermal and structural efficiencies and performance variations of radiators can therefore influence overall spacecraft configurations, weight, payload, and electrical power capability.

The system shown in Fig. 7.10 is based on the requirements of the Vapor Chamber Radiator Study (VCRS) by General Electric (GE) under contract number NAS9-744 by NASA for NASA/MSC.

The VCRS illustrates the role the radiator plays in serving as structure and shell for system components. The relative size of future radiators is even more dramatically illustrated by Fig. 7.11, in the study of a Low Acceleration Space Transportation System (LAST).

For this study, the vapor chamber (heat pipe) radiator is defined and evaluated as a potential candidate for rejecting waste heat from a radioisotope Brayton cycle space power system. A comparison in this study was made with an operationally equivalent conduction fin radiator. Both radiators employed DC-200 heat transfer fluid within the primary ducts and aluminum as the basic structural material. Vapor

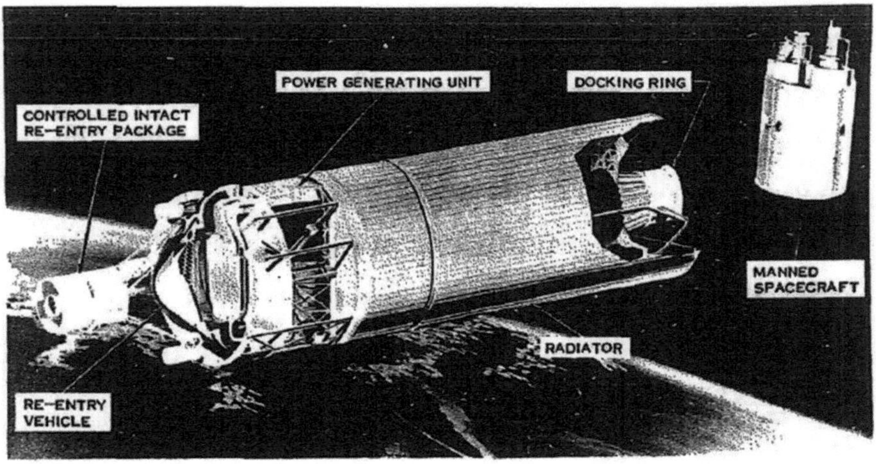

Fig. 7.10 Power module configuration

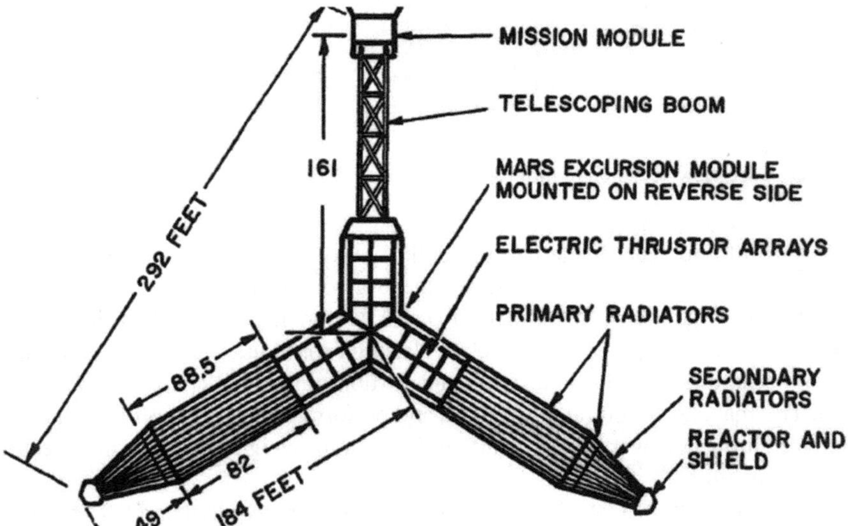

Fig. 7.11 Eight MW vehicle concept interplanetary configuration

chamber fluids are evaluated and selected for thermal performance and containment within the radiator. Vapor chamber compatibility and performance tests are made for a number of candidate fluids. Preliminary designs are developed for both conduction fin and vapor chamber radiator concepts. A comparison shows no significant advantages attributable to the Brayton cycle vapor chamber radiator where reliability and meteoroid criteria specify 0.99–0.999 probability of survival over a 5-year lifetime.

Table 7.1 Basic requirements summary

REFERENCE CONDITIONS AND SPECIFICATIONS FOR THE BRAYTON CYCLE VAPOR CHAMBER RADIATOR	
Thermal Heat Rejection	12.39 kWt Primary, 2.19 kWt Secondary
Radiator Fluid Inlet Temperature	288 °F Primary, 118 °F Secondary
Radiator Fluid Outlet Temperature	64 °F Primary and Secondary
Effective Radiator Sink Temperature	−10 °F
Radiator Surface Thermal Emissivity	0.85
Radiator Fluid	Dow Corning 200, 2 Centistokes at 77 °F
Primary Fluid Pressure Drop	25 PSI maximum
Reliability	0.99 or 0.999 for 5 years on Vapor Chambers and on Primary Fluid Loops
Supported Load	6000 Pounds including Heal Rejection System

At the end of the study, there was a report provided under NASA CR-1677 containing a description and results of the work performed toward the identification of a promising Brayton cycle vapor chamber fin (heat pipe) radiator. The specifications and requirements on which the radiator design is based were provided by NASA. A summary of reference specifications is contained in Table 7.1.

The reference Brayton cycle space power plant uses a separate radiator loop with a heat rejection about 15 kWt. A compact heat exchanger transfers waste heat from the power conversion loop to a liquid coolant. This coolant is then circulated through a radiator where the waste heat is rejected to space. This radiator is called the primary radiator. In addition, an auxiliary circuit and radiator rejected heat lost by cooling the power plant electrical and other components. This radiator is called the secondary radiator.

The general configuration visualized for these radiators is an array of tubes through which coolant flows and to which are attached solid, conducting fins. It was anticipated that a significant reduction in radiator weight and area might be achieved by using "vapor chamber" fins.

Is summary, heat pipe is a simple device of very high thermal conductivity with no moving parts that can transport large quantities of heat efficiently over large distances fundamentally at an invariable heat pipe is essentially a conserved slender tube containing a wick structure lined on the inner surface and a small amount of fluid such as water at the saturated state. It is composed of three sections: the evaporator section at one end, where heat is absorbed and the fluid is vaporized; a condenser section at the other end, where the vapor is condensed and heat is rejected; and the adiabatic section in between, where the vapor and the liquid phases of the fluid flow in opposite directions through the core and the wick, respectively, to complete the cycle with no significant heat transfer between the fluid and the surrounding medium. The operating pressure and the type of fluid inside the heat pipe depend largely on the operating temperature of the heat pipe.

References

1. Zohuri, B. (2015). *Dimensional analysis and self-similarity methods for engineers and scientists* (1st ed.). New York: Springer.
2. Zohuri, B. (2016). *Heat pipe design and technology: Modern applications for practical thermal management* (2nd ed.). New York: Springer.
3. Zohuri, B. (2017). *Thermal-hydraulic analysis of nuclear reactors* (2nd ed.). New York: Springer.
4. El-Genk, M. (2008). Space nuclear reactor power system concepts with static and dynamic energy conversion. *Energy Conversion and Management, 49*(3), 402–411.

Chapter 8
Application of Heat Pipes to Fissionable Nuclear Reactor

Heat pipes are often proposed as cooling system components for small fission reactors. Heat transport in heat pipe reactors is complex and highly system dependent. Nevertheless, in general terms it relies on heat flowing from the fuel pins through the heat pipe, to the heat exchanger, and then ultimately into the power conversion system and heat sink. Heat pipes have been used in reactors to cool components within radiation tests; however, no reactor has been built or tested that uses heat pipes solely as the primary cooling system. Heat pipe cooled reactors will likely require the development of a test reactor to determine the main differences in operational behavior from forced cooled reactors.

8.1 Introduction

As it was described in Chap. 6 of this book, the heat pipe is a vapor-liquid phase-change device that transfers heat from a hot reservoir to a cold reservoir using capillary forces generated by a wick or porous material and a working fluid.

The heat pipe is a device normally used to obtain a comparatively high transfer of thermal energy. It consists of a containment vessel (the pipe), a wicking material, and a fluid in two phases. The most common geometry is that shown schematically in Fig. 8.1.

It consists of a long narrow tube which forms a closed outer shell. The wick is a porous material which is in firm contact with the inside wall of the tube. The liquid phase of the fluid is in the wick and vapor phase is in the core or center of the tube.

The heat pipe works in the following manner. Heat is added to the evaporator end of the heat pipe. This causes liquid contained in the wick of the evaporator end to vaporize and results In a slight pressure drop in the vapor phase from the evaporator to condenser. Vapor then flows to the condenser end where it condenses on the wick. The evaporation of liquid in the evaporator section causes tiny menisci to form on the wick surface. These supports a pressure drop between the liquid and vapor

B. Zohuri, *Heat Pipe Applications in Fission Driven Nuclear Power Plants*,
https://doi.org/10.1007/978-3-030-05882-1_8

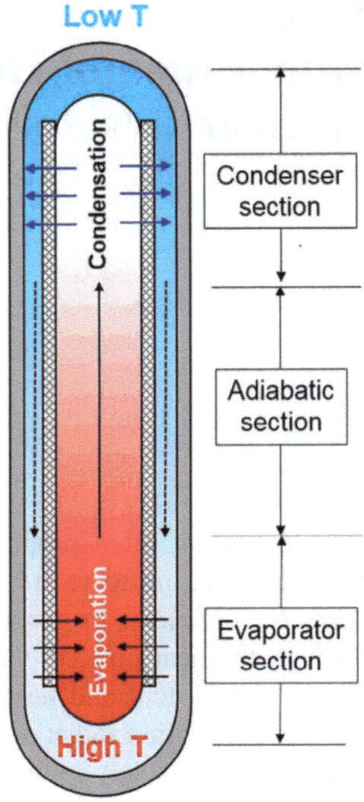

Fig. 8.1 Schema of the vertical constant heat pipe with adiabatic section

phases of the evaporator section; liquid flows through the wick from the condenser to the evaporator and completes the cycle [1].

The heat pipe is composed of a container lined with a wick that is filled with liquid near its saturation temperature. The vapor-liquid interface, usually found near the inner edge of the wick, separates the liquid in the wick from an open vapor core. Heat flowing into the evaporator is transferred through the container to the liquid-filled wicking material, causing the liquid to evaporate and vapor to flow into the open core portion of the evaporator. The capillary forces generated by the evaporating interface increase the pressure difference between the vapor and liquid. The vapor in the open core flows out of the evaporator through the adiabatic region and into the condenser. The vapor then condenses, generating capillary forces similar, although much less in magnitude, to those in the evaporator. The heat released in the condenser passes through the wet wicking material and container out into the cold reservoir. The condensed liquid is then pumped, by the liquid pressure difference due to the net capillary force between the evaporator and condenser, out of the condenser back into the evaporator. Proper selection and design of the pipe container, working fluid, and wick structure are essential to the successful operation of a heat pipe. The heat transfer limitations, effective thermal conductivity, and axial temperature difference define the operational characteristics of the heat pipe.

In Fig. 8.2 a schematic of a heat pipe aligned at angle ψ relative to the vertical axis (gravity vector) is shown.

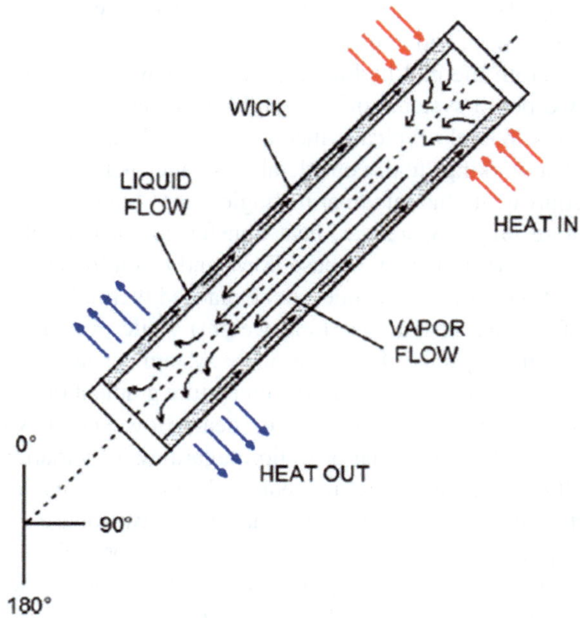

Fig. 8.2 Schema of the wick heat pipe

Due to the two-phase characteristics, the heat pipe is ideal for transferring heat over long distances with a very small temperature drop and for creating a nearly isothermal surface for temperature stabilization. As the working fluid operates in a thermodynamic saturated state, heat is transported using the latent heat of vaporization instead of sensible heat or conduction where the heat pipe then operates in a nearly isothermal condition. This nearly isothermal condition offers benefits of transporting large amounts of heat efficiently, decreasing the overall heat transfer area and saving system weight [2].

The amount of heat that can be transported through the use of latent heat is typically several orders of magnitude greater than transported by sensible heat for a geometrically equivalent system. Additionally, no mechanical pumping systems are required due to the capillary-driven working fluid. Given the wide range of operating temperatures for working fluids, the high efficiencies, the low relative weights, and the absence of external pumps in heat pipes, these systems are seen as attractive options in a wide range of heat transfer applications [3].

The wick provides a means for the flow of liquid from the condenser to the evaporator section of the heat pipe. It also provides surface pores that are required at the liquid-vapor interface for development of the required capillary pressure. The wick structure also has an impact on the radial temperature drop at the evaporator end between the inner heat pipe surface and the liquid-vapor surface. Thus, an effective wick requires large internal pores in a direction normal to the heat flow path. This will minimize liquid flow resistance. In addition, small surface pores are required for the development of high capillary pressure and a highly conductive heat flow path for minimization of the radial surface to liquid-vapor surface temperature drop. To satisfy these requirements, two types of wick structure have been developed. See Chap. 6 of this book.

From thermal hydraulic point of view, we can easily see that capillary-driven two-phase systems offer significant advantages over traditional single-phase systems. With the typically increased thermal capacity associated with the phase change of a working fluid, considerably smaller mass flow rates are required to transport equivalent amounts than in single-phase liquid or gas systems for a given temperature range. Moreover, heat transfer coefficients of two-phase systems are much greater than in single-phase flows and result in enhanced heat transfer.

Lower mass flow rates and enhanced thermal characteristics provide the benefits of smaller system size (and weight) while providing increased performance. The thermal capacity of a single-phase system depends on the temperature change of the working fluid; thus, a large temperature gradient or a high mass flow rate is required to transfer a large amount of heat. However, a two-phase system can provide essentially isothermal operation regardless of variations in the heat load. Additionally, single-phase systems require the use of mechanical pumps and fans to circulate the working fluid, while capillary-driven two-phase systems have no external power requirements, which make such systems more reliable and freer of vibration [3].

The best known capillary-driven two-phase system is the heat pipe, where a schematic of a conventional heat pipe is shown in Fig. 8.3. The concept of the heat pipe was first presented by Gaugler [4] and Trefethen [5] but was not widely publicized until an independent development by Grover et al. [6] at the Los Alamos Scientific Laboratories. Heat pipes are passive devices that transport heat from a heat source (evaporator) to a heat sink (condenser) over relatively long distances via the latent heat of vaporization of a working fluid. As shown, a heat pipe generally has three sections: an evaporator section, an adiabatic (or transport) section, and a condenser section. The major components of a heat pipe are a sealed container, a wick structure, and a working fluid. The wick structure is placed on the inner surface of the heat pipe wall and is saturated with the liquid working fluid and provides the structure to develop the capillary action for liquid returning from the condenser to the evaporator section.

With evaporator heat addition, the working fluid is evaporated as it absorbs an amount of heat equivalent to the latent heat of vaporization, while in the condenser section, the working fluid vapor is condensed. The mass addition in the vapor core of the evaporator section and mass rejection in the condenser end results in a pressure gradient along the vapor channel which drives the corresponding vapor flow. Return of the liquid to the evaporator from the condenser is provided by the wick structure.

As vaporization occurs in the evaporator, the liquid meniscus recedes correspondingly into the wick structure, as shown in Fig. 8.3. Similarly, as vapor condenses in the condenser region, the mass addition results in an advanced meniscus. The difference between the capillary radii in the evaporator and condenser ends of the wick structure results in a net pressure difference in the liquid-saturated wick. This pressure difference drives the liquid from the condenser through the wick structure to the evaporator region, thus allowing the overall process to be continuous [3].

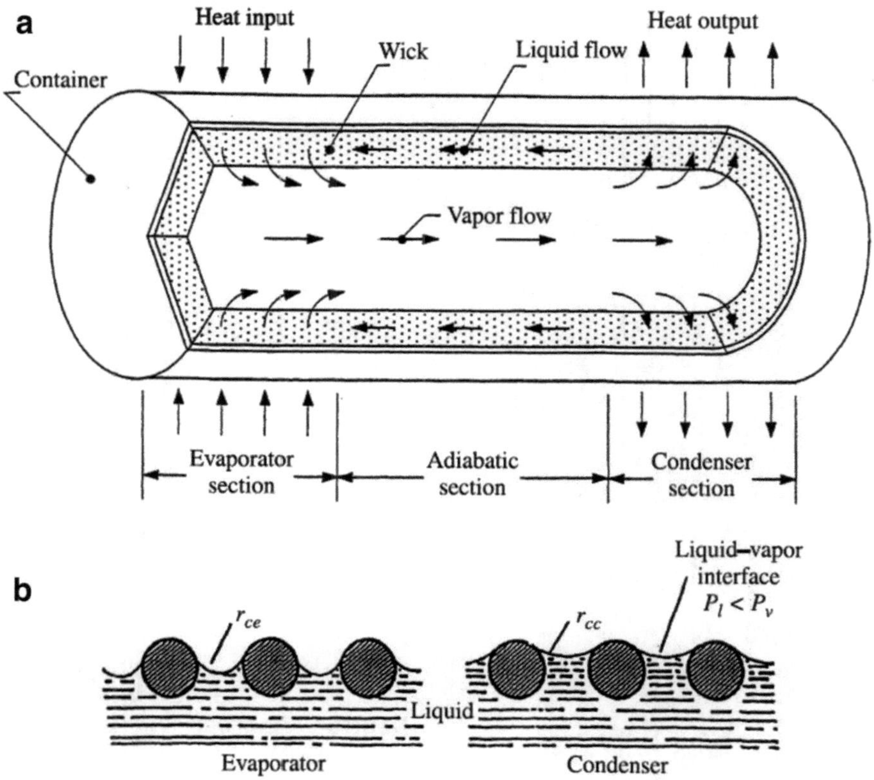

Fig. 8.3 (**a**) Typical heat pipe construction and operation; (**b**) radii of curvature of the liquid-vapor interface in the evaporator and condenser

Due to the two-phase characteristics, the heat pipe is ideal for transferring heat over long distances with a very small temperature drop and for creating a nearly isothermal surface for temperature stabilization. As the working fluid operates in a thermodynamic saturated state, heat is transported using the latent heat of vaporization instead of sensible heat or conduction where the heat pipe then operates in a nearly isothermal condition. This nearly isothermal condition offers benefits of transporting large amounts of heat efficiently, decreasing the overall heat transfer area and saving system weight. The amount of heat that can be transported through the use of latent heat is typically several orders of magnitude greater than transported by sensible heat for a geometrically equivalent system. Additionally, no mechanical pumping systems are required due to the capillary-driven working fluid. Given the wide range of operating temperatures for working fluids, the high efficiencies, the low relative weights, and the absence of external pumps in heat pipes, these systems are seen as attractive options in a wide range of heat transfer applications [3].

As part of heat pipe design and choice of working fluid as heat transfer media within the pipe is concerned, the geometry of working fluid selection and wick

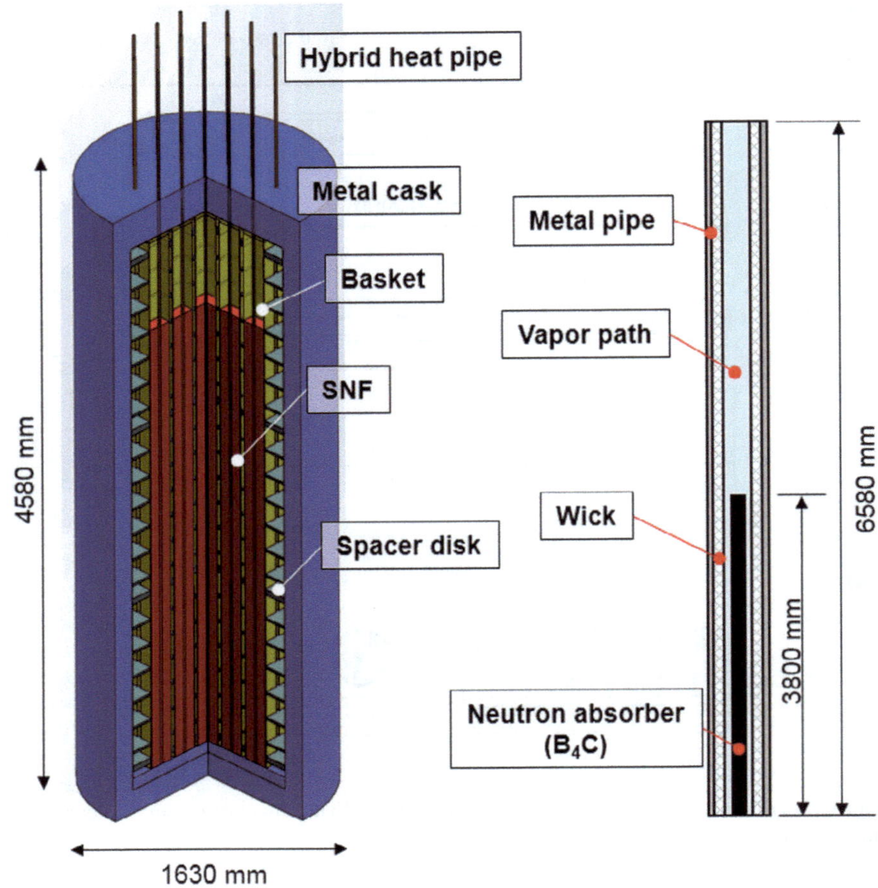

Fig. 8.4 Hybrid heat pipe equipped-metal cask for dry storage and hybrid heat pipe

structure should be properly selected in the heat pipe design for each specific application area such as Spent Nuclear Fuel (SNF) [7].

In case of SFN, there are recommendations that for heat pipe of hybrid type. Hybrid heat pipe, which is a heat pipe containing neutron absorber materials, was suggested for passive cooling in nuclear applications by UNIST thermal-hydraulics and reactor safety laboratory in Korea [8].With the heat transfer characteristics of a heat pipe, a hybrid heat pipe has the capability of both removing the decay heat and controlling reactivity with neutron absorber materials.

Hybrid heat pipe can be used in nuclear applications such as a passive in-core cooling system substituting control rod for an advanced nuclear power plant, wet storage pool, and dry storage cask for SNF. In this study, hybrid heat pipe was applied to a metal cask-type dry storage facility. The illustration in Fig. 8.4 shows the hybrid heat pipe-equipped Dual-Purpose metal Cask (DPC) for dry storage of spent fuel.

For a dry cask, hybrid heat pipes can be installed vertically at the guide tube or at the center instrumentation tube with only a simple change in the upper structure to assure the metal cask seal.

The lengths of the evaporator section and the condenser section were determined to be 3.8 m and 2 m, respectively, to reflect the active heated length of the SNF assemblies and the area exposed to the outside for sufficient cooling.

To evaluate the thermal performance and heat removal capacity of the heat pipe-based cooling device for a dry cask, a two-step Computational Fluid Dynamics (CFD) analysis is needed to be performed, which consisted of a single fuel assembly model and a full-scope dry cask model. Both steps possible should be calculated in steady state to start with. In the single fuel assembly simulation, heat transfer capacity of hybrid heat pipe was studied with obtaining detailed temperature and velocity field in single fuel assembly.

Based on the results of the single fuel assembly analysis, a full-scope dry cask simulation then can be conducted to examine the effects of the overall cooling performance of the hybrid heat pipe on the dry storage cask. DPC developed by Korea, which has 21 pressurized water reactor (PWR) Spent Nuclear Fuel (SNF) [7] assemblies that are 16 × 16 CE-type assemblies cooled in a pool for 10 years, was considered. Hybrid heat pipe can be installed at the guide tube and instrumental tube space of the spent fuel assembly. In [7] provided at the end of this chapter in the Reference section, two cases of hybrid heat pipe applications were analyzed: one is a case of single hybrid heat pipe per SNF assembly at the center instrumental guide tube and the other is a case of five hybrid heat pipes per SNF assembly at four guide tubes and one instrumentation tube. In this case, heat transferred from a dry cask embedded hybrid heat pipe is accumulated in a storage tank and can be used in another area as necessary. Figure 8.5 shows schematics of the thermoelectric module, the Stirling engine, and the PCM tank applied hybrid heat pipe cooling device.

In summary, as stated in Chap. 6 and at the beginning of the present chapter, there is a wide variety of working fluids depending on the function and temperature of operation of the heat pipe.

The working fluid is chosen so that its normal boiling temperature is somewhat below the temperature of operation. This temperature can vary from the cryogenic range to any upper range that can be tolerated by the containment.

Wick structures as illustrated in Fig. 8.2, also vary widely in design depending on the use. They can be made from any porous material or from channels machined in the wall. These channels can be uncovered or covered with screen. The wick structure can also be merely an annular space with a fine mesh screen separating the liquid and vapor phases with the liquid flowing down the annular space.

Heat pipes have been used in many applications where high thermal energy transfer is needed or constant temperature or heat flux is desired. This includes such things as cooling devices for electronic components, heat exchangers, and temperature control devices. Although they have been used to some extent in radioisotope [9] power sources, their application for cooling reactors is only in the preliminary design stages [10–12].

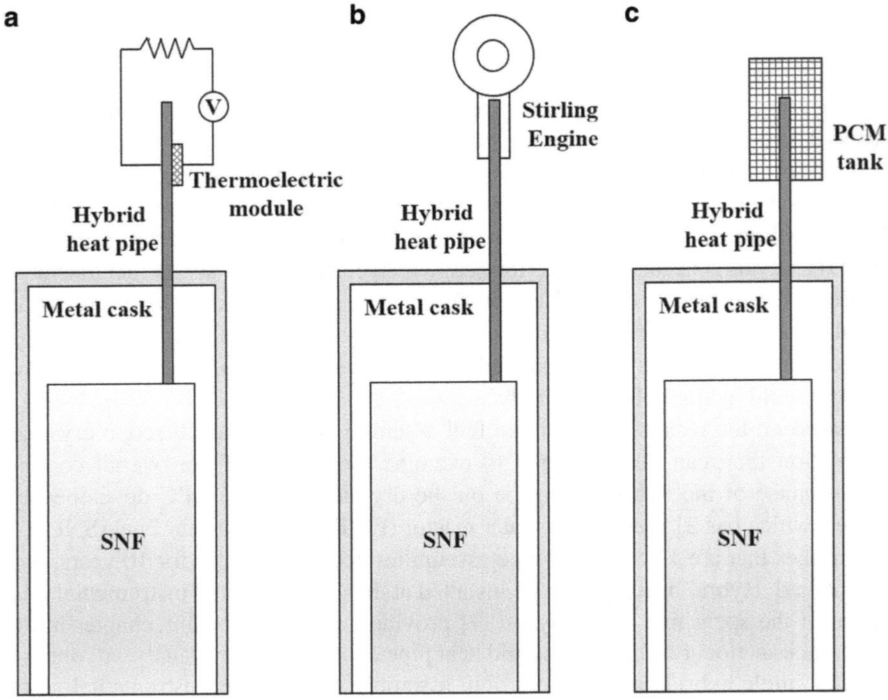

Fig. 8.5 Schematics of the (**a**) thermoelectric module, (**b**) Stirling engine, and (**c**) PCM tank applied with hybrid heat pipe

8.2 Operational Limits of Heat Pipes

As with any other system, the performance and operation of a heat pipe is limited by various parameters. Physical phenomena that might limit heat transport in heat pipes include capillary forces, choked flow, interfacial shear, and incipient boiling. The heat transfer limitations depend on the size and shape of the pipe, working fluid, wick parameters, and operating temperature. The lowest limit among these constraints defines the maximum heat transport limitation of a heat pipe at a given temperature [1].

The operation limits of a heat pipe as illustrated in Fig. 8.6, in the order of increasing power throughput and temperature, are the viscous, sonic, wicking or capillary, entrainment, boiling, and heat rejection. The latter is dictated by the length of the condenser section, surface emissivity, and available area for heat rejection in the radiators for space nuclear reactor power systems.

In the design of heat pipes, consideration must be given not only to the internal structure and fluid dynamics of the pipe but also to the external conditions imposed upon it. By *fully operational steady characteristics* of heat pipe up to now, we have assumed a steady-state heat pipe, with heat being added to and removed from the heat pipe at a constant rate. Under this condition we mean that the heat pipe is

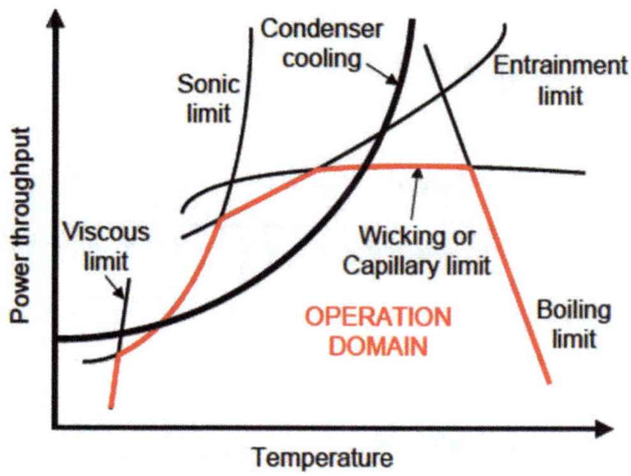

Fig. 8.6 Operation limits of a heat pipe

relatively *isothermal,* and heat is being dissipated over the entire length of the condenser. If the heat pipe input and output rates are then equal, the heat pipe will be functioning in the steady-state condition. If an imbalance between the heat input and output rates takes place, then the temperature of the fully operational heat pipe will continue to change with time to a level at which the balance between heat input and output rates is restored [1].

The viscous limit dominates at low temperature, near the melting point of the working fluid. The high liquid pressure losses in the wick, due to the high viscosity and low permeability, limit liquid flow from the condenser to the evaporator section. Avoiding this limit requires operating at a relatively low input power until the heat pipe temperature is high enough to decrease the liquid viscosity and, hence, the pressure losses in the wick.

The sonic limit, also dominant at low temperatures, should be avoided. The vapor pressure of the working fluid is a good indicator of reaching this limit [1]. The vapor pressure and physical state of the heat pipe liquid at ambient temperature, as well as the thermal resistance between the condenser and the adjacent heat sink, have significant influence on the start-up behavior of a heat pipe.

Prior to start up, the temperature of a heat pipe is equal to the ambient temperature, and its internal pressure is equal to the vapor pressure of the heat pipe liquid at ambient temperature. Also, depending on its freezing point, the heat pipe liquid may be in the liquid or the solid state. The transient behavior and problems of heat pipe start-up have been studied by Cotter [13] and Deverall et al. [14]. Tests results reported by the latter indicate that the transient behavior of a heat pipe depends on the circumstances mentioned above.

As liquid and vapor move in opposite directions, the vapor exerts a shearing force on the liquid at the liquid-vapor interface. If this shear force exceeds the surface tension of the liquid, liquid droplets are entrained into the vapor flow and are carried toward the condenser section as it is illustrated in Fig. 8.7. The magnitude of this shear force depends on the thermo-physical properties of the vapor and its velocity, and if it becomes large enough, it causes dry out of the evaporator [1]. Entrainment Limit at high vapor velocities, droplets of liquid in the wick are torn from the wick and sent into the vapor, which results in dry out. An abrupt wick dry out will take

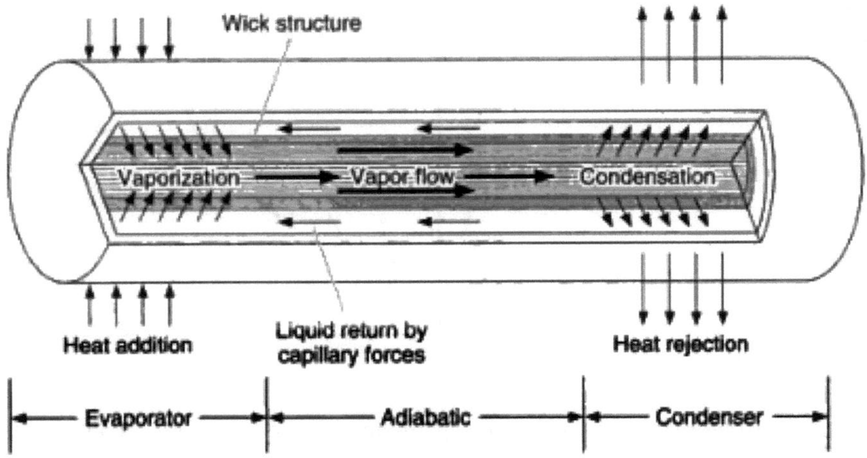

Fig. 8.7 Schematic of heat pipe

place when entrainment begins, and there is a sudden substantial increase in fluid circulation to the point that return liquid system cannot accommodate this flow increase [14]. This limit was identified by Kemme [14] when he discovered the sound sounds that were made by droplets from media liquid within heat pipe striking the condenser end of heat pipe and through the abrupt overheating of the evaporator [1]. The entrainment limit also is known as an axial heat flux, the heat transport rate per unit of vapor space cross-sectional area. Under this condition, the fluid velocities increase so as drag force as the heat transport rate through the heat pipe increases. The drag force on the heat pipe liquid is proportional to the liquid surface area in the wick pores, whereas the resisting surface tension force is proportional to the pore width normal to the drag force. Consequently, the ratio of the drag force to the surface tension force is proportional to the pore size and decreases as the pore size diminishes [1].

The entrainment limit is typically encountered during a heat pipe start-up, when the vapor flow at the evaporator section exit is chocked (velocity is near sonic). The induced interfacial shear stress at the surface of the liquid saturated wick by the vapor countercurrent flow could not only slow down the liquid flow to the evaporator section but also break up and entrain tiny liquid droplets back to the condenser. The reduced replenishing of the wick in the evaporator section with liquid could result in a local dry out. The entrainment limit could be raised by employing a small pore-size wick and/or increasing the cross-sectional flow area for the vapor in the heat pipe to lower its velocity at the exit of the evaporator section.

The “wicking limit” or “capillary limit” is the best understood. This condition is occurring when an applied heat flux causes the liquid in the wick structure to evaporate faster than it can be supplied by capillary pumping power of the wick. Once this event takes place, the meniscus at the liquid-vapor interface continues to withdraw and move back into the wick until all of the liquid has been depleted. This action will cause the wick to become dry, and heat pipe container temperature may continue to rise at the evaporator until a "burnout" condition is reached [15]. The difference in the capillary pressure across the liquid-vapor interfaces governs the operation of the heat pipes. This is one of the most important parameters that affect the performance and operation of a heat pipe. It is usually a major limiting factor in the working of low-temperature or cryogenic heat pipes [1].

The capillary limit is encountered when the capillary pressure is not sufficient to pump the liquid back to evaporator causing the dry out of the wick of the evaporator end. The physical structure of the wick is one of the most important reasons for this limit, and the type of working fluid affects it. Once limit is encountered, any further increase in heat input may cause serious damage to the heat pipe [11].

The performance and operational characteristics for a given heat pipe and thermosyphons as a function of the mean adiabatic or operating temperature and envelop of these operating limits have been discussed in various sections of this book and is depicted here as well (see Fig. 2.18).

Any design of heat pipe that falls within operation envelop of its function (red color) essentially considered as a good design and will work within specific function of operating temperature that is defined for that design [1].

Thus, the capillary (or wicking) limit is encountered when the net capillary pressure head is less than the combined pressure losses of the liquid flow in the wick and of the countercurrent vapor flow in the heat pipe. The capillary pressure head for circulating the heat pipes working fluid increases with increasing the liquid surface tension and decreasing radius of curvature of the liquid-vapor meniscus in the surface pores in the wick, as illustrated in term of R_c in Fig. 8.8 [1].

A boiling at the inside surface of the heat pipe wall in the evaporator section is likely when the local liquid superheat exceeds that for incipient nucleate boiling. The ensuing nucleation and growth of vapor bubbles blocks the flow of returning liquid to the evaporator section. In alkali-metal heat pipes, the boiling limit is typically encountered at high wall temperatures, beyond those selected for nominal operation.

An additional operation limit, often overlooked, is that of heat rejection in the condenser section. In space reactor power systems, the condenser section of the radiator heat pipes is cooled by thermal.

radiation into outer space. Therefore, the heat removal rate is proportional to the effective surface emissivity and area in the condenser section and the average surface

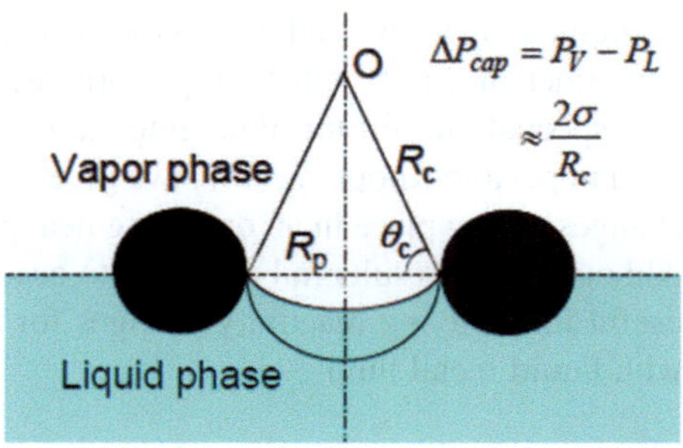

Fig. 8.8 Capillary pumping of the working fluid in heat pipe

temperature to the fourth power. Metallic surfaces typically have low emissivity and thus are treated with a black coating or paint to improve heat rejection. This operation limit might be encountered late in life, due to the accumulation of non-condensable gases, generated by neutron activation, at the end of the condenser, gradually decreasing the rate of heat rejection.

8.3 Heat Pipe as Control Cooling Device for Nuclear Reactor

A new concept involving the use of heat pipes as control devices for nuclear reactors will be investigated in this thesis. The feature of the concept is that the beat pipe will contain a fissionable material as the working fluid. The primary purpose of the heat pipe will be to change the amount of fuel within a reactor instead of the usual purpose of transferring heat.

This change in the amount of fuel results because of the relationship of the heat transferred in the heat pipe to the amount of liquid in the evaporator section. Section 8.3.1 gives a description of the system used for this study which is a model based on the nuclear rocket engines used in a Nerva rocket. The equations to determine this relationship are derived, and more details can be found in a book by Zohuri [1] as well as PhD thesis work by Monte Bryce Parker [16].

As it was stated previously, in various chapters of this book, historically, the use of heat pipes as passive heat transfer technique and mechanisms has been as recent as 1964, where Grove et al. [6] at Los Alamos National Laboratory (LANL) published their first results of such device experimental work using the heat pipe as a passive heat transfer device. Thus, since then, the interest in heat pipe as a means of passive heat transfer apparatus has grown greatly, and more and more experiment and theoretical work were established by others as well, yet the bulk of it was done by scientists at both Sandia National Laboratory (SNL) and LANL on the performance of heat pipe and its related technologies [17–19].

As result of such activities by these scientists, in 1965, Cotter published the comprehensive theoretical treatment of steady state heat transfer operational of heat along with dynamical startup of heat pipe following with theoretical treatment of it in 1967 [13, 20].

Kemme [21–23] and his associates at Los Alamos have done much of the experimental work with high-performance heat pipes. The results from his work are especially useful for high temperatures and liquid metals.

Hampel and Koopman (I6) have given a theoretical treatment of some of the mass changes taking place in an operating heat pipe for a number of liquid metal systems. Although their results fail to take into account some important parameters, they are useful for studying reactivity changes for cases of solid fuels cooled by heat pipes with liquid metal fluid.

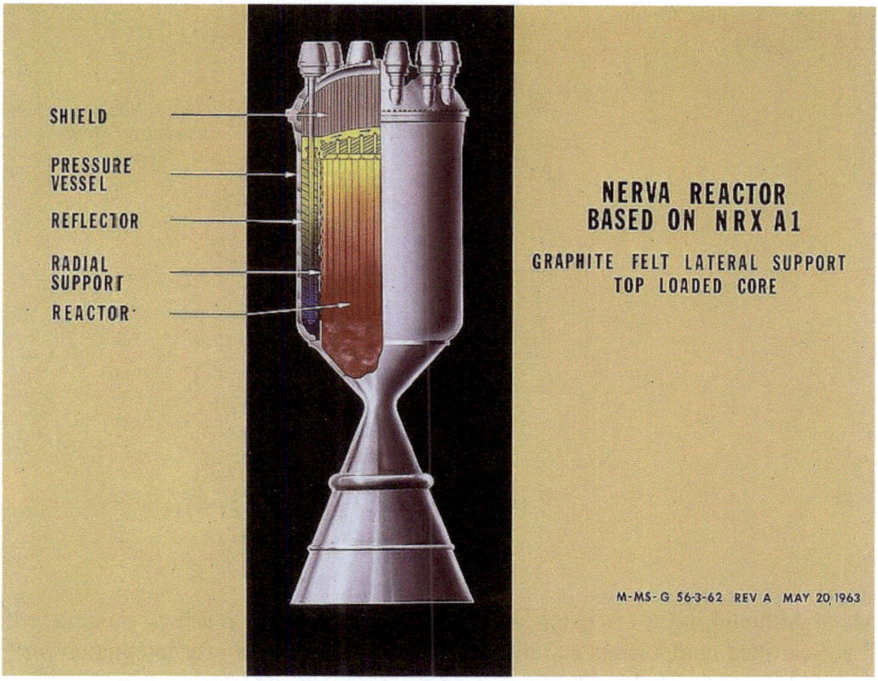

Fig. 8.9 Diagram of the NERVA nuclear rocket engine

Cheung [24] has published a comprehensive general review of heat pipe theory and applications up through the middle of 1968 and Peldman and Whiting [25, 26] have presented a review of many novel applications.

The theoretical study of rocket reactor control has been presented by Smith [27], Smith and Stenning [28], Jansen and Buckner [29], and Mohler and Perry [30].

8.3.1 Investigation of Heat Pipe in Nuclear Reactor

We can study the visibility of utilization of heat pipes containing liquid fissionable materials used in space-based nuclear rocket reactor such as Nuclear Engine for Rocket Vehicle Application (NERVA) as it was suggested by LANL and companies like Westinghouse or General Electric as it is shown in Figs. 8.9 and 8.10, under Atomic Energy Commission (AEC) and National Aeronautics and Space Administration (NASA) program managed by the Space Nuclear Propulsion Office (SNPO) until both the program and the office ended at the end of 1972.

NERVA demonstrated that nuclear thermal rocket engines were a feasible and reliable tool for space exploration, and at the end of 1968 SNPO certified that the latest NERVA engine, the NRX/XE, met the requirements for a human mission to

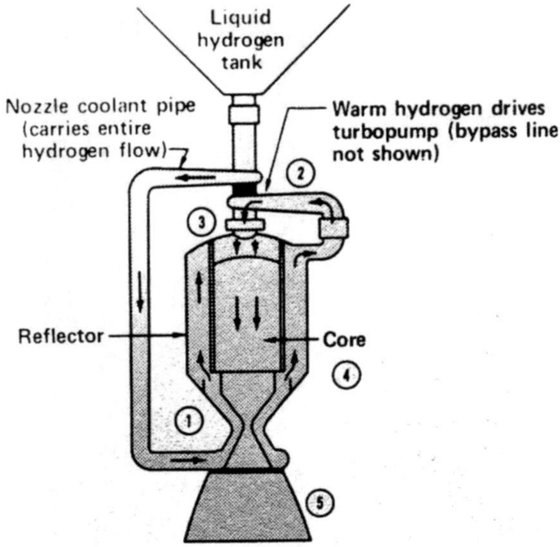

Fig. 8.10 Flow diagram for NERVA engine

Mars. Although NERVA engines were built and tested as much as possible with flight-certified components and the engine was deemed ready for integration into a spacecraft, much of the US space program was canceled by Congress before a manned mission to Mars could take place.

NERVA was considered by the AEC, SNPO, and NASA to be a highly successful program; it met or exceeded its program goals. Its principal objective was to "establish a technology base for nuclear rocket engine systems to be utilized in the design and development of propulsion systems for space mission application" [31]. Virtually all space mission plans that use nuclear thermal rockets use derivative designs from the NERVA NRX or Pewee.

Los Alamos Scientific Laboratory began researching nuclear rockets in 1952, accelerating into Project Rover in 1955 when the deputy director of Lawrence Livermore National Laboratory, Herbert York, postulated a way to shrink reactor weights considerably. By 1961, after unexpectedly fast-paced progress on the part of Project Rover, NASA's Marshall Space Flight Center began to use nuclear thermal rockets in their mission plans. Marshall planned to use a nuclear-powered rocket from Los Alamos to power a Reactor-In-Flight-Test (RIFT) nuclear stage to be launched as early as 1964, and the need for planning and oversight led to the formation of the Space Nuclear Propulsion Office. SNPO was formed so that the AEC and NASA could work together, and H. B. "Harry" Finger was selected as its first director. Finger made a decision to delay Reactor In-Flight-Test (RIFT) system, and he defined strict objectives for nuclear rocket engines to achieve before RIFT would be allowed.

Almost all of the NERVA research, design, and fabrication were done at Los Alamos Scientific Laboratory. Testing was done at a large installation specially built by SNPO on the Nevada Test Site. Although Los Alamos tested several KIWI and

Phoebus engines during the 1960s, testing of NASA's NERVA NRX/EST (Engine System Test) contractor engine didn't begin until February 1966.

The objectives were:

1. Demonstrate the feasibility of starting and restarting the engine without an external power source.
2. Evaluate the control system characteristics (stability and control mode) during start-up, shutdown, cooldown, and restart for a variety of initial conditions.
3. Investigate the system stability over a broad operating range.
4. Investigate the endurance capability of the engine components, especially the reactor, during transient and steady-state operation with multiple restarts.

All test objectives were successfully accomplished, and the first NERVA NRX operated for nearly 2 h, including 28 min at full power. It exceeded the operating time of previous KIWI reactors by nearly a factor of two.

Although the harsh space environment imposes many severe challenges to space pioneers, space exploration is a realistic and profitable goal for long-term humanity survival. One of the viable and promising options to overcome the harsh environment of space is nuclear propulsion. Particularly, the nuclear thermal rocket (NTR) is a leading candidate for near-term human missions to Mars and beyond due to its relatively high thrust and efficiency. Traditional NTR designs use typically high power reactors with fast or epithermal neutron spectrums to simplify core design and to maximize thrust. In parallel there are a series of new NTR designs with lower thrust and higher efficiency, designed to enhance mission versatility and safety through the use of redundant engines (when used in a clustered engine arrangement) for future commercialization. This paper proposes a new NTR design of the second design philosophy, for future space application as illustrated in Fig. 8.11.

The reactor of a nuclear thermal reactor (NTR), which could be named as an Extremely High-Temperature Gas-Cooled Reactor (EHTGR), is heated to the maximum temperature without fuel melting by fission energy and is cooled by the H^2 propellant. The propellant absorbs enthalpy from the EHTGR and is expanded out through a converging-diverging nozzle at very high temperature and exhaust velocity V_e as depicted in Fig. 8.11. The main components of an NTR engine are a propellant feeding system (PFS) housing a turbopump assembly (TPA), an EHTGR, and a thermodynamic nozzle.

The liquid H^2 is stored in insulated propellant tanks and is drawn through a pump of the TPA. The pump then sends the pressurized H^2 flow to coolant channels in the secondary reactor components (the nozzle, moderator, reflector, etc.) to extract heat for the propellant feeding power. After cooling the secondary components, the accessorily heated H^2 flows up to the TPA to drive its turbine. The H^2 then flows into the primary fuel zone in the reactor core and is exhausted through the thermodynamic nozzle at 2500 K to over 3000 K to generate rocket thrust.

Choosing this rocket reactor as the reactor model which will be used as we indicated is based on NERVA system. The heat pipes are inserted into the reactor as shown in Fig. 8.12. The evaporator end of the heat pipe projects into the end of the reactor where the hydrogen gas leaves the reactor and enters the rocket chamber. The

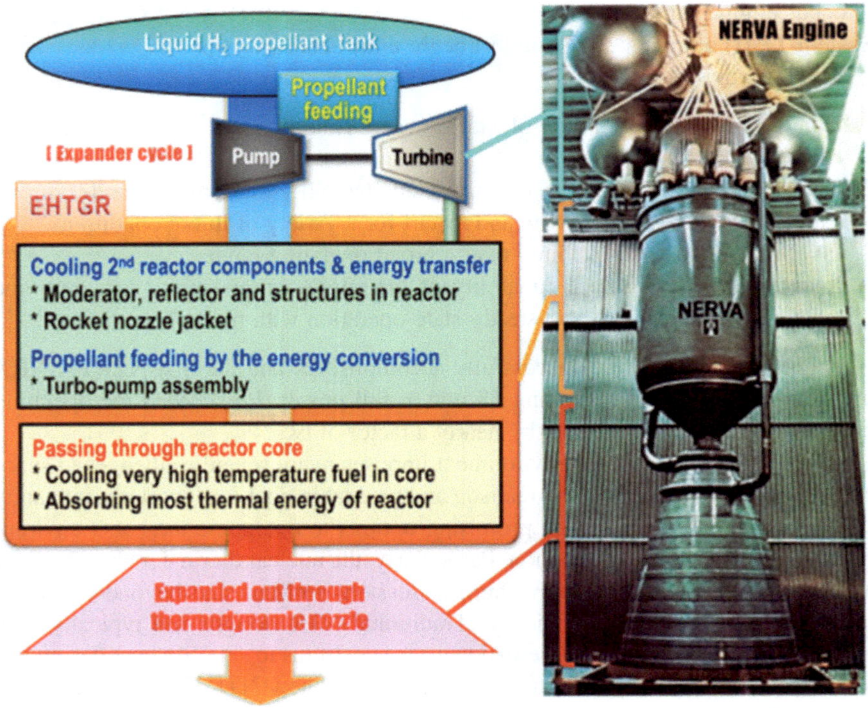

Fig. 8.11 Typical operating principle of a nuclear thermal rocket

Hg is used to cool the reactor and propel the rocket. The condenser end of the heat pipe will be in the chamber where it will be cooled by the exiting H^2 gas.

The working liquid of the heat pipe is UF^4, and this fluid will fission in the evaporator end causing the heat pipe to circulate UF^ vapor from evaporator to condenser and liquid UF^4 from the condenser to evaporator in the normal fashion of a heat pipe.

The heat pipe will affect the reactivity because the amount of UP^4 in the evaporator will vary with the heat flux and temperature of operation. Instead of a wick of porous material, an annular channel for liquid flow with a fine mesh screen separating the two phases as shown in Fig. 8.13 will be used. This type of liquid return channel was chosen because of the function of this heat pipe. The heat will be generated within the liquid channel, and it is therefore not necessary to have as good thermal conduction between the fluid and containment wall as channels would provide. It is also necessary to have a screen because a variation in the amount of fluid in the evaporator section of the heat pipe will be due to the change in the depth of the meniscus within each screen opening.

This meniscus is illustrated in Fig. 8.14 where r_s is the radius of the screen opening which has been approximated with a circle. This type of wick has been tested experimentally by Kemme [21] and has high-performance characteristics. The

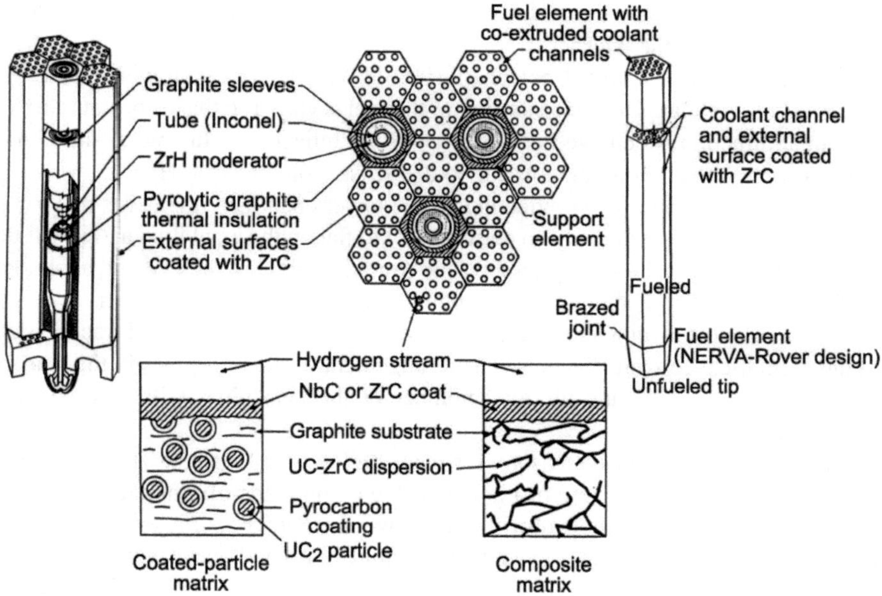

Fig. 8.12 NERVA reactor core with heat pipe

Fig. 8.13 Heat pipe cross section

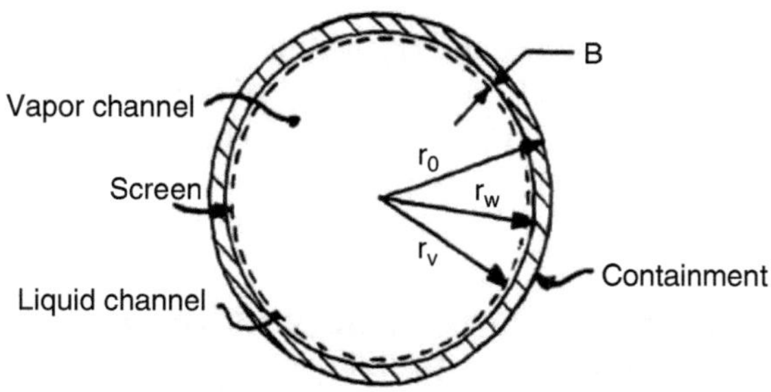

Fig. 8.14 Meniscus in screen opening

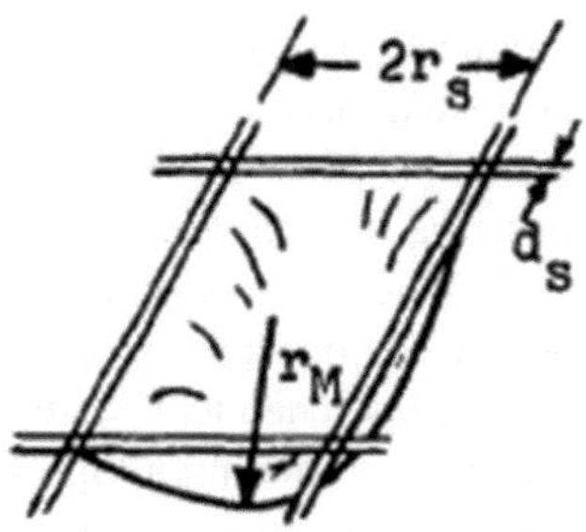

reason for this is it allows easier flow of liquid than would be allowed by using a porous material for the wick.

The reactor has a nominal power of 1000 MW. The core is graphite with UC dispersed throughout. The concentration of the fuel is varied so that a constant flux in the radial direction is achieved. It has a beryllium reflector on the radial sides, and there is some reflection on the top due to the dense, cold entering H^2.

There is assumed to be no reflection on the lower end where the exists due to its low density of the gas. There is no reason to try to flatten the axial flux, and it will be assumed to be one-half of a chopped cosine from the center to the exit end as shown by Cooper [32]. The maximum flux at nominal power is 2×10^5 n/cm^2 s.

In this case, the reactor size is approximately 3.0 m^3 with a *33%* void fraction due to the cylindrical coolant channels. The arrangement of the heat pipes in the reactor is shown in Fig. 8.11, and the total number of the heat pipes will be used as a parameter.

There were a number of reasons for using this type of system. The reason for choosing UF^ is that it is one of the few stable compounds of uranium which is a liquid at low enough temperatures and on which there is sufficient data on its properties. UF^4 is stable at elevated temperatures and will react very little with graphite [33]. Another compound which, would possibly work be UCl^4 with a normal boiling point of 792 °C. However, very limited data are available on the properties of this compound.

The reason for using a reactor model design based on a rocket reactor is that this heat pipe must be at very high temperatures. The normal boiling point of UF^4 is 1723 °K, and efficient operation of the heat pipe dictates that a high-temperature reactor be used. The nominal exit temperature of the H^2 is chosen to be 2200 °K.

8.4 The Role of Nuclear Power and Nuclear Rocket Propulsion in Space Exploration

Application of nuclear energy in rocket propulsion appears essential to extensive space exploration, and the solid-core reactor has reached a prototype engine demonstration after a decade of development. The purpose of this section review is to describe the major characteristics of solid-core nuclear rocket engines, to indicate their state of development and probable application, and to discuss a variety of advanced concepts for high-thrust propulsion systems. See Fig. 8.15.

In most chemical systems, the propulsive energy is derived from the heat of reaction, producing a high-temperature gas that is expanded and ejected through a nozzle, converting the thermal energy to directed energy. An energy balance can be used to derive the gas exhaust velocity υ_s, which is an important measure of engine performance. Assuming all the thermal energy to be converted to kinetic energy gives the trust.

Fig. 8.15 Artistic image deep space exploration

A goal of planetary scientists is a systematic survey of the outer planets by a combination of flyby, orbiter and atmospheric entry probe missions. To accomplish such an ambitious scientific objective, probe components must maintain functional capabilities during prolonged exposure to deep space environment as part of mission for the role of nuclear power and nuclear propulsion in the peaceful exploration of space.

Nuclear power can be used for a rocket propulsion system. The reactor power is used to heat a propellant that is forced through the rocket nozzle to provide motion in the opposite direction. Figure 8.16 shows a typical nuclear rocket propulsion module.

Two parameters that provide a measure of the efficiency of a rocket propulsion energy source are the theoretical specific impulse and the ratio of the take-off mass to the final mass in orbit. As shown in Fig 8.17, chemical reaction using hydrogen, oxygen, or fluorine can achieve a specific impulse of 4300 s with a mass ratio for earth escape of 15. However, hydrogen heated by a fission reactor instead of a chemical reaction achieves twice the specific impulse with a solid core while at the same time having a mass ratio of 3:2. With different cores, the specific impulse can be as much as seven times greater again with a mass ratio of only 1:2.

The fundamental advantage of nuclear thermal propulsion (NTP) units over liquid propellant rocket engines lies in being able to use a single component working fluid with the minimum molecular weight, which, all other factors being equal, provides a maximum specific impulse. Thus, when using hydrogen, the specific impulse produced by NTP can be more than twice as high as that of chemical engines.

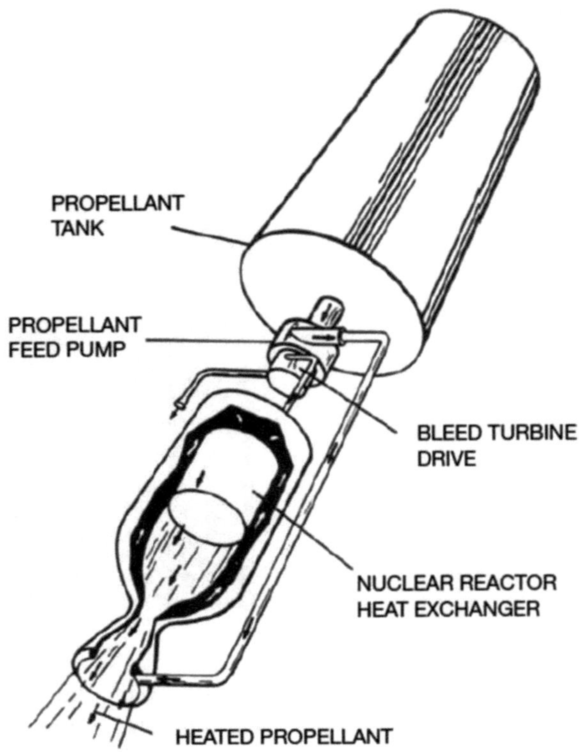

Fig. 8.16 Typical nuclear rocket propulsion module (Source: NASA/US Department of Energy)

Owing to its higher specific impulse, NTP can perform the same space mission with a smaller mass of propellant than a chemical engine. On a purely theoretical basis, it is possible to do even better with direct fission, or thermonuclear fusion, with theoretical impulses rising to 36 × 106 s and a mass ratio that is barely larger than unity.

8.4.1 US Directions in Deep Space Exploration Driven by Nuclear Program

The basic improvement in mass ratio to be obtained from a nuclear fission propulsion system was recognized very early on. In the United States, a test program named Rover was conducted from 1955 to 1973 using various designs. Table 8.1 lists the achievements that culminated in 1969 in the XE-prime engine (see Fig. 8.18), a first down firing prototype operating at 1100 MW. In 1972, a 44 MW nuclear furnace demonstrated peak fuel power densities of 4500 MW/m^3 with temperatures of up to 2500 K for 109 min. Figure 8.18 is illustration of the ground experimental engine

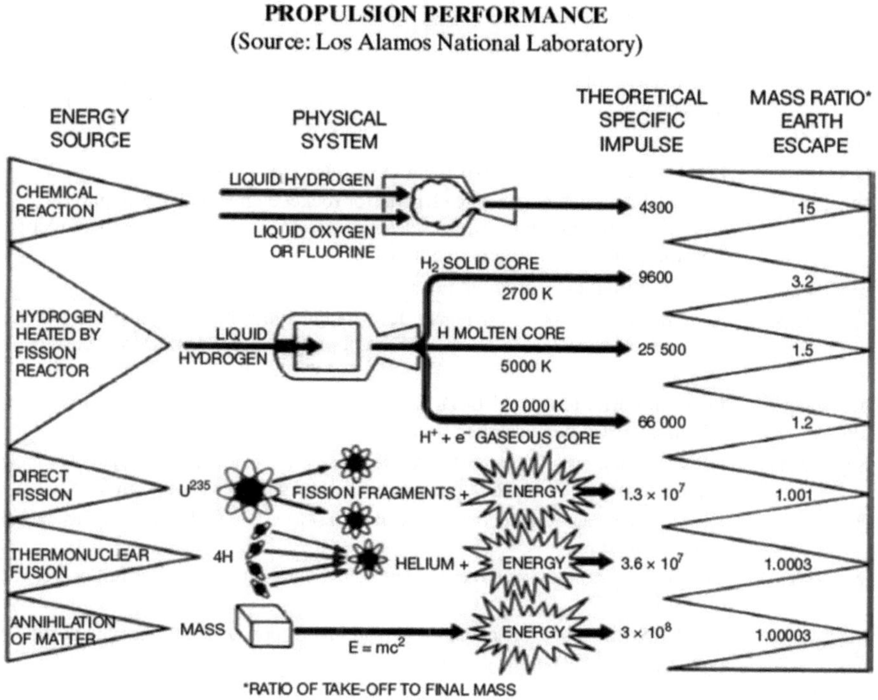

Fig. 8.17 Propulsion performance

(XE)-Prime) installed in Engine Test Stand No. 1 at the Nuclear Rocket Development Station in Nevada.

Figure 8.19 shows a comparison of the sizes of reactors tested in the Rover program.

The nuclear rocket program, which costs \$1.7 billion (\$7 billion in current dollars), was considered a technical success, but it was terminated, and the US program has since used chemical propellants.

However, nuclear propulsion may be the option of choice for a manned mission to Mars according to recent conference announcements. Indeed, a 1990 assessment of the program concluded that a Mars mission could be performed with Nuclear Engine for Rocket Vehicle Application (NERVA) technology [34].

8.4.2 Safe Affordable Fission Engine

Another aspect and concept of Safe Affordable Fission Engine (SAFE) is a propulsive heat pipe (HP) power system on which work is being done at the Los Alamos National Laboratory and the Marshall Space Flight Center. SAFE-400 (see Fig. 8.20) is designed to provide 400 kW of thermal power for more than 10 years

Table 8.1 The Rover program

1955	Following several years of nuclear rocket studies, nuclear rocket programme initiated as Project Rover at Los Alamos National Laboratory. Concept to be pursued was solid core. H_2 cooled, reactor expanding gas through a rocket nozzle
Jul. 1959	First reactor test, Kiwi-A, performed at 70 MW for 5 min
Oct. 1960	Proof-of-principle tests (Kiwi-A series of three reactors completed)
Jul. 1961	Industrial contractors (Aerojet General for rocket engine and Westinghouse Electric Corporation for reactor) selected to perform rocket development phase. Reactor in-flight test programme initiated
1963	Reactor in-flight test programme cancelled
1961–1964	Kiwi-B series of 1000 MW reactor tests included five reactors plus several cold flow unfueled reactors to resolve vibration problems and demonstrate design power
May–Sep. 1964	First full power test, Kiwi-B4D, at design power with no indications of core vibrations. Also demonstrated restart capability
Sep. 1964	NRX-A2, first tests of the NERVA (NERVA = nuclear engine for rocket vehicle application) reactor, reached full power of 1100 MW for about 5 min
Jan. 1965	Kiwi-B type reactor deliberately placed on fast transient to destroy itself as part of safety programme
Jun. 1965	The prototype of a new class of reactors. Phoebus-1A, was run at full power for 10.5 min
Mar. 1966	The NRX/EST, first rocket engine 'bread board' power plant, operated at full power (1100 MW) for 13.5 min
Dec. 1967	The fifth fuelled NRX reactor in the NERVA series exceeded the design goal of 60 min at 1100 MW
Jun. 1968	The Phoebus-2A—the most powerful nuclear rocket reactor ever built—ran for 12 min above 4000 MW
Dec. 1968	Set records in power density and temperature, operating at 503 MW for 40 min at 2550 K and core power density of 2340 MW/m^3
Mar. 1969	The first down firing prototype nuclear rocket engine, XE-prime, was successfully operated at 1100 MW.
1969	Saturn V production suspended (prime launch vehicle for NERVA)
Jun. 1972	In the 44 MW Nuclear Furance (NF-1), fuel was demonstrated at peak power densities of approximately 4500 MW/m^3 and temperatures up to 2500 K for 109 min.
Jan. 1973	Nuclear rocket programme terminated. Judged a technical success but changing national priorities resulted in cancellation decision

through two independent Brayton power systems, the reactor heat being deposited into the gas (He 72%, Xe 28%) flow via two independent heat pipes to gas heat exchangers. This provides 100 kW(e) for a 25% efficiency.

The SAFE-400 reactor contains 127 identical modules made of niobium-zirconium (1 wt%) alloy. Each module contains a Nb1Zr-Na heat pipe at its center surrounded by three niobium-zirconium tubes each of which contains a rhenium clad uranium nitride fuel sleeve. The wick of the heat pipe is fabricated from Nb1Zr mesh; the 60% void being filled with sodium during operation. The heat pipes extend

Fig. 8.18 The ground experimental engine (XE-Prime) (Source: NASA)

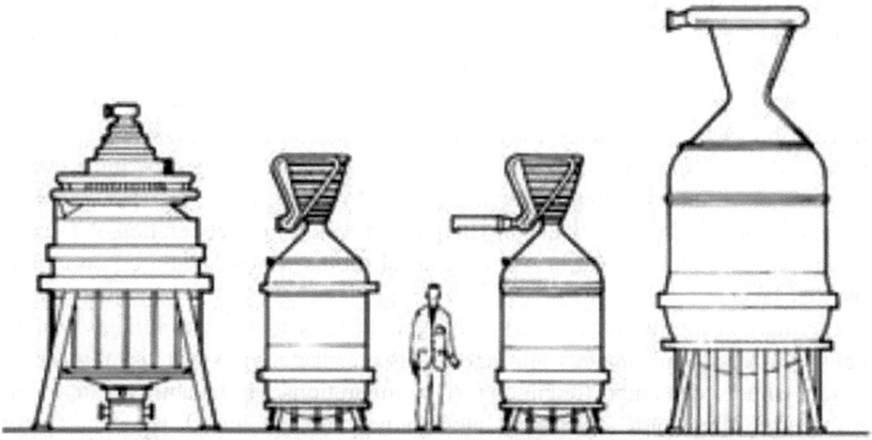

Fig. 8.19 Comparison of reactor sizes (Source: NASA)

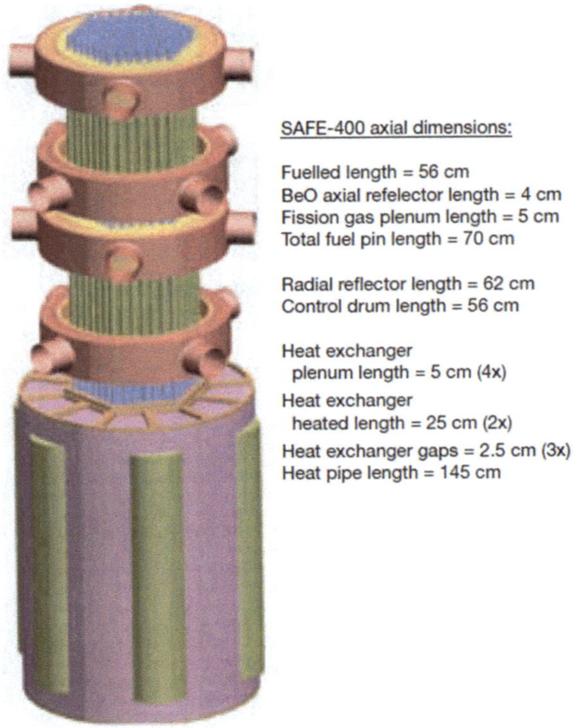

Fig. 8.20 The SAFE-400 reactor (Source: Nuclear News)

75 cm outside the core. The fission power is transferred to the heat pipes at a vapor temperature of 1200 K and, thence, to the Brayton cycle heat exchangers [35, 36].

The system uses existing technology and can be tested with electrical heating in existing facilities, so development time is short. It is also flexible since it can be used with Stirling or Brayton cycles. Furthermore, it is designed to be passively safe in all credible launch or reentry accident scenarios. For example, it is subcritical even if fully immersed and surrounded by wet sand. Moreover, it is designed so that no operations are required after launch to prepare it for start-up. In operation, the reactor is controlled by Nb1Zr clad beryllium control drums which have a boron carbide absorber layer.

The mass of the reactor is 512 kg. This could be reduced to as little as 80 kg if changes in design parameters and operations are made. However, in all cases of reduced dimensions and different control configurations, the reliability, safety margins, ease of fabrication, and ease of integration would be reduced. Thus, the mass is a compromise between greater safety and increased reliability.

The technology has been tested in heat pipe demonstrations and with a 12 module SAFE-30 core. Also, the SAFE-400 has been tested through the 19 module SAFE-100 program, which is of similar design but where the Nb1Zr has been replaced by stainless steel for economy. The program also tested fabrication techniques. The integrated core and heat exchanger were tested in 2003 in a new facility at the Marshall Space Flight Center.

8.4.3 Heat Pipe Operated Mars Exploration Reactor

Heat Pipe Operated Mars Exploration Reactor (HOMER) is driving the mission of getting to Mars and maybe is the attainment of a primary objective, but for a person to survive on the surface a source of electrical energy is needed. Approximately 3–20 kW(e) are required, which just takes the task beyond the capabilities of a radioisotope-powered TEG (or RTG) because of the mass of plutonium required. Solar power is impractical because of the distance of Mars from the sun and for seasonal and geographic sunlight issues. Thus, nuclear fission is the remaining option. HOMER is shown in Fig. 8.21.

Note that radioisotope-powered TEGs (or RTGs) of milliwatt electrical power for space application, such as the Angel RTG and its modifications, are compact and reliable in operation and have low mass and size, which makes them convenient for probes. The RTG waste heat is enough to maintain design temperatures for equipment working in deep space environments.

HOMER fulfils the need for a small power source. It is designed specifically for producing electricity on the surface of the planet. The low power requirement means that the reactor operates within well-understood regimes of power density, burnup,

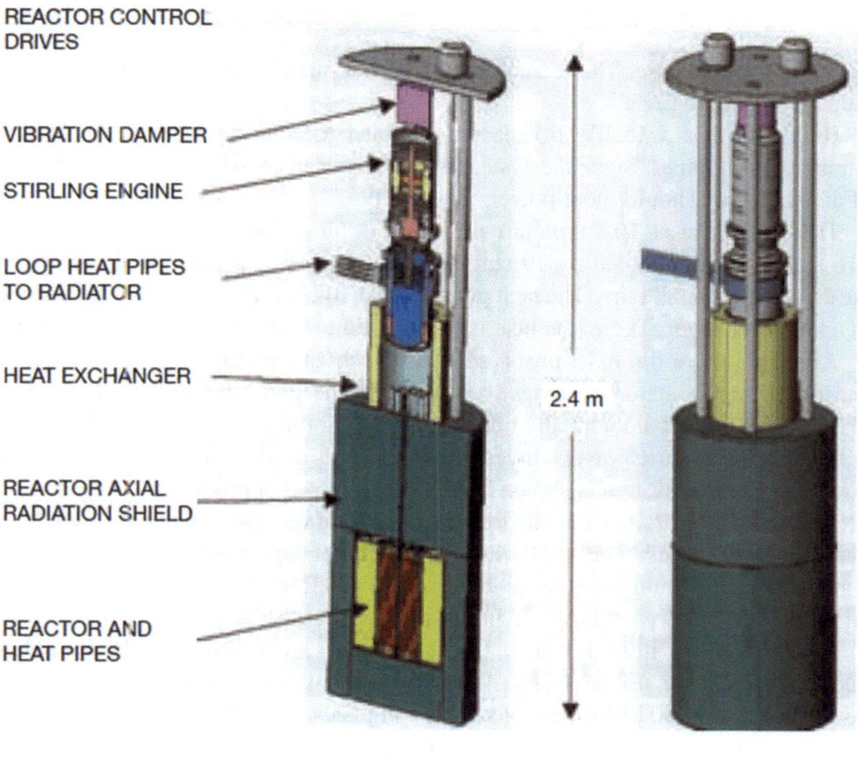

Fig. 8.21 The HOMER power source (Source: Nuclear News)

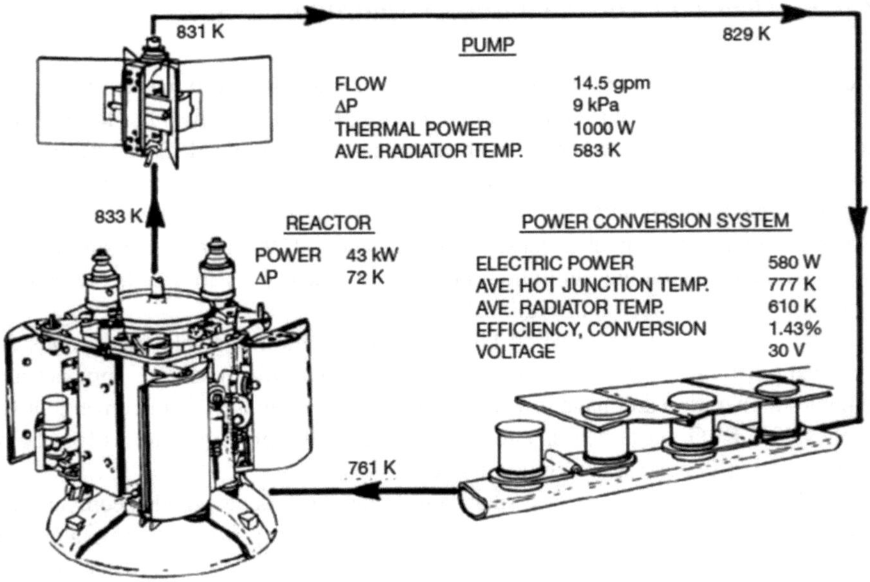

Fig. 8.22 SNAP-10A power conversion system (Source: Atomics International)

and fission gas release. The fluence is so low that there is no significant irradiation damage to core materials.

HOMER-15 is a 15 kW(th) reactor designed to couple with a 3 kW Stirling engine via heat pipes. Since the system is low power, it can be considered a module of a larger array should more power be needed.

The reactor uses 102 uranium nitride fuel pins, each 44 cm long, clad in 316 stainless steel, cooled by 19 stainless steel-sodium heat pipes and assisted by the 0.38 g Martian gravity. The heat pipes extend 40 cm beyond the core axial shield to a heat exchanger. There, the heat is transmitted to a Stirling cycle engine [37].

Although, there did exist unsolved issues such as mercury corrosion and crud, and protection of stator windings, bearings and the pump made the reliability of SNAP-2 and SNAP-8 potentially poor and it was for this reason that the SNAP-10A with its thermoelectric power conversion system was used in flight. The SNAP-10A power conversion system is shown in Fig. 8.22.

In case of SNAP-10A reactor for space exploration, the requirement would be that the reactor needs to be small, restrained, and not dependent upon gravity for its control, which would be normal on earth. Therefore, the design uses rotating beryllium control drums that have boron carbide absorber segments.

Furthermore, as it is shown in Fig. 8.22, there are many possible designs of nuclear reactors for use in space. Advanced space mission requirements for high power levels (25,500 kW(e)) coupled with compact size and long lifetimes favor the use of the fast reactor spectrum with highly enriched fuel. One design for a liquid metal cooled space reactor, which is still a major contender for the future,

However, this design is heavily dependent upon the designs of terrestrial liquid metal cooled fast reactors but is adapted for spacecraft in which the mission is power production rather than breeding or waste reduction.

8.4.4 On Board Nuclear Reactor Studies for Space Exploration

There are many possible designs of nuclear reactors for use in space. Advanced space mission requirements for high power levels (25,500 kW(e)) coupled with compact size and long lifetimes favor the use of the fast reactor spectrum with highly enriched fuel. One design for a liquid metal cooled space reactor, which is still a major contender for the future, is shown in Fig. 8.23.

The design depicted in Fig. 8.23 is heavily dependent upon the designs of terrestrial liquid metal cooled fast reactors but is adapted for spacecraft in which the mission is power production rather than breeding or waste reduction.

The reactor needs to be small, restrained, and not dependent upon gravity for its control, which would be normal on earth. Therefore, the design uses rotating beryllium control drums that have boron carbide absorber segments.

However, the materials technology and proof of that technology have been completed in the non-space liquid metal fast breeder reactor program. This example

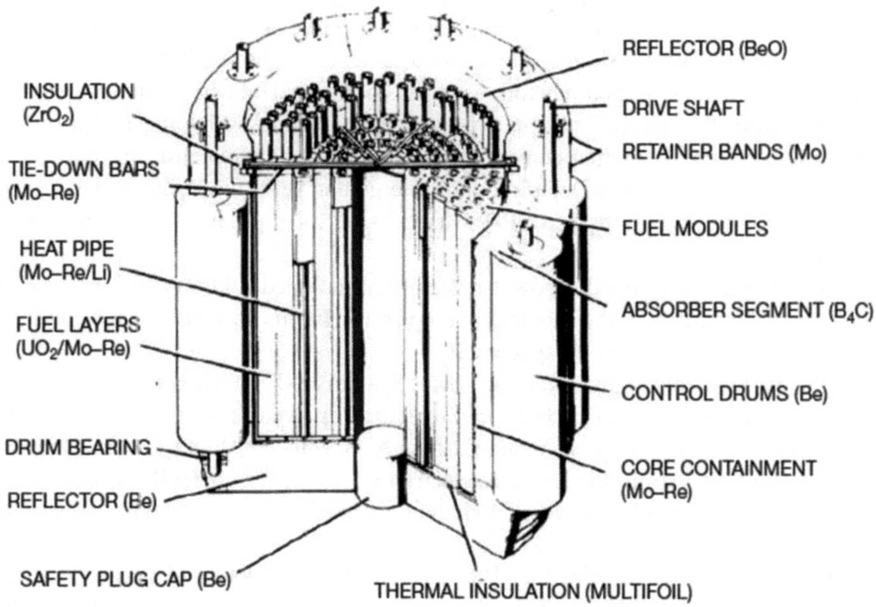

Fig. 8.23 Distributed cooled liquid metal space reactor (Source: Los Alamos National Laboratory)

Table 8.2 Examples of solid-core nuclear reactor systems

Solid-core type	Variant 1	Variant 2	Variant 3
Integral heat transport reactor	Matrix fuel, gaseous (He) coolant	Pin fuel, Na-Li coolant	In-core cylindrical thermionics, Na-K coolant
Distributed heat transport reactor	Heat pipe wafer or coated particle fuel with heat pipe	Liquid metal wafer fuel with electromagnetic pumps	In-core thermionics wafer or coated particle fuel with either electromagnetic pumps or heat pipe

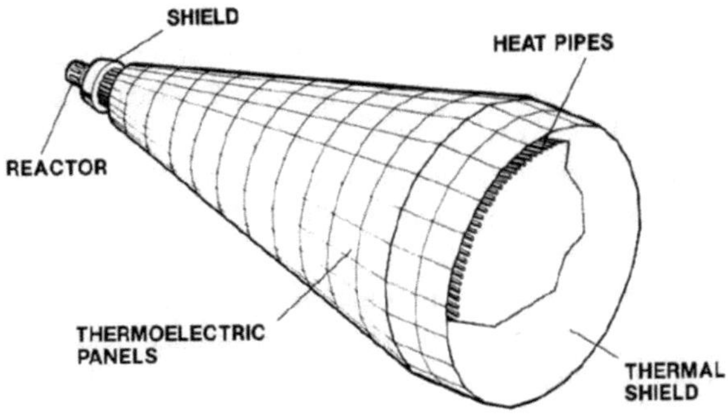

Fig. 8.24 SP-100 Nuclear Power System (radioactively coupled system design) (Source: Los Alamos National Laboratory)

of a distributed liquid metal cooled reactor is merely one of many candidate systems that include several variants of solid-core reactors (see Table 8.2).

For a mass density of 30 kg/kW(e) in a small reactor, outlet temperatures must be of the order of 1200–1500 K. This temperature objective defines both the form of the fuel and the coolant. For higher power requirements, in the 0.5–5.0 MW(e) range, fluidized bed and pellet bed reactors with gas cooling have been studied.

Apart from the nuclear reactor, a power plant includes shielding and a power conversion system, including converters and an excess heat rejection system.

In 1983, NASA, the US Department of Energy and several other agencies agreed to fund a joint program, named SP-100 (see Fig. 8.24), to develop reactor system technology.

This program developed a power system that included a lithium-cooled reactor coupled by heat pipes to thermoelectric converters. In this way the reactor could be used remotely from a manned spacecraft.

The development of space Nuclear Power Systems (NPSs) with direct conversion of nuclear fission thermal power into electrical power started in the early to mid-1950s. The former Soviet Union's first NPS with the direct (thermoelectric) conversion of nuclear fission heat into electricity was the terrestrial Romashka NPS. This NPS first operated in August 1964 and generated about 6100 kWh of electrical energy over 15,000 h.

The BUK space thermoelectric NPS was created in the 1960s and has an electrical power output of about 3 kW. After the conclusion of tests in the early 1970s, this Nuclear Power System (NPS) was put into operation in Near Earth Orbits. From

Table 8.3 Soviet Nuclear Power Systems in space

No.	Space vehicle (NPS)	Launch date	Time operated (day)
1.	Cosmos-367 (BUK)	3 Oct. 1970	For one orbit
2.	Cosmos-402 (BUK)	1 Apr. 1971	For two orbits
3.	Cosmos-469 (BUK)	25 Dec. 1971	9
4.	Cosmos-516 (BUK)	21 Aug. 1972	32
5.	Cosmos-626 (BUK)	27 Dec. 1973	45
6.	Cosmos-651 (BUK)	15 May 1974	71
7.	Cosmos-654 (BUK)	17 May 1974	74
8.	Cosmos-723 (BUK)	2 Apr. 1975	43
9.	Cosmos-724 (BUK)	7 Apr. 1975	65
10.	Cosmos-785 (BUK)	12 Dec. 1975	For three orbits
11.	Cosmos-860 (BUK)	17 Oct. 1976	24
12.	Cosmos-861 (BUK)	21 Oct. 1976	60
13.	Cosmos-952 (BUK)	16 Sep. 1977	21
14.	Cosmos-954 (BUK)	18 Sep. 1977	43
15.	Cosmos-1176 (BUK)	29 Apr. 1980	134
16.	Cosmos-1249 (BUK)	5 Mar. 1981	105
17.	Cosmos-1266 (BUK)	21 Apr. 1981	8
18.	Cosmos-1299 (BUK)	24 Aug 1981	12
19.	Cosmos-1365 (BUK)	14 May 1982	135
20.	Cosmos-1372 (BUK)	1 Jun. 1982	70
21.	Cosmos-1402 (BUK)	30 Aug 1982	120
22.	Cosmos-1412 (BUK)	2 Oct. 1982	39
23.	Cosmos-1579 (BUK)	29 Jun. 1984	90
24.	Cosmos-1607 (BUK)	31 Oct. 1984	93
25.	Cosmos-1670 (BUK)	1 Aug. 1985	83
26.	Cosmos-1677 (BUK)	23 Aug. 1985	60
27.	Cosmos-1736 (BUK)	21 Mar. 1986	92
28.	Cosmos-1771 (BUK)	20 Aug. 1986	56
29.	Cosmos-1818 (TOPAZ)	2 Feb. 1987	142
30.	Cosmos-1860 (BUK)	18 Jun. 1987	40

1970 to 1988, there were 32 launches of these power systems (reactors) as a component of the Cosmos series of spacecraft. See Table 8.3.

Note that Romashka Nuclear Power System (NPS), the main unit of the Romashka NPS is shown in Fig. 8.25. The Romashka NPS is a converter based on a fast reactor, in which the heat generated in the reactor core is conducted to a coaxially arranged radioisotope power system known as Thermoelectric General (TEG) or Radioisotope Thermoelectric General (RTG) as shown in Fig. 8.26, located on the radial reflector outer surface. The reactor core comprises a stack of 11 fuel elements; the segmented fuel elements consisting of discs of uranium bicarbide with 90% enriched ^{235}U. This is located within a graphite package, so

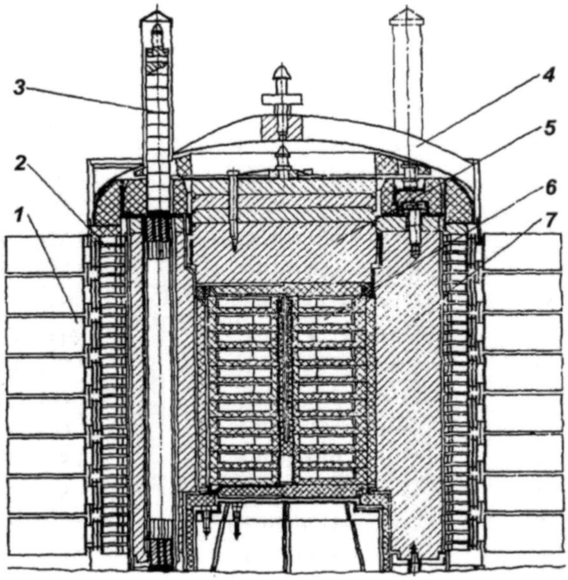

Fig. 8.25 The Romashka NPS reactor convertor layout (Source: Kurchatov Institute)

Fig. 8.26 The Mars science laboratory rover (Source: NASA)

built that a significant part of the heat from the core goes through the package body, thus reducing the temperature drop in the uranium bicarbide.

A radial beryllium reflector encloses the reactor. Graphite bushings are located between the core and the reflector to prevent reflector deformation at the high operating temperatures. The bushings are coated with silicon carbide and beryllium oxide to protect them from chemical interaction with beryllium.

In Fig. 8.25, where the Romashka Nuclear Power System (NPS) reactor is depicted, the following remarks do apply as:

1. Radiator ribs.
2. Thermoelectric elements.
3. Control rod.
4. Reactor vessel.
5. Upper reflector.
6. Reactor core.
7. Radial reflector.

The reactor end reflectors are also made of beryllium. The high-temperature heat insulation made of foam graphite and multilayer graphitized fabric is mounted on the reactor end walls to reduce the heat transfer. The combination allows the reactor to operate with a temperature of up to 2173 K in the center of the core and between 1273 and 1373 K on the reflector outer surfaces.

The reactor control system consists of four rods located in the radial reflector and in the lower end reflector. Two rods are used for automatic and manual control, whereas the other two, together with the movable end reflector, are used for reactor protection in case of emergency.

Figure 8.26 is the NASA Mars science laboratory rover that via RTGs system provides electrical power for spacecraft by converting the heat generated by the decay of plutonium-238 (^{238}Pu) fuel into electricity using devices called thermocouples. Since they have no moving parts that can fail or wear out, RTGs have historically been viewed as a highly reliable power option. Thermocouples have been used in RTGs for a total combined time of over 300 years, and a not a single thermocouple has ever ceased producing power.

High-temperature semiconductor-grade silicon-germanium alloy (Si, 85% by mass; Ge, 15% by mass) is used in the TEGs. These are mounted inside the hermetically sealed steel vessel in four groups, each group having an independent power outlet. The cell comprises two thermopiles with the n- and p-conductivity joined together on the hot side by the molybdenum keyboard. On the cold side, separate pairs are joined with each other in series by a copper bridge onto a common arm running the height of the generator.

To prevent the thermoelectric converters shorting, insulating plates of beryllium oxide are used on the hot and cold sides. To reduce heat loss, all clearances between the thermoelectric cell and hollows in the TEG structure are filled with a cotton-like quartz and a helium atmosphere. A total of 192 enamel coated fins, with an emissivity of at least 0.9, reject excess heat. Basic details of the Romashka NPS reactor converter characteristics are presented in Table 8.4.

The reactor converter generated an electric power of 460–475 W at a constant optimum outer load at start-of-life. By the end of testing (after approximately 15,000 h), the reactor converter electrical power had decreased to 80% of its initial power. This electrical power loss was mainly accounted for by an increase in the thermoelectric converter inner resistance owing to the diffusion processes operating at the graphite disc/silicon-germanium alloy interface resulting in the formation of a high-resistance silicon carbide layer and, partly, to the failure of contacts on the hot side.

Table 8.4 The Romashka NPS reactor converter characteristics

Characteristic	Value
Reactor core diameter/height (by package) (mm)	241/351
Radial reflector outer diameter/height (mm)	483/553
Reactor load mass by uranium-235 (kg)	49
Total mass of the TEG (with the casing and radiator) and reactor (without drives and control rods) (kg)	635
Reactor converter effective thermal power (without taking into account the end wall spread of heat) (kW)	28.2
Reactor converter electrical output (at start-of-life) (W)	460–475
Electrical power reduction over a lifetime of 15,000 h	80%
Reactor converter terminal operating voltage (four groups of thermoelectric converters connected in series) (V)	21
Number of thermoelectric converters in a TEG	3072

The BUK NPS includes the reactor, the shielding, and the conic/cylindrical radiator located in series along the axis. The radiator comprises a system of ribbed pipes for coolant flow united by input and output collectors. It is located on the load-bearing frame structure that is joined to the spacecraft.

The BUK NPS uses a small fast reactor which contains 37 fuel rods. The fuel is a highly enriched uranium-molybdenum alloy; the 235 U load being about 30 kg. Longitudinally movable control rods are placed in the beryllium side reflector. A two-loop liquid metal heat removal system uses a eutectic alloy of sodium and potassium as coolant. The first loop's coolant heated to about 973 K is supplied to the outer casing of the TEG. The TEG, the inner cavities of which are hermetically sealed and filled with inert gas, is located under the radiator, behind the reactor shielding. The second circuit coolant removes the excess heat to the radiator with the coolant maximum temperature at the radiator inlet being about 623 K. The TEG has two independent sections: one for the spacecraft users and an auxiliary one for the power-to-conduction type electromagnetic pumps (electromagnetic pumps) (see Appendix B for more information on electromagnetic pumps) used for both coolant loops. The BUK NPS layout is shown in Fig. 8.27.

In consideration for using nuclear reactors in Near Earth Orbits (NEOs) is the final stage of reentry to the earth's atmosphere of the reactor and its accumulated driven fission products. The BUK NPS was designed to separate from the spacecraft at altitude.

The following are the main components in the BUK layout as illustrated in Fig. 8.27:

1. Nuclear reactor
2. Liquid metal circuit pipeline.
3. Reactor shielding.

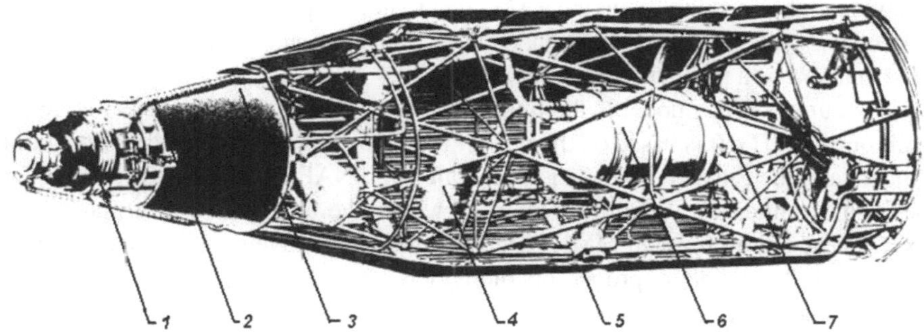

Fig. 8.27 The BUK NPS layout (Source: Kurchatov Institute)

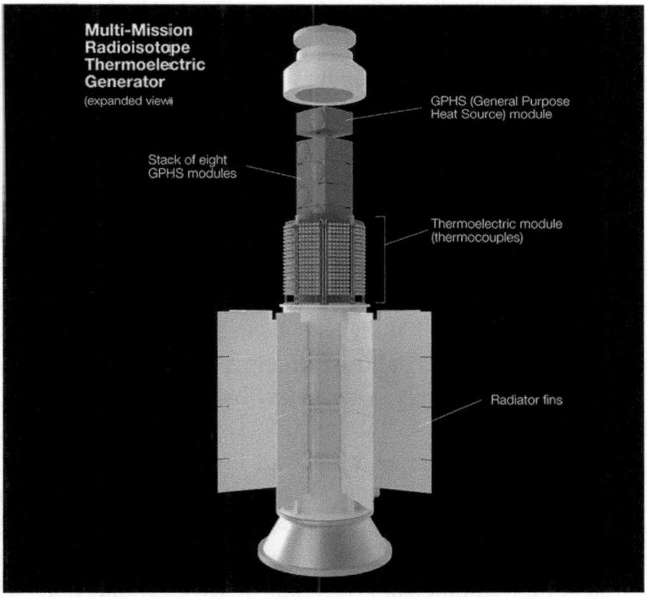

Fig. 8.28 Picture of multi-mission radioisotope thermoelectric generator (Source: NASA)

4. Liquid metal circuit expansion tanks.
5. Radiator
6. Thermoelectric generator (TEG)
7. Load-bearing frame structure.

As it is illustrated in Fig. 8.28, thermocouples are common in everyday items that must monitor or regulate their temperature, such as air conditioners, refrigerators, and medical thermometers. The principle of a thermocouple involves two plates, each made of a different metal that conducts electricity. Joining these two plates to form a closed electrical circuit while keeping the two junctions at different temperatures produces an electric current. Each of these pairs of junctions forms an individual thermocouple. In an RTG, the radioisotope fuel heats one of these junctions, while the other junction remains unheated and is cooled by the space environment or a planetary atmosphere.

The current RTG model is the multi-mission radioisotope thermoelectric generator, or MMRTG. It is based on the type of RTG flown previously on the two Viking landers and the Pioneer 10 and 11 spacecraft (the SNAP-19 RTG). It is designed to be used in either the vacuum of space or within the atmosphere of a planet. The excess heat energy from an MMRTG can be used as a convenient and steady source of warmth to maintain proper operating temperatures for a spacecraft and its instruments in cold environments.

Figure 8.28 is presentation of labeled pull-apart view showing the major components of the MMRTG, or multi-mission radioisotope thermoelectric generator. Image credit: NASA.

The multi-mission radioisotope thermoelectric generator (MMRTG) is a type of radioisotope thermoelectric generator developed for NASA space missions such as the Mars Science Laboratory (MSL), under the jurisdiction of the US Department of Energy's Office of Space and Defense Power Systems within the Office of Nuclear Energy. The MMRTG was developed by an industry team of Aerojet Rocketdyne and Teledyne Energy Systems. Figure 8.29 shows depiction of layout of MMRTG with more details and description each accompanied infrastructure of this satellite.

Furthermore, a reactor placed in such Near Earth Orbit (NEO) takes several hundred years to fall back into the atmosphere, by which time the radioactivity of the fission products will have decayed to acceptable levels. However, this reactor disposal system failed for the COSMOS-954 satellite, which fell out of orbit over Canada in 1978 and scattered its radioactive debris over a large portion of Canada's Northwest Territory. Subsequently, a secondary backup safety system was incorporated into the BUK Nuclear Power System NPS. This secondary system can eject the fuel assembly from the NPS, either while still in Low-Earth Orbit (LEO) or just before reentry, so that the fuel is widely dispersed into minuscule particles in the upper atmosphere as a result of its thermal destruction upon reentry. Such dispersion prevents resulting doses at any place on the earth's surface exceeding the International Commission on Radiological Protection (ICRP) recommended level.

A United Nations (UN) report issued after the COSMOS-954 failure conducted that nuclear reactors "can be used safely in space provided that all necessary safety requirements are met [38]."

Space exploration missions require safe, reliable, long-lived power systems to provide electricity and heat to spacecraft and their science instruments. A uniquely capable source of power is the radioisotope thermoelectric generator (RTG)—essentially a nuclear battery that reliably converts heat into electricity. Radioisotope power has been used on eight Earth orbiting missions, eight missions traveling to each of the outer planets, as well as each of Apollo missions following 11 more to the Earth's moon. Some of the outer solar system missions are the Pioneer, Voyager, Ulysses, Galileo, Cassini, and New Horizons missions. The RTGs on Voyager 1 and Voyager 2 have been operating since 1977. Similarly, Radioisotope Heat Units (RHUs) were used to provide heat to critical components on Apollo 11 as well as

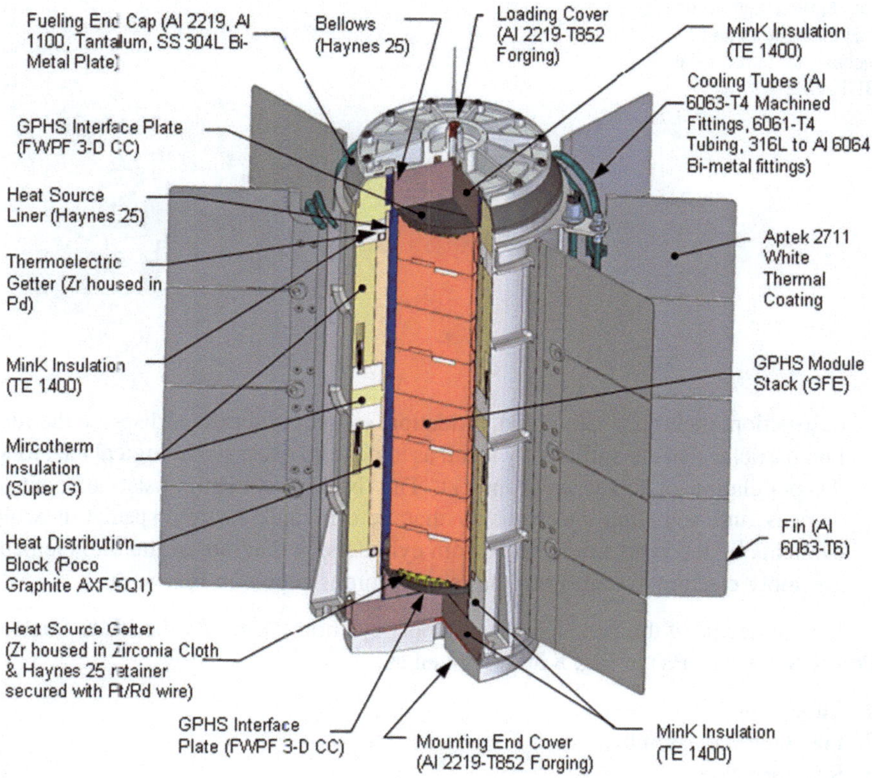

Fig. 8.29 Details oriented MMRTG

the first two generations of Mars rovers. In total, over the last four decades, 26 missions and 45 RTGs have been launched by the United States.

Radiation safety is provided by two diverse systems:

1. The basic safety system, the spacecraft component, relied on moving the spacecraft into a long-term burial orbit, close to circular, at a height of more than 850 km. There, nuclear reactor fission products can decay safely to the level of natural radioactivity. The orbit change system is located in the spacecraft module and is mechanically joined to the nuclear power unit and separated from the spacecraft service module in low operational orbit. The orbit change system includes an off-line propulsion system with its own control systems and an off-line source of electrical power.
2. The backup emergency system provides for the dispersion of fuel, fission products, and other materials with induced activity into the upper layers of the earth's atmosphere. This system ejects the fuel element assembly either in the operational orbit or when the object, which includes the nuclear reactor, enters denser atmospheric layers. During the descent, aerodynamic heating, thermal

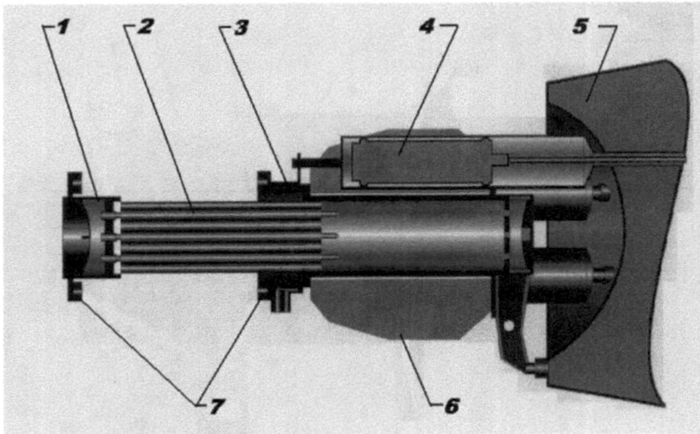

Fig. 8.30 Diagram of the fuel element assembly ejection system for the BUK NPS

destruction, melting, evaporation, oxidation, etc. are expected to disperse the fuel into particles that are sufficiently small as to pose no excess radiological hazard to the population or to the environment. This backup system consists of control devices and actuating mechanisms that deform and destroy special flexible elements by the pressure of gases from cylinders. A diagram of the fuel element assembly ejection system from the reactor core is shown in Fig. 8.30.

Layout details of the fuel element assembly ejection system for the BUK Nuclear Power System (NPS) in Fig. 8.30 are given as:

1. Tube plate
2. Fuel element assembly.
3. Reactor vessel.
4. Control rod.
5. Reactor shielding.
6. Side reflector.
7. Actuating mechanism (Source: Kurchatov Institute)

The backup safety system was introduced into the BUK NPS after the failure of Cosmos-954 spacecraft's change of orbit system.

The spacecraft's descent resulted in large radioactive fragments of wreckage being strewn in a line across Northern Canada in 1978. Characteristics of the BUK NPS are shown in Table 8.5.

An advanced Nuclear Power System (NPS) known as TOPAZ was begun in the early 1970s. The TOPAZ NPS includes a thermionic reactor converter with a cesium vapor supply system and control drum drive unit, the reactor shielding, the radiator, and the frame by which the system is joined to the spacecraft service module as illustrated in Fig. 8.31 with detailed oriented marked components.

These components are listed as below:

1. Cesium vapor supply system and control drum drive unit
2. Thermionic reactor converter.
3. Liquid metal circuit pipeline.
4. Reactor shielding.

Table 8.5 Characteristics of the BUK Nuclear Power System (NPS)

Characteristic	Value
Power (kW(e))	<3
Design lifetime (*a*)	1
Reactor power (kW)	<100
Reactor outlet temperature (K)	973
Fuel and spectrum	U–(90% enriched)–Mo, fast
Coolant	Na–K eutectic
Power conversion	Two cascade thermoelectric converter (Si–Ge)
Hot junction temperature (K)	623
Unshielded weight (kg)	900

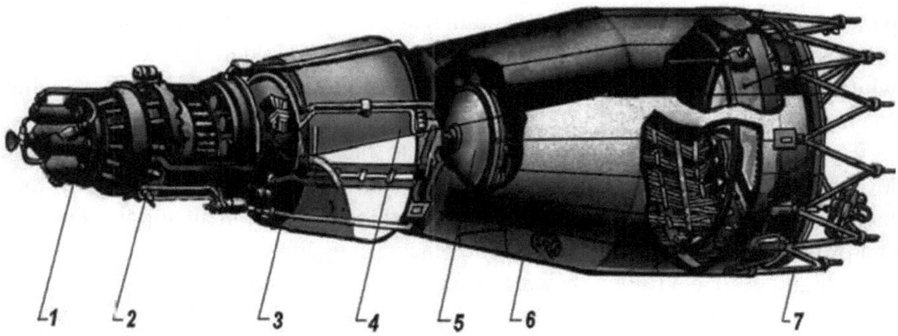

Fig. 8.31 The TOPAZ Nuclear Power System (Source: Kurchatov Institute)

5. Liquid metal circuit expansion tank.
6. Radiator
7. Frame structure.

The automatic control system is placed in the hermetically sealed service module and connected to the related nuclear power unit systems by electrical service lines.

The core consists of 79 thermionic fuel elements (TFEs) and four zirconium hydride moderator discs. The TFEs and cooling channels are located in the moderator disc openings and form a system of five concentric rows. Five-cell TFEs with a three layer collector stack are used, with fission gas vented from emitter assemblies to the interelectrode gap. The TFEs are electrically connected so that they form the working section of 62 TFEs and the pump section of 17 TFEs. The pump section, where the TFEs are connected in parallel, is intended to energize the conduction type electromagnetic pump of the nuclear power unit's heat removal system. The TFEs within this section are connected at both ends in the cesium vapor. The operating section terminal electrical output is about 6 kW at a voltage of ~32 V. The pumping section current is ~1200 A at a voltage of 1.1 V. Before the reactor converter is brought to the rated electrical power level, the electromagnetic pump is fed from the start-up unit by a high-current storage battery located behind the radiation shielding.

Twelve rotating cylinders (drums) located in the side reflector provide thermal power control, reactivity compensation, and emergency shutdown. These beryllium cylinders have sector cover plates of boron carbide and are divided into four groups of three cylinders. Each group is controlled by its own drive.

The cesium vapor supply system pumps vapor through the thermionic fuel element (TFE) interelectrode space at a flow rate of about 10 g/day. A pyrolyzed graphite trap absorbs used cesium and non-condensing impurities are ejected into space.

The NPS uses a lithium hydride reactor shield located in a hermetically sealed steel container with the inner load-bearing elements. The single circuit sodium-potassium heat removal system includes a radiator that has load carrying capacity and also serves as a structural member.

The radiator is designed as a system of D-shaped tubes placed hydraulically in parallel. The tubes are welded into the radiator O-ring collectors and supported by the load-bearing elements. The tubes' plane surface is soldered to a steel radiator which has a high emissivity coating. The area of the radiator is about 7 m^2, which ensures rejection of at least 170 kW(th) at a coolant temperature of 880 K.

The automatic system controls the NPS to rated thermal and electrical power levels, maintains the working section current or coolant temperature at the rated level, maintains the voltage of ~28 V for the onboard equipment supply lines, and can provide shutdown of the thermionic reactor converter.

A high-speed controller that redistributes the direct current of the thermionic reactor converter section between the spacecraft and ballast loads controls the voltage. In the nominal operating mode, the rated current of the operating section and, consequently, its electrical power are sustained by correcting thermal power. As the efficiency degrades the coolant temperature rises to 880 K. After that, instead of maintaining the current, the automatic control system limits the coolant temperature. The thermal power then remains practically constant, whereas the operating section current will fall to values at which the onboard network voltage exceeds allowable limits, requiring NPS shutdown. NPS shutdown is also provided for specific emergency situations, as well as by radio command from earth.

The TOPAZ Nuclear Power System (NPS) generates approximately 6 kW at a start-of-life efficiency of about 5.5%. Its mass, including the nuclear power unit, the automatic control system, and the coupling service lines, is about 1200 kg, and it has a design lifetime of 4400 h. The nuclear power unit is 4.7 m long with a maximum diameter of 1.3 m.

Keep in your mind that TOPAZ NPS, thermionic energy conversion is used, and these thermionic cells were made an integral part of each fuel rod. Seventy-nine thermionic fuel elements of 90% enriched uranium oxide formed the TOPAZ reactor core, which was cooled by Na-K and waste heated rejected into space by a radiator module. To increase the efficiency of thermionic conversion, a once-through cesium vapor supply system was incorporated, and with a reactor thermal power of 150 kW, 5–6 kW of electrical power was produced. However, several TOPAZ systems were ground tested and refinements made, leading to the flight testing of two TOPAZ units in around 1987 through 1988 years. The COMOS-11818 test lasted 142 days,

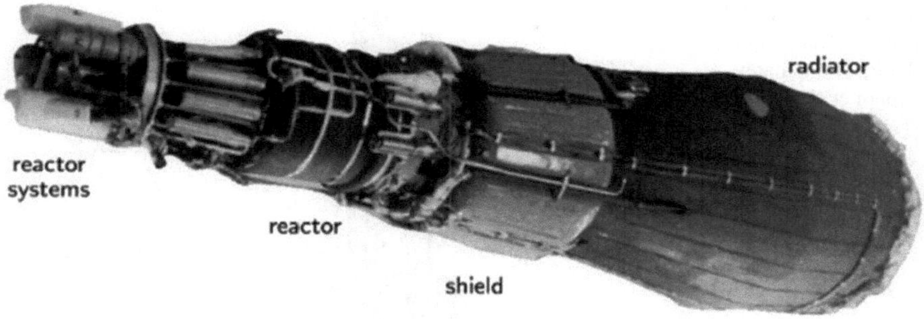

Fig. 8.32 The TOPAZ Nuclear Power System [39] (Source: Ponomarev-Stepnoi et al. 2000)

while the testing for COSMOS-1867 did last 342 days. In both cases, the tests were ended once the cesium vapor generator ran out of cesium. Nevertheless, these tests confirmed the reliable operation of thermionic conversion in a space NPS as a TOPAZ illustrated in Fig. 8.32.

As we stated the TOPAZ reactor illustrated in Fig. 8.32 has thermal output of 150 kW producing 5 kW of electrical power. The mass of the system is much as 980 kg.

A more advanced NPS known as ENISEY or TOPAZ-2 has also been developed by Russian as illustrated in Fig. 8.33.

This NPS originally designed for powering television relay satellites, also uses thermionic conversion technology and produces 5.5 kW(e) from a reactor power of about 135 kWt. Although this unit has yet to be flown in space, it represents the most advanced space reactor power system yet to be developed. In 1991 the United States and Russia began collaborating work on testing the advanced TOPAZ-2 system for use as an alternative to solar power systems for civil, commercial, and scientific applications. Six TOPAZ-2 units have been delivered to the United States and tested under this cooperative program.

8.5 Nuclear Power Assessment Study: Final

The objective of this study is to "discuss a sustainable strategy and present findings for the provisioning of safe, reliable, and affordable nuclear power systems that enable NASA Science Mission Directorate (SMD) missions and is extensible to Human Exploration and Operations Mission Directorate (HEOMD) needs in the next 20 years." See Fig. 8.34.

The context of this study is set by the confluence of ongoing requirements and continuing budget pressures. NASA has pursued different approaches for provisioning nuclear power systems. In recent history, radioisotope power systems (RPS) have been provisioned in support of the Science Mission Directorate (SMD) for robotic exploration. Fission power systems (FPS) have been in development in

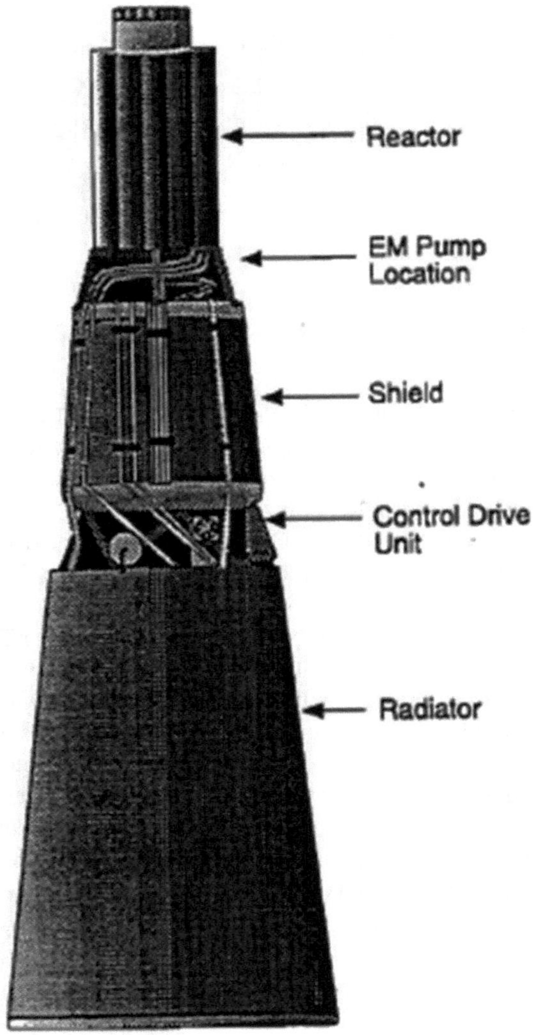

Fig. 8.33 The TOPAZ-II space nuclear reactor [39] (Source: Ponomarev-Stepnoi et al. 2000)

support of Human Exploration and Operations Mission Directorate (HEOMD) goals.

Nevertheless, fission and radioisotope power systems have traveled down parallel development paths, requiring separate resources. SMD is considering the possibility of using both RPS and FPS for future missions. This potential approach along with the current budget scenario presents an opportunity to explore development of common power system technologies that feed both FPS and RPS as an alternate provisioning strategy. This strategy may hold the possibility of furthering exploration goals for several mission directorates, while reducing technology risk associated with new systems development.

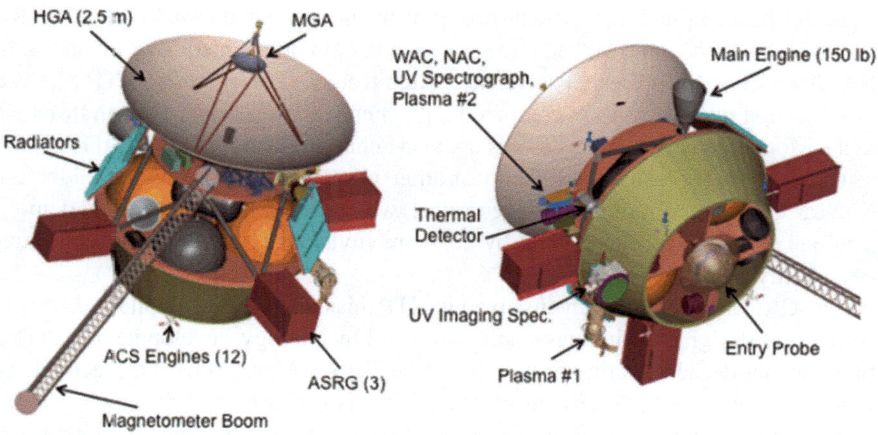

Fig. 8.34 Decadal survey UOP conceptual spacecraft layout

NASA's need for RPS to enable robotic scientific missions for planetary exploration has been a "given" for over four decades. The continuing need for planetary missions has been articulated clearly during the last decade from the NRC report of 2009 through the Planetary Decadal Survey of 2011, "Vision and Voyages."

Requirements across the other Divisions within the Science Mission Directorate and across other Directorates have not been as clear. Implementation of these systems has never been an easy task. An ongoing difficulty is the inherent difference between the development cycle of RPS and the shorter time from mission approval to launch.

Joining these two different sets of requirements—and their associated different ways of "doing business"—remains a challenge in the partnership between DOE and NASA to develop and fly these power systems.

8.6 Summary

The National Aeronautics and Space Administration (NASA) Glenn Research Center (GRC) has been actively involved in Nuclear Thermal Propulsion (NTP) technology development, mission, engine, and vehicle design dating back to the Rover and Nuclear Engine for Rocket Vehicle Applications programs. This technology was successfully developed in over 20 rocket/reactor tests, which demonstrated a wide range of thrust levels, high-temperature fuel, sustained engine operation, accumulated time at full power, and restart capability—everything required for a human mission to Mars. Furthermore, NTP requires no large technology scale-up. The smallest engine tested during the Rover program—the Pewee engine—is sufficient for this when used in a clustered engine arrangement. The GRC has led every major study involving NTP since the late 1980s and has helped quantify the evolution and growth potential of the nuclear thermal rocket (NTR), which includes the bimodal and liquid-oxygen (LOX)-augmented NTR concepts. In NASA's recent Mars Design Reference Architecture (DRA) study, NTP reduced total mission mass over 400 t compared with chemical propulsion. Human missions to the Moon and near-Earth asteroids are also enhanced using NTP. In 2011, NASA restarted an NTP technology demonstration effort that is continuing under the Nuclear Cryogenic Propulsion Stage project, which began in 2012. Ground demonstrations of a small, scalable NTR by 2020 are envisioned, with a flight demonstration shortly thereafter.

The GRC has been actively involved in NTP mission analysis, engine and vehicle conceptual design, requirements definition, and technology development planning for over four decades, dating back to the Rover/NERVA programs. This technology was developed to a high TRL in more than 20 rocket/reactor tests, which demonstrated a wide range of thrust levels, high-temperature fuel, sustained engine operation, accumulated time at full power, and restart capability—everything required for a human mission to Mars. Compared with other advanced propulsion systems, NTP requires no large technology or performance scale-ups. In fact, the smallest engine tested during the Rover program—the 111.2-kN Pewee engine—is sufficient when used in a clustered engine arrangement.

The NTR also has significant evolution and growth capability that the GRC has helped quantify and demonstrate through numerous collaborative efforts with the industry. Configured as a bimodal system, it can generate its own electrical power to support spacecraft operational needs. Adding an O^2 afterburner nozzle introduces a variable thrust and capability as well as allowing bipropellant operation. For human missions to Mars, study after study has identified NTP as the preferred propulsion option because of its proven technology, higher performance, lower launch mass, simple assembly, and growth potential. In NASA's recent Mars DRA 5.0 study, when compared with chemical propulsion, the use of NTP helped reduce the total mission IMLEO by more than 400 t—the equivalent mass of the International Space Station. This same proven performance can be used to enhance human precursor missions to the Moon, NEAs, and Mars orbit prior to a Mars landing mission.

Finally, and most importantly, NASA restarted an NTP technology development and demonstration effort in 2011, which includes foundational technology development work in the five key task areas identified by the GRC. The results from these tasks will provide the basis for continuing work in these same areas under the NCPS project over the next 3 years. If successful, this effort could be followed by system-level technology demonstrations that include ground testing a small, scalable NTR by 2020 with a flight test shortly thereafter.

We have concluded that nuclear power is an important source of energy on Earth and that it is also needed in space to provide the electricity to power both propulsion systems of various types and all of the other mission electronic functions. We have found that ideas initially developed for space applications have also stimulated a new vision for Earth-based power systems, both large and small. These systems include new ion plasma propulsion systems, and new high-efficiency, gas-cooled reactors. This new vision also includes a re-examination of high-efficiency generation cycles perhaps involving fluids other than steam and the use of heat pipes for compact reactors for very specialized and localized usage.

A nuclear reactor space power system using out-of-pile thermionic diodes, heat pipes, and a dual central absorber rod type of reactivity control has been studied past decades. Emphasis is placed on the neutronic aspects and general feasibility of the concept. Comparison is made between uranium-233 and -235 nitride and plutonium-239 nitride fuels. From a neutronic standpoint, plutonium-239 nitride is found to be a slightly better fuel than uranium-233 nitride for this fast-spectrum reactor. In this concept, heat is transferred from the reactor core to the thermionic diodes by layers of radial heat pipes stacked alternately with slabs of fuel. For this out-of-pile concept, which would supply ~130 kW(e), the reactor can be considerably smaller than the equivalent reactor with in-pile diodes. It would be particularly adaptable to a shadow-shielded type of application.

The merited applications of an NTR include solar system exploration, transportation, and power support for the International Space Station (ISS) and Near Earth Orbit (NEO) missions. The first one includes human exploration missions to Mars and beyond, because an NTR is more reliable and flexible for long-distance space missions at lower costs than Commercial Resupply Service (CRS). On the basis of launch cost, two NTR missions could be conducted with the same price of one CR mission. Second, a bimodal NTR is able to provide not only propulsion for transportation but also electric power to the ISS. By using a bimodal NTR, the ISS can be placed into a lunar cycle orbit and be electrically powered for lunar exploitation [40]. In addition, an NTR is suited to conduct the missions for threat investigation or interception of NEOs, whether they are asteroids or comets. An NTR enables flight velocities over two times those achievable of CRS with the same amount of propellant, allowing interaction with a potential NEO threat in a shorter time or at a longer range [41].

Note that, Commercial Resupply Services (CRS) are a series of contracts awarded by **NASA** from 2008 to 2016 for delivery of cargo and supplies to the International Space Station (ISS) on commercially operated spacecraft. The first CRS contracts were signed in 2008 and awarded $1.6 billion to SpaceX for 12 cargo transport missions and $1.9 billion to Orbital Sciences for eight missions, covering deliveries to 2016. In 2015, NASA extended the Phase 1 contracts by ordering an additional three resupply flights from SpaceX and one from Orbital Sciences [42]. After additional extensions late in 2015, SpaceX is currently scheduled to have a total of 20 missions and Orbital 10 [15].

In assessing the status and reviewing the role of nuclear power in the peaceful exploration of space, it also aims to initiate a discussion on the potential benefits of space-related nuclear power technology research and development for the research and development of innovative terrestrial nuclear systems.

Active space exploration started at the beginning of the twentieth century when amateur and semiprofessionals engaged in rocket science, conducting trial and error experiments to leave the ground, if not Earth. They made steady but not dramatic progress.

However, just as flight had received government support during World War I, so rocket development received the same support during World War II, and spacecraft development received government backing during the Cold War. The advances in each case were dramatic. Fortunately, since the Cold War ended, space exploration has matured in a healthier environment. International cooperation is the order of the day, including the building and manning of the Earth's first space station.

As close up of this chapter, we should see that suggestions for future study of heat pipe drive fissionable reactors as passive cooling system can be stated as follow.

There are numerous applications of this type of heat pipe on which further study could be done. Experimental studies, where facilities permit, are especially needed. A few suggestions for both experimental and analytical work are listed here:

1. One of the areas where experimental work is needed is in the area of high-temperature properties of compounds of fissionable material. More data is needed on uranium, thorium, and plutonium chlorides, bromides, and fluorides.
2. These compounds appear to have the best stability and high-temperature properties for this application.
3. Another possible investigation is the use of this heat pipe in other types of reactors such as liquid metal fast breeder reactors. It should be possible to use a natural uranium compound in this reactor, with the heat pipes in different areas of the reactor such as in reflectors.
4. One other possible design is to use heat pipes in a fast reactor with all of the fuel in the heat pipes. This could be used as an auxiliary power supply for space travel.
5. The heat pipes could be used as heat sources for thermoelectric or thermionic energy converters.
6. Finally, it could be speculated that two component heat pipes could be used. These would have two different working fluids within the heat pipe. One of these fluids would have a slightly higher boiling point than the other, and it would stay in the evaporator section, while the other fluid stays in the condenser. The interface of these fluids would move back and forth short distances depending on the heat flux in the heat pipe. If the fluid in the evaporator were a fissionable fuel and the fluid in the condenser a poison, it would conceivably be possible to get very high reactivity changes with flux changes.

In conclusion, it appears that the overall concept is a good method of increasing reactivity feedback if a heat pipe working fluid is used which has optimum temperature operation that matches the operating temperature of the reactor and that it is better for larger reactors.

References

1. Zohuri, B. (2016). *Heat pipe design and technology: Modern applications for practical thermal management* (2nd ed.). New York: Springer.

2. Nemec, P., Caja, A., & Malcho, M. (2013). Mathematical model for heat transfer limitations of heat pipe. *Mathematical and Computer Modelling, 57*, 126–136.
3. Ochterbeck, J. M. (2003). Heat pipes. In *Heat transfer handbook* (1st ed.). Hoboken: Wiley.
4. Gaugler, R. S. (1944). *Heat transfer devices*. U.S. patent 2,350,348.
5. Trefethen, L. (1962). *On the surface tension pumping of liquids or a possible role of the candlewick in space exploration*. GE Tech. Int. Ser. No. G15-D114. Schenectady, NY: General Electric Co.
6. Grover, G. M., Cotter, T. P., & Erikson, G. F. (1964). Structures of very high thermal conductivity. *Journal of Applied Physics, 218*, 1190–1191.
7. Jeong, Y. S. (2016). Hybrid heat pipe based passive cooling device for spent nuclear fuel dry storage cask. *Applied Thermal Engineering, 96*, 277–285.
8. Jeong, Y. S., Kim, K. M., Kim, I. G., & Bang, I. C. (2015). Hybrid heat pipe with control rod as passive in-core cooling system for advanced nuclear power plant. *Applied Thermal Engineering, 90*, 609–618.
9. Deverall, J. E., & Kemme, J. E. (1965). *Satellite heat pipe*. U.S. Atomic Energy Commission Report LA-3278-MC. Los Alamos, NM: Los Alamos Scientific Laboratory.
10. Anderson, J. L., & Lantz, E. (1968). A nuclear thermionic space power concept using rod control and heat pipes. *Nuclear Applications, 5*(6), 424–436.
11. Croke, E. J., & Roberts, J. J. (1968). Compact power concept features a fast reactor, heat pipes, and direct converters. *Reactor and Fuel-Processing Technology, 11*(4), 187–200.
12. Lubarsky, B., & Shure, L. I. (1966). Applications of power systems to specific missions. In *Space Power Systems Advanced Technology Conference NASA-SP-131* (pp. 269–277). National Aeronautics and Space Administration.
13. Cotter, T. P. (1967). Heat pipe startup dynamic. In *Proceedings of SAE Thermionic Conversion Specialist Conference, Palo Alto, CA*.
14. Deverall, J. E., Kemme, J. E., & Florschuetz, L. W. (1970, September). *Sonic limitations and startup problems of heat pipes*. Los Alamos Scientific Laboratory Report No. LA-4578.
15. de Selding, P. B. (2016, July 17). SpaceX wins 5 new space station cargo missions in NASA contract estimated at $700 million. *Space News*.
16. Parker, M. B. (1970). *Application of heat pipes to nuclear reactor engineering*. Ames: Iowa State University.
17. Anand, D. K. (1966). On the performance of a heat pipe. *Journal of Spacecraft and Rockets, 3* (5), 763–765.
18. Carnesale, A., Cosgrove, J. H., & Ferrill, J. K. (1966). *Operating limits of the heat pipe*. U.S. Atomic Energy Report SC-M-66-623. Albuquerque, NM: Sandia Laboratories.
19. Cotter, T. P., Deverall, J., Erickson, Q. F., Qrover, Q. M, Keddy, E. S., Kemme, J. E., & Salmi, E. W. (1965). *Status report on theory and experiments on heat pipes at Los Alamos*. U.S, Atomic Energy Commission Report LA-DC-7206. Los Alamos, NM: Los Alamos Scientific The Laboratory.
20. Cotter, T. P. (1965). *Theory of heat pipes*. U.S. Atomic Energy Commission Report LA-324o-MC. Los Alamos, NM: Los Alamos Scientific Laboratory.
21. Kemme, J. E. (1966). *Heat pipe capability experiments*. U.S. Atomic Energy Report LA-3585-MS. Los Alamos, NM: Los Alamos Scientific Laboratory.
22. Kemme, J. E. (1969). *Heat pipe design considerations*. Unpublished mimeographed paper presented at National Heat Transfer Conference, Minneapolis, Minnesota. Los Alamos, NM: Los Alamos Scientific Laboratory.
23. Kemme, J. E. (1967). *High performance heat pipes*. U.S. Atomic Energy Commission Report LA-DC-9027. Los Alamos, NM: Los Alamos Scientific Laboratory.
24. Cheung, H. (1968). *A critical review of heat pipe theory and applications*. U.S. Atomic Energy Commission Report UCRL-50453. Livermore, CA: Lawrence Radiation Laboratory, University of California.
25. Peldman, K. T., & Whiting, G. H. (1968). Applications of the heat pipe. *Mechanical Engineering, 90*(11), 48–53.

26. Peldman, K. T., & Whiting, G. H. (1967). The heat pipe. *Mechanical Engineering, 89*(2), 30–33.
27. Smith, H. P. (1962). Closed loop dynamics of nuclear rocket engines with bleed turbine drive. *Nuclear Science and Engineering, 14*(4), 371–379.
28. Smith, H. P., & Stenning, A. H. (1961). Open loop stability and response of nuclear rocket engines. *Nuclear Science and Engineering, 11*(1), 76–84.
29. Jansen, W., & Buckner, J. K. (1963). Starting and control characteristics of nuclear rocket engines. *AIAA Journal, 1*(3), 563–573.
30. Mohler, R. R., & Perry, J. E. (1961). Nuclear rocket engine control. *Nucleonics, 19*(4), 80–84.
31. Robbins, W. H., & Finger, H. B. (1991, July). *An historical perspective of the NERVA nuclear rocket engine technology program.* NASA Contractor Report 187154/AIAA-91-3451. NASA Lewis Research Center, NASA.
32. Cooper, R. S. (1968). Nuclear propulsion for space vehicles. *Annual Review of Nuclear Science, 18*, 203–228.
33. Valfells, A. (1962). *Preliminary design of a power generating system using a fissile gas.* Unpublished PhD thesis, Library, Iowa State University of Science and Technology, Ames.
34. Robbins, W. H. (1991). *An historical perspective of the NERVA nuclear rocket engine technology program.* NASA Contractor Report 187154 AIAA-91-3451.
35. Zohuri, B., & McDaniel, P. (2017). *Combined cycle driven efficiency for next generation nuclear power plants: An innovative design approach* (2nd ed.). New York: Springer.
36. Zohuri, B. (2016). *Compact heat exchangers: Selection, application, design and evaluation.* New York: Springer.
37. Zohuri, B., & McDaniel, P. (2019). *Thermodynamics in nuclear power plant systems* (2nd ed.). Springer.
38. UN, United Nations Committee on the Peaceful Users of Outer Space. (1981, February). *Report of the Working Group of the Uses of Nuclear Power Sources in Outer Space, Annex II to the report designated as Report of the Scientific and Technical Subcommittee on the Work of the Eighteenth Session.* UN Document A/AC, 105/287.
39. Ponomarev-Stepnoi, N. N., Talyzin, V. M., & Usov, V. A. (2000). Russian space nuclear power and nuclear thermal propulsion system. *Nuclear News.*
40. Paniagua, J., Maise, G., & Powell, J. (2008). Converting the ISS to an Earth-Moon transport system using nuclear thermal propulsion. In: *Space Technology and Applications International Forum (STAIF 2008)* (pp. 492–502). Albuquerque, NM: AIP.
41. Powell, J., Maise, G., Ludewig, H., & Todosow, M. (1997). Highperformance ultra-light nuclear rockets for near-earth objects interaction missions. *Annals of the New York Academy of Sciences, 822*, 447–467.
42. Bergin, C. (2015). *NASA lines up four additional CRS missions for Dragon and Cygnus.* Retrieved April 19, 2015, from nasaspaceflight.com.

Chapter 9
Design Guide and Heat Pipe Selection

Heat pipes and its closed two-phase thermosyphons are highly efficient heat transfer devices utilizing the continuous evaporation/condensation of suitable working fluid for two-phase heat transport in a closed system. Due to a variety of advantageous features, these devices have found a number of applications in space, terrestrial, nuclear power plant, and electronics technology. The operational principles and the performance characteristics of the different types of heat pipes are described. For the heat pipe designs, which have found the widest application, versus the classical capillary-wick heat pipes and the wickless heat pipes or closed two-phase thermosyphons, mathematical schemes are given to calculate the performance and performance limits in Chap. 2. Here in this chapter we will discuss the design criteria and steps for a heat pipe depending on its application. In this chapter also, few design examples are gathered from different authors or researchers in this field to provide a better guideline and procedures to the readers. Heat pipes are being used very often in particular applications when conventional cooling methods are not suitable. Once the need for heat pipe arises, the most appropriate heat pipe needs to be selected.

9.1 Introduction

It was recognized rather early in the history of the heat pipe research [1] that techniques could be developed which would provide for control of the effective thermal conductance of the heat pipe. This was first envisioned as blocking a portion of the condenser by a non-condensable gas. More recently several other types of control have been developed including liquid blockage and liquid and vapor modulation. Such techniques enable the device to be operated at a fixed temperature independent of source and sink conditions [1].

The methodology of arriving at the optimum heat pipe design is a complex one, not only because of the mathematical analysis involved, but more particularly the many qualitative judgments that must be considered and introduced. Figure 9.1 is an

B. Zohuri, *Heat Pipe Applications in Fission Driven Nuclear Power Plants*,
https://doi.org/10.1007/978-3-030-05882-1_9

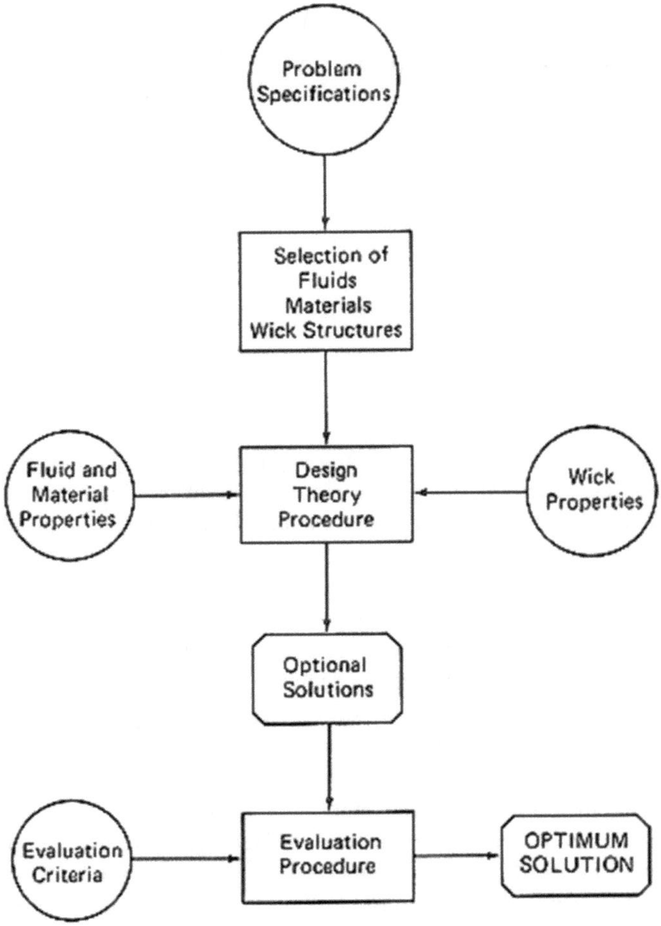

Fig. 9.1 Schematic of a heat pipe design procedure [2]

illustration by Chi [3] that schematically provides a high-level approach and design procedure. Physical properties of several metals and fluids are given in Appendices B and C reference by Zohuri [1].

Figure 3.4 shows temperature ranges for several fluids from melting point to critical temperatures. Because of overlapping of these temperature ranges for different fluids, several fluids can often be used for a given operating temperature. Heat pipe design guidelines in this chapter will provide information on the selection of the pipe working fluids, wick structures, and container materials along with some data that are gathered from various references and researchers reports in this filed.

A simple approximation theory is developed around a simple cylindrical shape heat pipe operation in a zero-gravity field. The following simplified assumptions also are made as follows [2]:

1. The pipe is capillary limited.
2. Vapor pressure losses are negligible.
3. Wick thickness of t_w is much smaller than the vapor core radius.
4. Heat flux density is uniform at the evaporator or condenser surface.
5. Thermal conductivity of the liquid-saturated wick is proportional to that of the liquid.

The heat transport factor $(QL)_{\text{capillary}_{\max}}$ under above assumptions can then be written as follows using 2.77 of Chap. 2 of Zohuri book [1]:

$$(QL)_{\text{capillary}_{\max}} = 2\left(\frac{\sigma \rho_l \lambda}{\mu_l}\right)\left(\frac{K}{r_c}\right)(2\pi r_v t_w) \tag{9.1}$$

where

Q = Heat flow rate
L = Heat pipe length
σ = Surface tension coefficient
ρ_l = Liquid density
λ = Latent heat of vaporization
μ_l = Liquid dynamic viscosity
K = Wick permeability
r_c = Effective capability radius
r_v = Vapor core radius
t_w = Wick thickness

The first term in bracket represents the liquid property, while the second one is the wick property, and the third is the wick cross-sectional area, respectively (Fig. 9.2).

Equation 9.1 indicates that for a pipe of fixed wick structure and dimensions, its heat transport factor $(QL)_{\text{capillary}_{\max}}$ is proportional to the liquid transport factor N_l, which is defined as $\left(\frac{\sigma \rho_l \lambda}{\mu_l}\right)$. Figure 3.6 shows the values of the liquid transport factor N_l for several fluids. For a maximum temperature gradient of a heat pipe, the temperature drop across the liquid-saturated wick must be minimum. If we further to simplify the assumptions 3 and 4 that are described in above, the temperature drop across the wick is proportional to Qt_w/k_l and can be written as follows [2]:

$$\Delta T \propto \frac{Qt_w}{k_l} \tag{9.2}$$

where k_l is thermal conductivity of liquid (Fig. 9.3).

Also, from 9.1 we can deduce that the required wick thickness t_w for the same heat transport is inversely proportional to the liquid transport factor N_l so the 9.2 can be written as

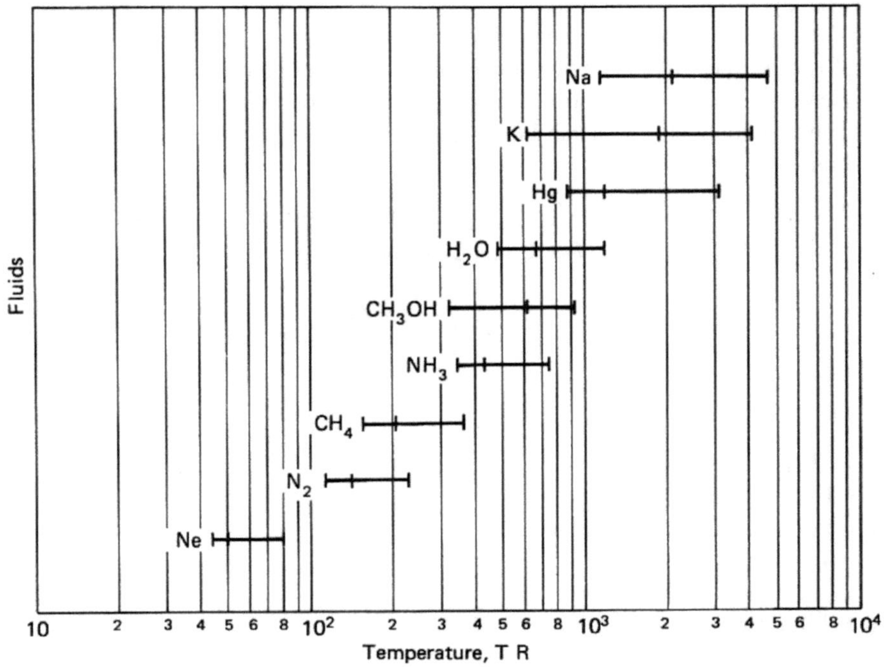

Fig. 9.2 Normal melting point, boiling point, and critical point temperatures for several heat pipe working fluids ($1R = 0.5556$ K) [2]

$$\Delta T \propto \frac{Q}{k_l N_l} \tag{9.3}$$

From this equation one can induce that the temperature drop across a wick is inversely proportional to the liquid property ($k_l N_l$), which has been called the liquid conductance factor. The values of the liquid conductance factor ($k_l N_l$) for several fluids are shown in Fig. 9.4.

Per Dunn and Kew [2], the design of a heat pipe or thermosyphon to fulfill a particular duty involves four broad processes:

1. Selection of appropriate type and geometry
2. Selection of candidate materials
3. Evaluation of performance limits
4. Evaluation of the actual performance

The background to each of these stages is covered in Chap. 2 of Zohuri [1]. In this chapter, the theoretical and practical aspects are discussed with reference to sample design calculations.

Per Reay and Kew [2], the design procedure for a heat pipe is outlined in Fig. 9.5. As with any design process, many of the decisions that must be taken are interrelated, and the process is iterative. For example, choice of the wick and case material

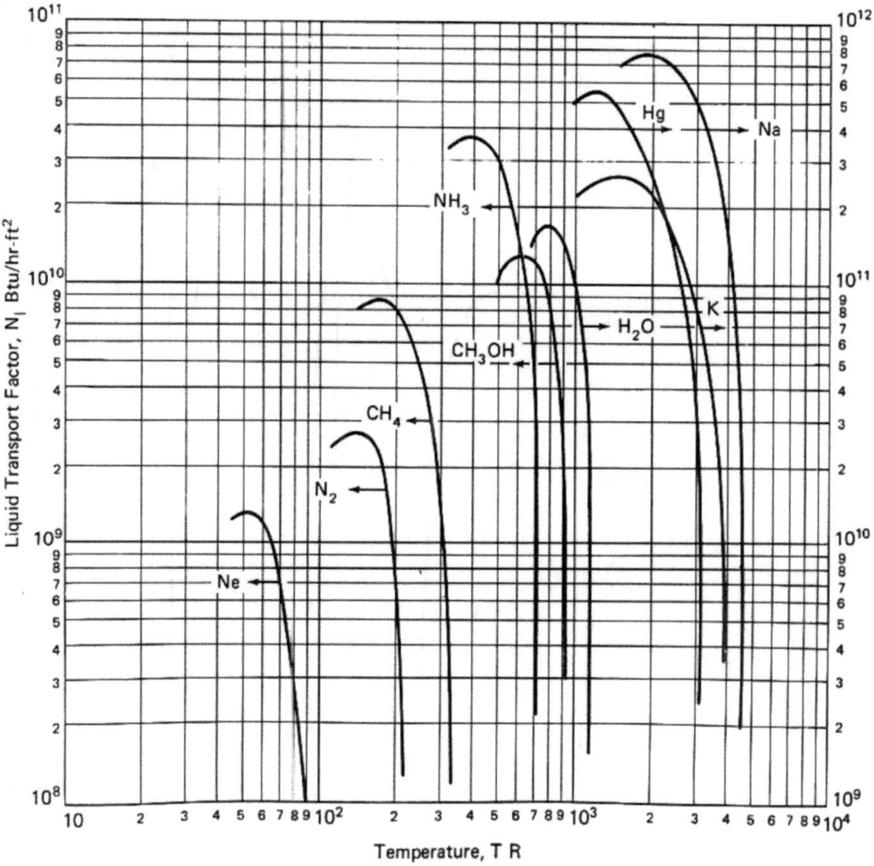

Fig. 9.3 Liquid transport factor versus temperature for several heat pipe working fluids (1 Btu/h-ft^2 = 3.153 W/m^2; 1R = 0.5556 K) [2]

eliminates many candidate working fluids (often including water) due to compatibility constraints. If the design then proves inadequate with the available fluids, it is necessary to reconsider the choice of construction materials.

Two aspects of practical design, which must also be taken into consideration, are the fluid inventory and start-up of the heat pipe.

9.1.1 Fluid Inventory

A feature of heat pipe design, which is important when considering small heat pipes and units for space use, is the working fluid inventory. It is a common practice to include a slight excess of working fluid over and above that required saturating the

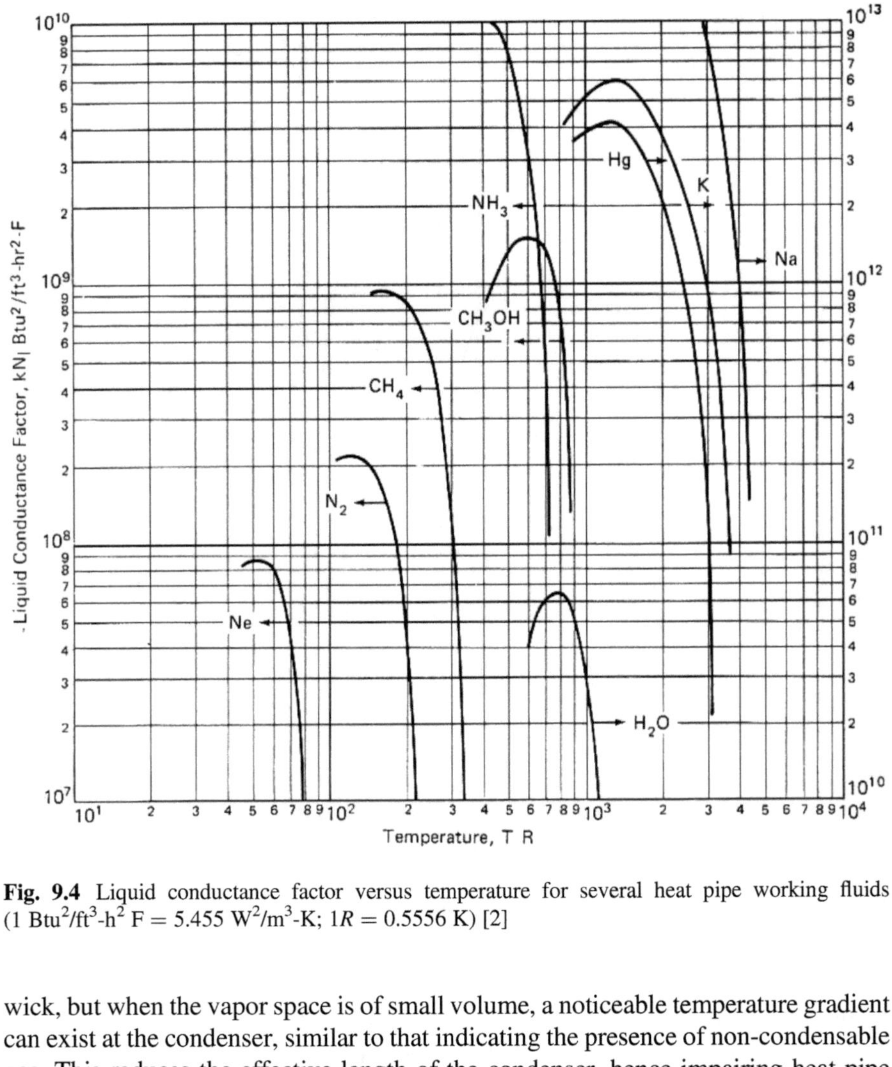

Fig. 9.4 Liquid conductance factor versus temperature for several heat pipe working fluids (1 Btu^2/ft^3-h^2 F = 5.455 W^2/m^3-K; 1R = 0.5556 K) [2]

wick, but when the vapor space is of small volume, a noticeable temperature gradient can exist at the condenser, similar to that indicating the presence of non-condensable gas. This reduces the effective length of the condenser, hence impairing heat pipe performance. Another drawback of excess fluid is peculiar to heat pipes in space, where in zero gravity the fluid can move about the vapor space, affecting the dynamics of the spacecraft. If there is a deficiency of working fluid, the heat pipe may fail because of the inability of any artery to fill. This is not as critical with homogeneous wicks as some of the pores will still be able to generate capillary action. Chi [3] discusses in detail these effects and the difficulties encountered in ensuring whether the correct amount of working fluid is injected into the heat pipe. One way of overcoming the problem is to provide an excess fluid reservoir, which behaves as a sponge, absorbing working fluid that is not required by the primary wick structure.

Fig. 9.5 Flow sheet for heat pipe design

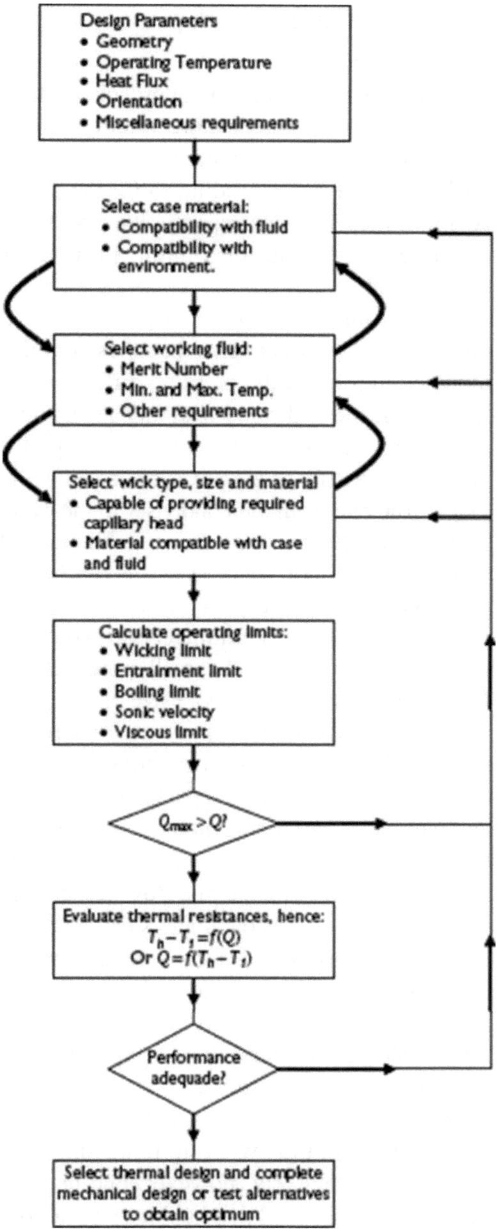

For the successful operation of heat pipe, its working fluid must be in liquid state, and selected fluid for the heat pipe must have a melting or evaporation point temperature *below* and its critical point temperature *above* the heat pipe operating temperature.

9.1.2 Priming Heat Pipe

The artery of a heat pipe includes a closed tube mounted at the evaporator end of the artery and having one or more venting pores located in a thin-walled section of the tube. The wall section containing the venting pore is so thin as to cause meniscus coalescence of liquid tending to fill the pore when it borders a gas bubble. By the action of meniscus coalescence, the pore remains open to vent any gas bubbles that are present within the artery during priming thereof. The wick must satisfy two main requirements.

First, it must be capable of generating a high capillary pressure. The capillary pressure of a wick is a term that is used to define the ability of the pore structure of the wick to tolerate the pressure difference between the vapor and the liquid in the wick that it can withstand without causing the liquid menisci in the pores to collapse.

Second the wick must have a low liquid flow resistance. The liquid flow resistance is a measure of the pressure drop that the liquid experiences when flowing at a given flow rate from the condenser to the evaporator. With increasing heat input, the liquid flow rate increases so as to increase the pressure drop in the liquid. The thermal capacity of the heat pipe is reached when the pressure drop is so great that the maximum capillary pressure of the wick pores in the evaporator region is exceeded and the wick ultimately dries out. The wick can be designed and fabricated in a tight structure with small pore size to achieve a high capillary pressure. However, the smaller the pore size of the wick, the greater will be its flow resistance. To overcome the above conflicting wick requirements of high capillary pressure and low liquid flow resistance, designers of heat pipes have restored to the use of arteries to handle very high heat loads. Generally, an artery is a closed tube filled with liquid and having at least a portion of its wall structure porous and in communication with the wick. In the artery, the maximum capillary pressure is determined by the pore size of the wall, whereas the flow resistance is determined by the diameter of the tube. Thus, these two parameters can be adjusted independently [1].

Although arteries can provide an order of magnitude increase in heat pipe capacity, they suffer a serious disadvantage in that they are extremely difficult to *prime* and *re-prime* reliably without entrapment of a bubble. An arterial bubble is intolerable because far before maximum capacity is reached, the bubble grows and the artery empties of liquid.

Bubbles present problems only if the heat pipe contains none-condensable gas as well as the heat pipe working fluid, since a pure vapor bubble will spontaneously collapse. It is unlikely, however, that heat pipe fluid can be maintained so free of none-condensable gas that bubbles will not be a problem. Also, there is an important class of heat pipes that are intentionally filled with some none-condensable gas for purpose of heat pipe control. The reason an arterial bubble is formed is that a liquid sheath on the artery wall prevents gas from venting during the priming process which usually proceeds from the condenser to the evaporator. An artery can de-prime when vapor bubbles become trapped in it. It may be necessary to reduce the heat load in such circumstances, to enable the artery to re-prime.

If some form of arterial wick is built into heat pipe, it is necessary to ensure that an artery becomes depleted of working fluid, and it should be able to refill automatically. It is possible to calculate the maximum diameter of an artery to ensure that it will be able to *re-prime*. The maximum priming height that can be achieved by a capillary is given by the equation [1]

$$h + h_c = \frac{\sigma_l \cos\theta}{(\rho_l - \rho_v)} \times \left(\frac{1}{r_{p1}} + \frac{1}{r_{p2}}\right) \tag{9.4}$$

where

h = The vertical height to the base of the artery
h_c = The vertical height to the top of the artery
ρ_l = Density of liquid
ρ_v = Density of vapor
σ_l = Liquid surface tension
r_{p1} = The first principal radius of curvature of the priming meniscus
r_{p2} = The second principal radius of curvature of the priming meniscus
θ = Contact angle

Per Reay and Kew [2], for the purpose of priming, the second principal radius of curvature of the meniscus is extremely large ($\approx 1 \ \sin\phi$). For a cylindrical artery,

$$h_c = d_a$$

and

$$r_{p1} = \frac{d_a}{2}$$

where d_a is the artery diameter.

Hence 9.4 can be rearranged and becomes as follows:

$$h + d_a = \frac{2\sigma_l \cos\theta}{(\rho_l - \rho_v)g \times d_a} \tag{9.5}$$

which produces a quadratic in d_a that may be solved, and 9.6 can be obtained:

$$d_a = \frac{1}{2}\left[\left(\sqrt{h^2 + \frac{8\sigma_l \cos\theta}{(\rho_l - \rho_v)}}\right) - h\right] \tag{9.6}$$

Several patents have been filed by various designers in regard to assembly and structure of different artery wicks such as helically grooved, trapezoidal grooves, and sinusoidal grooves, and artery, disposed within the tube, conducts condensed liquid by capillary action from the condenser to the evaporator and is provided with a channel formed on its upper surface. See the picture below (Fig. 9.6).

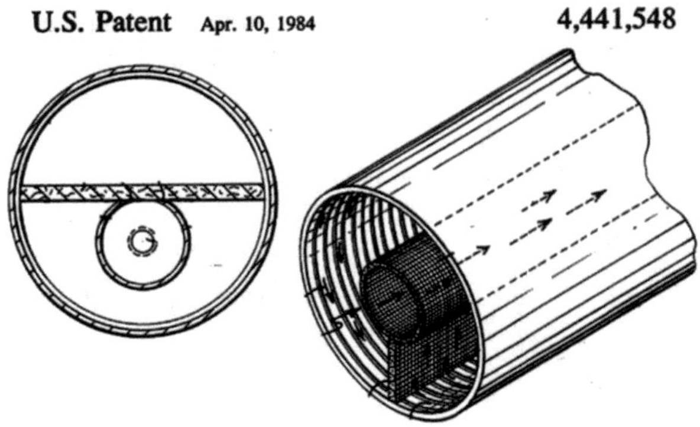

Fig. 9.6 Partial perspective schematic view of a conventional heat pipe having an arterial US patent 4,441,548 by Franklin et al.

9.2 How to Select a Heat Pipe

1. Investigate and determine the following operational parameters:
   (a) Heat load and geometry of the heat source
   (b) Possible heat sink location, the distance, and orientation relative to the heat source
   (c) Temperature profile of heat source, heat sink, and ambient
   (d) Environmental condition (such as existence of corrosive gas)
2. Select the pipe material, wick structure, and working fluid. (Consult with an expert engineer or original heat pipe manufacturer to select the most appropriate heat pipe designed by them. In Chap. 1 list of these manufactures is provided).
   (a) Determine the working fluid appropriate for your application.
   (b) Select pipe material compatible to the working fluid.
   (c) Select wick structure for the operating orientation.
   (d) Decide on the protective coating.
   (e) Determine the length, size, and shape of the heat pipe.

Figure 9.7 gives the performance of heat pipes with diameter from 3 to 22.23 mm. In this case the selection should be within this given range for diameter of chosen heat pipe. Note that the constraint in this selection is based on vertical orientation of operation environment for the heat pipe and plots are depicted for a copper water artery-grooved wick heat pipe. Similar plots can be obtained for different types of working fluid and wick selections.

The rate of vapor traveling from the evaporator to the condenser is governed by the difference in vapor pressure between them. It is also affected by the diameter and the length of the heat pipe. In the large-diameter heat pipe, the cross-sectional area will allow higher vapor volume to be transported from the evaporator to the condenser than in a small-diameter pipe. The cross-sectional area of a heat pipe is the direct function for both the sonic limit and entrainment limit. Figure 3.11 compares the heat transport of heat pipes with different diameters. Also, the operational temperature of a heat pipe affects the sonic limit. We can see, in Fig. 9.8, the heat pipes transport more heat at higher operational temperatures.

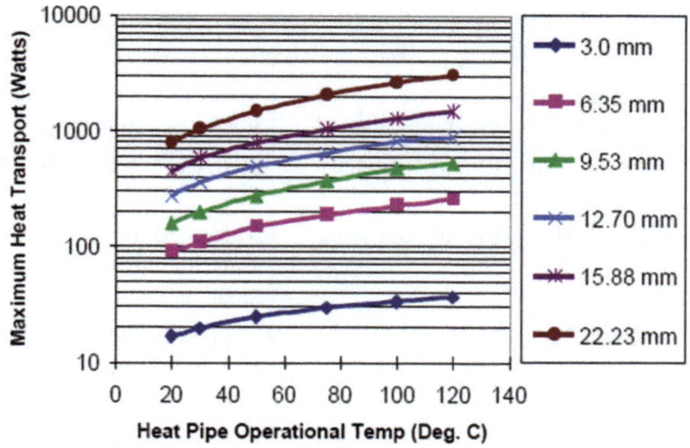

Fig. 9.7 Performance of copper water groove heat pipe at vertical orientation (gravity assist) (Enertron Corporation)

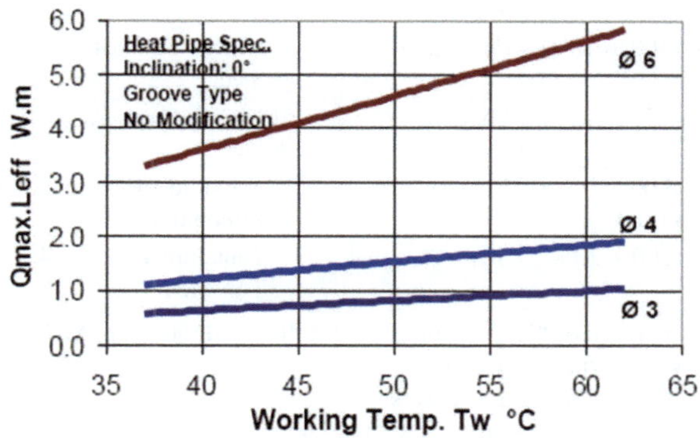

Fig. 9.8 The performance of various groove wick copper water heat pipes

The rate of working fluid return from the condenser to the evaporator is governed by capillary limit and is the reciprocal function of the heat pipe length. A longer heat pipe transports less heat versus the same heat pipe with a shorter length. In Fig. 9.8, the unit of the *Y*-axis is $Q_{max}L_{effective}$ (W-m) representing the amount of heat a pipe can carry per meter length. If the pipe is half a meter, it can carry twice the wattage.

As it can be seen, the selection of an appropriate heat pipe can be a complicated process.

9.3 What Materials Can Be Used to Construct a Heat Pipe?

A particular working fluid can only be functional at certain temperature ranges. Also, the particular working fluid needs a compatible vessel material to prevent corrosion or chemical reaction between the fluid and the vessel. Corrosion will damage the vessel and chemical reaction can produce a non-condensable gas. Refer to Table 9.1. For example, the liquid ammonia heat pipe has a temperature range from −70 to +60 °C and is compatible with aluminum, nickel, and stainless steel.

Table 9.1 Typical operating characteristics of heat pipes

Temperature range (°C)	Working fluid	Vessel material	Measured axial[a] heat flux (kW/cm^2)	Measured surface[a] heat flux (W/cm^2)
−200 to −80	Liquid Nitrogen	Stainless Steel	0.067 @ −163 °C	1.01 @ −163 °C
−70 to +60	Liquid Ammonia	Nickel, Aluminum, Stainless Steel	0.295	2.95
−45 to +120	Methanol	Copper, Nickel, Stainless Steel	0.45 @ 100 °C[b]	75.5 @ 100 °C
+5 to +230	Water	Copper, Nickel	0.67 @ 200 °C	146@ 170 °C
+190 to +550	Mercury[c] +0.02% Magnesium +0.001%	Stainless Steel	25.1 @ 360 °C[d]	131 @ 750°C
+400 to +800	Potassium[c]	Nickel, Stainless Steel	5.6 @ 750 °C	181 @ 750 °C
+500 to +900	Sodium[c]	Nickel, Stainless Steel	9.3 @ 850 °C	224 @ 760 °C
+900 to +1500	Lithium[c]	Niobium +1% Zirconium	2.0 @ 1250 °C	207 @ 1250 °C
1500 + 2000	Silver[c]	Tantalum +5% Tungsten	4.1	413

Reference of "Heat Transfer", 5th Edition, JP Holman, McGraw-Hill
[a]Varies with temperature
[b]Using threaded artery wick
[c]Tested at Los Alamos Scientific Laboratory
[d]Measured value based on reaching the sonic limit of mercury in the heat pipe

The liquid ammonia heat pipe has been widely used in space and only aluminum vessels are used due to lightweight. Water heat pipes, with a temperature range from 5 to 230 °C, are most effective for electronics cooling applications, and copper vessels are compatible with water.

Heat pipes are not functional when the temperature of the pipe is lower than the freezing point of the working fluid. Freezing and thawing is a design issue, which may destroy the sealed joint of a heat pipe when place vertically. Proper engineering and design can overcome this limitation.

9.4 When to Consider a Heat Pipe

Heat pipes are classified into two general types—"fixed conductance" and "variable conductance." A fixed conductance heat pipe is a device of very high thermal conductance with no fixed operating temperature. Its temperature rises or falls according to variations in the heat source or heat sink.

9.5 Things to Consider When Designing a Heat Pipe

As an effective heat conductor, heat pipe can be used in situations when a heat source and a heat sink need to be placed apart, to aid heat conduction of a solid, or to aid heat spreading of a plane. However, not every heat pipe is suitable for all applications. For that reason, the following need to be considered when designing with heat pipes:

1. Heat transport limitation of the heat pipe
2. Wick structure of the heat pipe
3. Length and diameter of the heat pipe
4. Heat pipe orientation
5. Effect of bending and flattening of the heat pipe
6. Heat pipe reliability

9.5.1 What are the Four Heat Transport Limitations of a Heat Pipe?

A heat pipe is a hermetically sealed evacuated tube normally containing a mesh or sintered powder wick and a working fluid in both the liquid and vapor phase. When one end of the tube is heated, the liquid turns to vapor absorbing the latent heat of vaporization. The hot vapor flows to the colder end of the tube where it condenses and gives out the latent heat. The recondensed liquid then flows back through the wick to the hot end of the tube. Since the latent heat of evaporation is usually very large, considerable quantities of heat can be transported with a very small temperature difference from one end to the other.

The vapor pressure drop between the evaporator and the condenser is very small; and, therefore, the boiling-condensing cycle is essentially an isothermal process. Furthermore, the temperature losses between the heat source and the vapor and between the vapor and the heat sink can be made small by proper design. Therefore, one feature of the heat pipe is that it can be designed to transport heat between the heat source and the heat sink with very small temperature losses.

The amount of heat that can be transported as latent heat of vaporization is usually several orders of magnitude larger than can be transported as sensible heat in a conventional convective system with an equivalent temperature difference. Therefore, a second feature of the heat pipe is that relatively large amounts of heat can be transported with small lightweight structures. The performance of a heat pipe is often expressed in terms of equivalent thermal conductivity. The huge effective thermal conductivity of the heat pipes can be illustrated by the following examples. A tubular heat pipe using water as the working fluid and operated at 150 °C would have a thermal conductivity several hundred times that of a copper bar of the same dimensions.

A heat pipe using lithium as the working fluid at a temperature of 1500 °C will carry an axial heat flux of 10–20 kW/cm^2. By suitable choice of working fluid and container materials, it is possible to manufacture heat pipes for use at temperatures ranging from −269 °C to in excess of 2300 °C.

The four heat transport limitations can be simplified as follows:

(a) ***Sonic limit***—The rate that vapor travels from evaporator to condenser.
(b) ***Entrainment limit***—Friction between working fluid and vapor that travel in opposite directions.
(c) ***Capillary limit***—The rate at which the working fluid travels from condenser to evaporator through the wick.
(d) ***Boiling limit***—The rate at which the working fluid vaporizes from the added heat to condenser side.

9.5.2 Heat Pipe Diameter

For this purpose, the round tubes and pipes of different materials are readily available, and they are the most efficient configuration from the stress point of view. The size of the pipe diameter necessary for a given application requirement should be analyzed so that vapor velocity is not excessive. Control of vapor velocity is required since at high *Mach* numbers, the flow compressibility of vapor contributes to a large axial temperature gradient. For that matter, the heat pipe can be designed so that its maximum *Mach* number in the vapor flow passage does not exceed value of 0.2. Given that value for consideration of first step of the design, the vapor can be considered incompressible which is the common theory approaches and heat pipe operating condition and available computer codes. Under this condition the axial temperature gradient is negligibly small as well.

For a heat pipe operation under this constraint whose heat transport mode requirements and consequently maximum axial heat flux $Q_{\max}$ are known and the required vapor core diameter d_v at vapor Mach number M_v equal to 0.2 can be determined by utilizing 2.59 in Chap. 2 of Zohuri [1] and as a result we get the following relationship:

$$d_v = \left(\frac{20 Q_{\max}}{\pi \rho_v \lambda \sqrt{\gamma_v R_v T_v}} \right)^{1/2} \tag{9.7}$$

where

d_v = Vapor core diameter
$Q_{\max}$ = Maximum axial heat flux
ρ_v = Vapor density
λ = Latent heat of vaporization
γ_v = Vapor specific heat ratio
R_v = Gas constant for the vapor
T_v = Vapor temperature

9.5.3 Heat Pipe Containers Design

The American Society of Mechanical Engineering (AMSE) standard for unfired pressure vessels [4] specifies that the maximum allowable stress at any temperature be one-quarter of the material's ultimate strength f_{tu} at that temperature. Metal properties including ultimate tensile strength for several metals can be found in Appendix B.

Per Chi [3] for round tubes in which the wall thickness is less than 10% of the diameter, the maximum pressure stress is closely approximated by the simple expression

$$f_{\max} = \frac{Pd_o}{2t} \tag{9.8}$$

where

$f_{\max}$ = Maximum hoop stress in the wall
P = Pressure differential across the wall
d_o = Tube outside diameter
t = Tube wall thickness

The maximum hoop stress in a thick-walled cylinder, subject to internal pressure, is given by the expression

$$f_{\max} = \frac{P\left(d_o^2 + d_i^2\right)}{d_o^2 - d_i^2} \tag{9.9}$$

where

$f_{\max}$ = Maximum hoop stress in the wall
P = Pressure differential across the wall
d_o = Tube outside diameter
d_i = Tube inside diameter

Ends of the heat pipe container can be capped off either with hemispherical, conical, or flat end caps. The maximum stress in a thick-walled hemispherical end cap is

$$f_{\max} = \frac{P\left(d_o^3 + 2d_i^3\right)}{2\left(d_o^3 - d_i^3\right)} \tag{9.10}$$

If the wall thickness of the hemispherical end cap is less than 10% of its diameter, 9.10 can be approximated by:

$$f_{\max} = \frac{Pd_o}{4t} \tag{9.11}$$

The maximum stress in a flat circular end cap can be calculated by the

$$f_{\max} = \frac{Pd_o^2}{8t^2} \tag{9.12}$$

where

$f_{\max}$ = Maximum stress
P = Pressure differential across the cap
d_o = End cap diameter
t = End cap thickness

For design calculation, the internal pressure of the pipe is equal to the saturation vapor pressure of the pipe working fluid at its operating temperature or its maximum cycle pressure whichever is larger. The pressure differential is equal to the vapor pressure minus the ambient pressure. As the vapor pressure is usually much larger than the ambient pressure, the vapor pressure is, therefore, approximately equal to the pressure differential. Figure 9.9 contains vapor pressure versus temperature information for several fluids. The maximum allowable stress is equal to one-quarter of the ultimate tensile stress (UTS).

Appendix B of Zohuri [1] is providing UTS for different materials. Having the knowledge and information about heat pipe outside diameter of tube, which will be equal to the vapor diameter plus allowances for the wick and wall thickness, one can use 9.8 through 9.12 to calculate the wall thickness for the pipe container and both end caps.

Figure 9.10 represents design curves, which can be utilized to determine the required tube size very quickly when the heat pipe operation pressure and the material UTS are known. Figure 9.11 also represents similar design curves for analyzing the required thickness of the flat end caps.

Additionally, Table 9.2 contains the dimensional data for commercial tubes ranging from ¼ to 1 in. outside diameter.

9.5.4 Heat Pipe Material Selection

The important consideration for material selection of heat pipe is its wick and container compatibility with working fluids since the heat pipes are subject to continuous performance degradation as a result of chemical reaction or decomposition of the working fluid and corrosion or erosion of the container or wick. Chemical reaction or decomposition of the working fluid may give rise to non-condensable gas evolution. A specific example of this is the hydrolysis of water yielding hydrogen

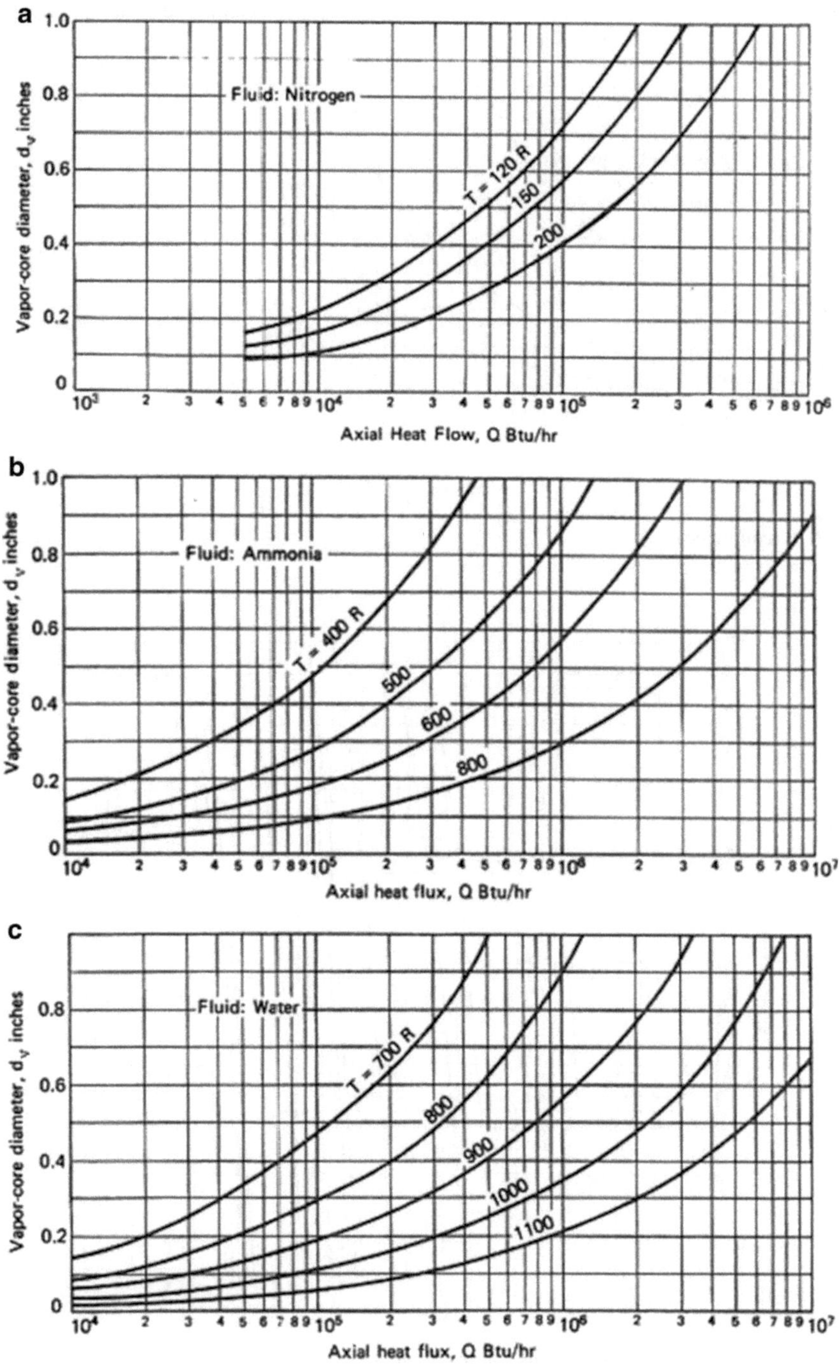

Fig. 9.9 (**a**) Vapor core diameter versus heat transfer rate for vapor Mach number of 0.2 for nitrogen [3] (Inch = 0.0254 m, 1 Btu/h = 0.2929 W, 1R = 0.5556 K). (**b**) Vapor core diameter versus heat transfer rate for vapor Mach number of 0.2 for ammonia [3]. (**c**) Vapor core diameter versus heat transfer rate for vapor Mach number of 0.2 for water [3]. (**d**) Vapor core diameter versus heat transfer rate for vapor Mach number of 0.2 for sodium [3] (Inch = 0.0254 m, 1 Btu/h = 0.2929 W, 1R = 0.5556 K)

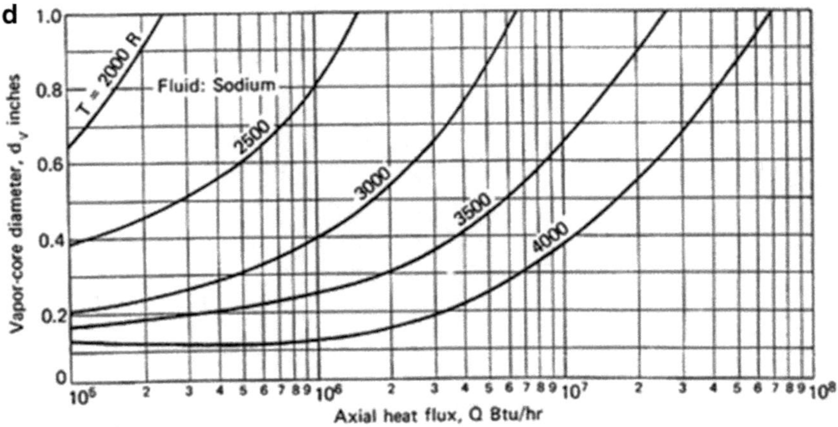

Fig. 9.9 (continued)

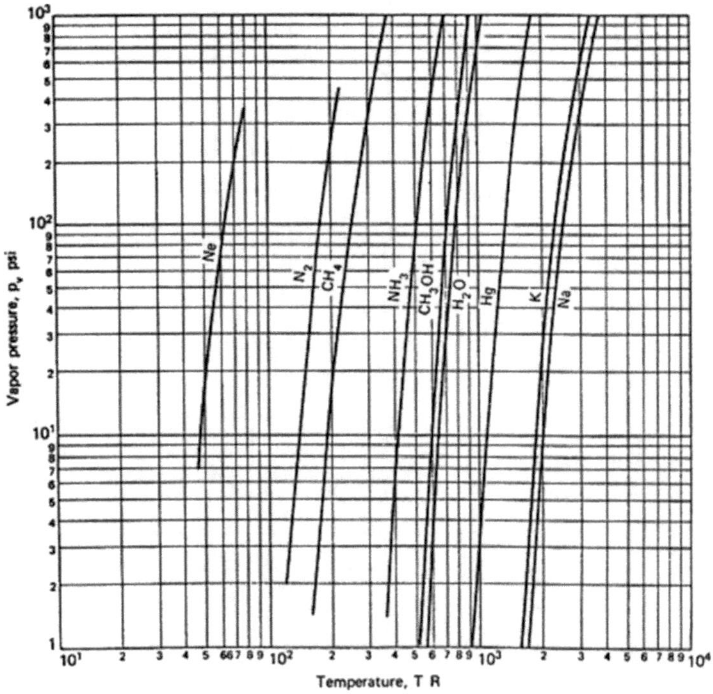

Fig. 9.10 Vapor pressure versus temperature for several heat pipe working fluids (1 psi = 6.895×10^3 N/m^2, 1*R* = 0.5556 K) [3]

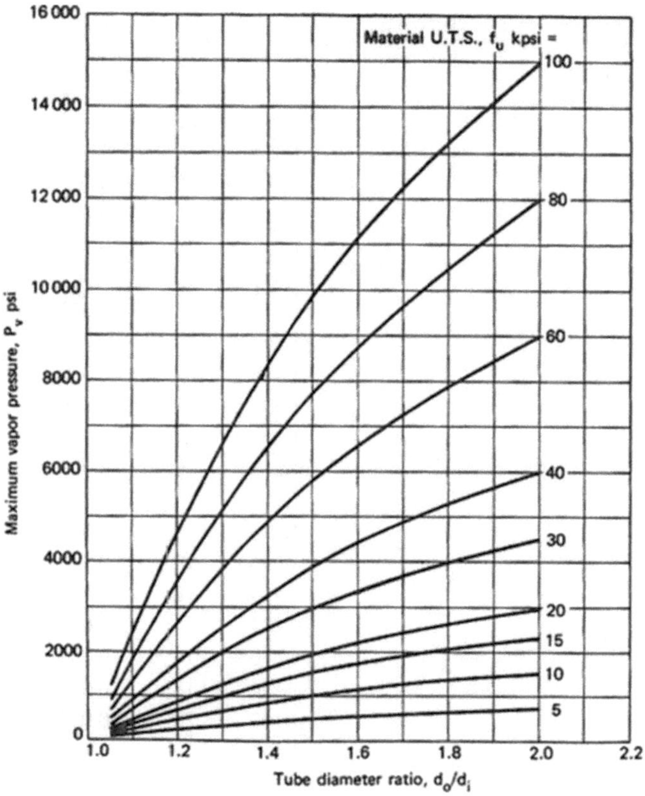

Fig. 9.11 Design chart for heat pipe containers tubes (1 psi = 6.895 × 10^3 N/M^2, 1 kpsi = 6.895 × 106 N/m^2) [3]

gas in a water-aluminum heat pipe. In a conventional heat pipe, all non-condensable gas is swept to the condenser end, thus, inactivating a portion of the condenser [3].

Corrosion and erosion of the container and wick may result in a change of the fluid wetting angle and the permeability or capillary pore size of the wick. As a result solid particles resulting from corrosion and erosion are transported by the flowing liquid to the evaporator region and deposited there. Table 9.3 is a compatibility compilation for fluid-metal combinations. It can be used for selection of wick and container materials.

Besides material compatibility, other factors, such as weight, temperature characteristic, and martial fabrication costs, may also be of importance.

Figure 9.12 shows the values of density divided by ultimate tensile stress versus temperature for several materials. For a minimum weight, the material with the smallest value of (ρ/f_u) should be chosen, where ρ is material density and f_u is ultimate tensile stress.

The temperature drop across the container wall is directly proportional to the wall thickness and inversely proportional to the thermal conductivity of the material.

Table 9.2 Data for sample commercial tubes (1 in. = 0.0245 m) [2]

Tube o.d., in.	o.d./i.d.	Bwg[a]	Thickness, in.	i.d., in.
$\frac{1}{4}$	1.289	22	0.028	0.194
	1.214	24	0.022	0.206
	1.168	26	0.018	0.214
$\frac{3}{8}$	1.354	18	0.049	0.277
	1.233	20	0.035	0.305
	1.176	22	0.028	0.319
	1.133	24	0.022	0.331
$\frac{1}{2}$	1.351	16	0.065	0.370
	1.244	18	0.049	0.402
	1.163	20	0.035	0.430
	1.126	22	0.028	0.444
$\frac{5}{8}$	1.536	12	0.109	0.407
	1.362	14	0.083	0.459
	1.263	16	0.065	0.495
	1.186	18	0.049	0.527
	1.126	20	0.035	0.555
$\frac{3}{4}$	1.556	10	0.134	0.482
	1.410	12	0.109	0.532
	1.284	14	0.083	0.584
	1.210	16	0.065	0.620
	1.150	18	0.049	0.652
	1.103	20	0.035	0.680
$\frac{7}{8}$	1.441	10	0.134	0.607
	1.332	12	0.109	0.657
	1.234	14	0.083	0.709
	1.174	16	0.065	0.745
	1.126	18	0.049	0.777
	1.087	20	0.035	0.805
1	1.493	8	0.165	0.670
	1.366	10	0.134	0.732
	1.279	12	0.109	0.782
	1.199	14	0.083	0.834
	1.149	16	0.065	0.870
	1.109	18	0.049	0.902
	1.075	20	0.035	0.930

[a]Birmingham wire gauge

Hence, for small temperature drop, the materials chosen must have a large product for thermal conductivity times the ultimate tensile strength (kf_u). Values of (kf_u) for several materials are plotted in Fig. 9.13 and 9.14.

Table 9.3 Fluid-solid compatibility

	Solids					
Fluids	Al	Cu	Fe	Ni	SS[a] 304	Ti
Nitrogen	C[b]	C	C	C	C	
Methane	C	C			C	
Ammonia	C		C	C	C	
Methanol	I	C	C	C	C	
Water	I	C		C	C[c]	C
Potassium				C		I
Sodium				C	C	I

[a]*SS* stainless steel
[b]*C* compatible; *I* incompatible; *blank* data not available
[c]Possible hydrogen generation

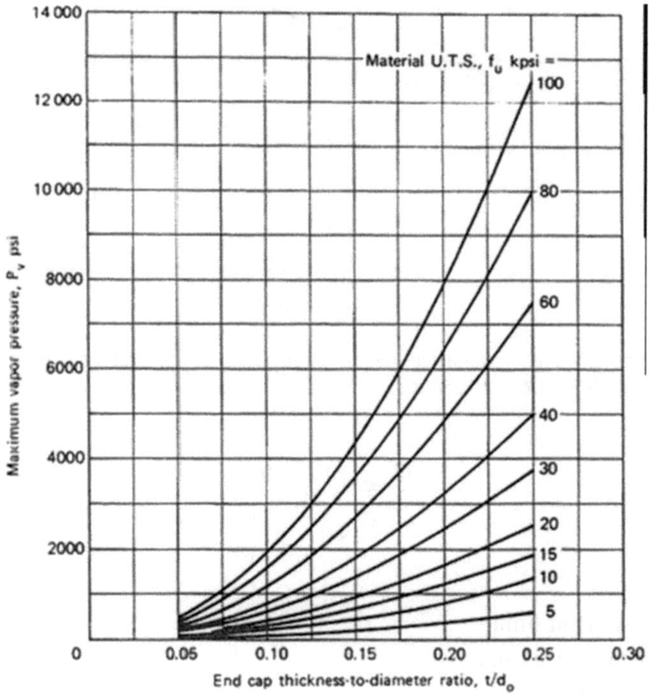

Fig. 9.12 End cap thickness-to-diameter ratio t/d_o

9.6 Entrainment and Boiling Limitations

Equation 2.34 of Chap. 2 (Zohuri [1]) was developed by Chi [3] to show the entrainment limit on the axial heat flux and can be shown as follows:

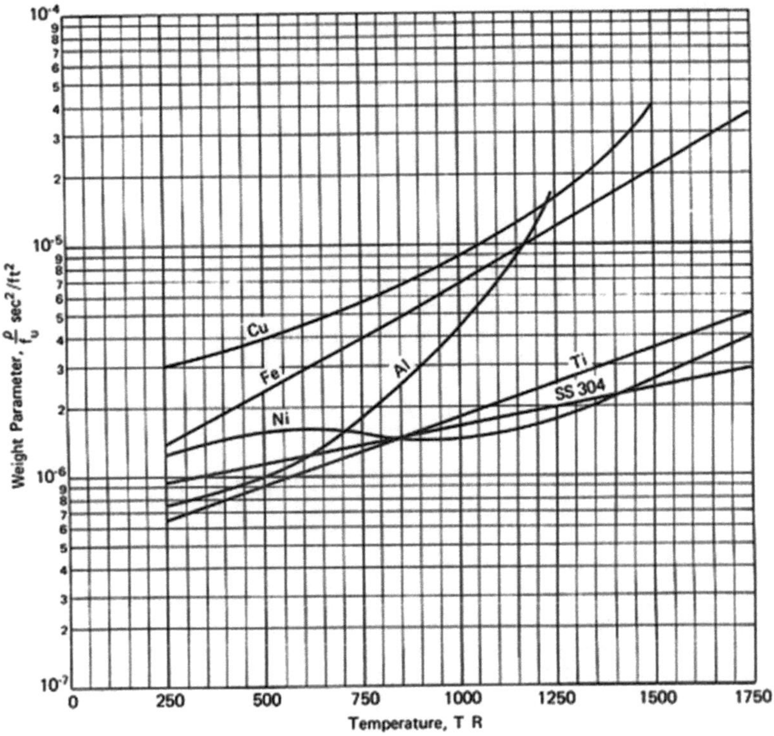

Fig. 9.13 Material weight parameter versus temperature for several heat pipe materials [2] (1 s^2/ft^2 = 10.76 s^2/m^2; 1R = 0.5556 K)

$$\frac{Q_{e_{\max}}}{A_v} = \lambda\sqrt{\frac{\sigma\rho_v}{2r_{h,s}}} \tag{9.13}$$

where

$Q_{e_{\max}}$ = Entrainment limit
λ = Latent heat of vaporization
ρ_v = Vapor density
$r_{h,s}$ = Hydraulic radius of the surface pores of the wick

The value of $r_{h,s}$ is equal to half of wire screen for any kind of wrapped-screen wick. Figure 9.15a through 9.15d from Chi [3] shows values for entrainment limit on the heat flux density, which is in alignment with 9.13 with the pipe fluid, screen mesh, and operating temperature of heat pipe as parameters. One can use this figure to easily read the value of $Q_{e_{\max}}/A_v$ for design of a heat pipe. This value should be larger than the actual value of Q/A_v at which the heat pipe is going to operate.

The boiling limits and its theory have been discussed in Sect. 2.8 of Chap. 2, and 2.90 (Zohuri [1]) has been developed and another version of it is written as follows:

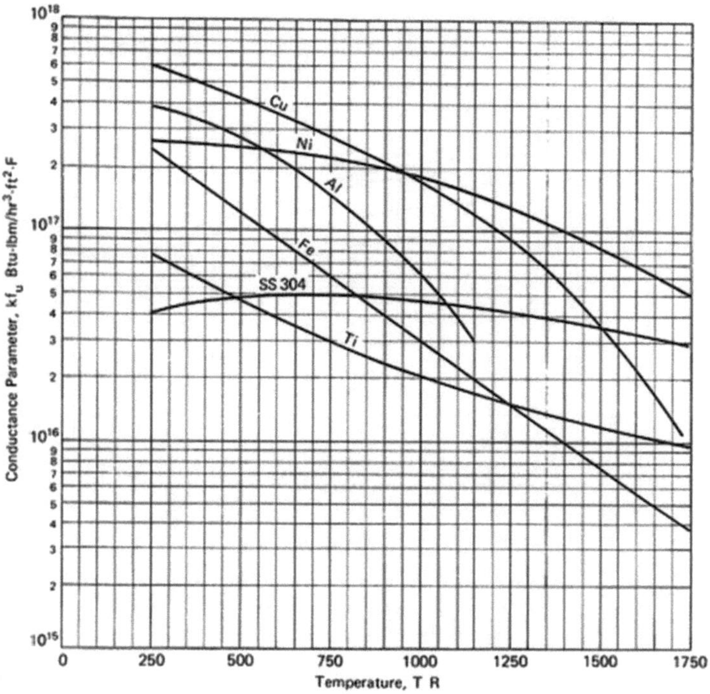

Fig. 9.14 Material conductance parameter versus temperature for several heat pipe materials [2] (1 Btu-lbm/h^3-ft^2-F = 1.986 × 10^{-7} W-kg/s^2-m^2-K; 1R = 0.5556 K)

$$\frac{Q_{\text{Boiling}_{\text{Maximum}}}}{L_{\text{e}}} = \frac{2\pi k_{\text{e}} T_{\text{v}}}{\lambda \rho_{\text{v}} \ln\left(\frac{r_{\text{i}}}{r_{\text{v}}}\right)} \left(\frac{2\sigma}{r_{\text{n}}} - P_{\text{c}}\right) \tag{9.14}$$

where

$\frac{Q_{\text{Boiling}_{\text{Maximum}}}}{L_{\text{e}}}$ = Boiling heat transfer limit per unit length of the evaporator
L_{e} = Effective thermal conductivity of the liquid-saturated wick
T_{v} = Vapor temperature
λ = Latent heat of vaporization
ρ_{v} = Vapor density
σ = Surface tension coefficient
P_{c} = Capillary pressure
r_{i} = Inside radius of pipe container
r_{v} = Vapor core radius of the pipe
r_{n} = Critical radius for nucleate boiling

For wick structure of wrapped screen, the value of k_{e} can be calculated by 9.15, namely,

a

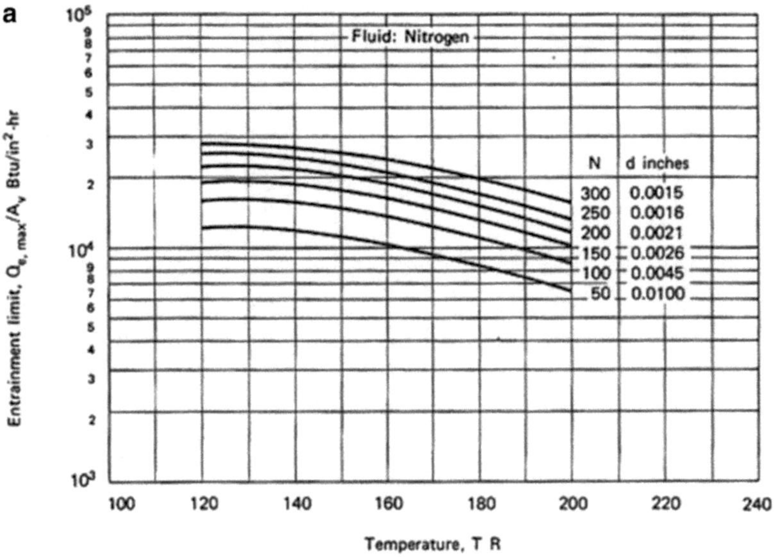

b

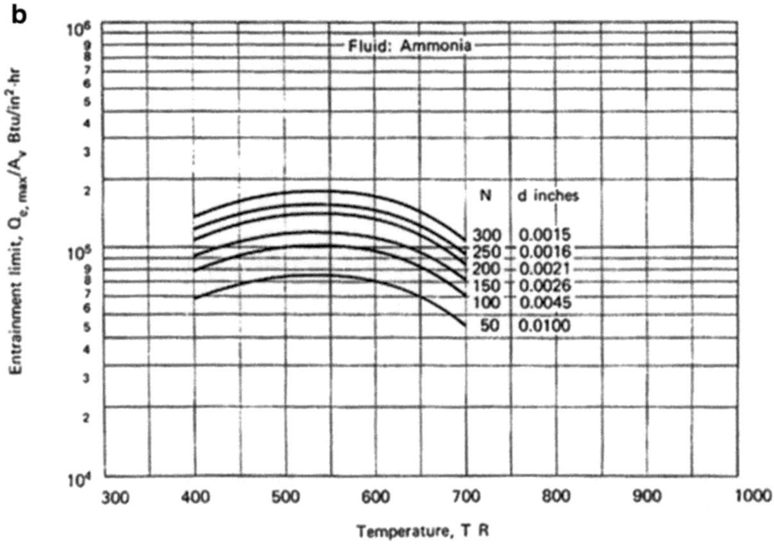

Fig. 9.15 (**a**) Entrainment limit on axial heat flux density for heat pipe with screen wicks for nitrogen as working fluid [9] (1 Btu/in.^{2}-h = 454 W/m^2, 1R = 0.5556 K, 1 in. = 0.0254 m). (**b**) Entrainment limit on axial heat flux density for heat pipe with screen wicks for ammonia as working fluid [9] (1 Btu/in.^{2}-h = 454 W/m^2, 1R = 0.5556 K, 1 in. = 0.0254 m). (**c**) Entrainment limit on axial heat flux density for heat pipe with screen wicks for water as working fluid [3] (1 Btu/in.^{2}-h = 454 W/m^2, 1 R = 0.5556 K, 1 in. = 0.0254 m). (**d**) Entrainment limit on axial heat flux density for heat pipe with screen wicks for sodium as working fluid [3] (1 Btu/in.^{2}-h = 454 W/m^2, 1R = 0.5556 K, 1 in. = 0.0254 m)

c

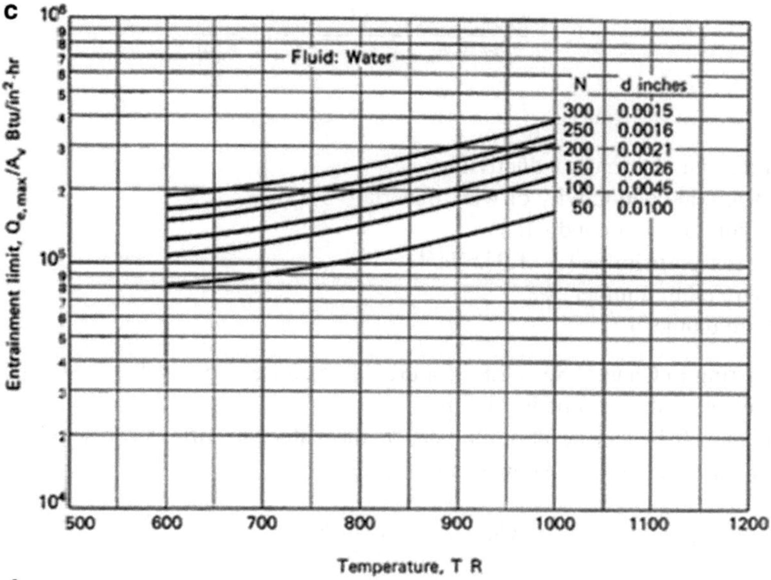

d

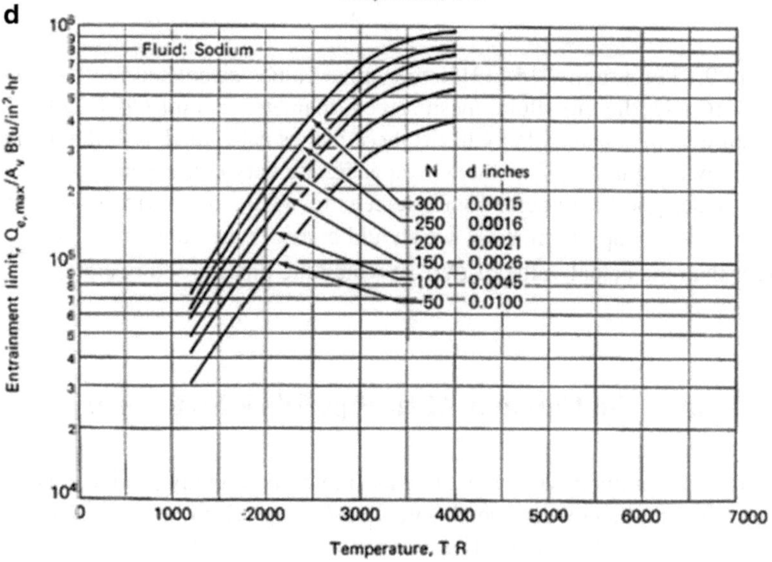

Fig. 9.15 (continued)

$$k_e = \frac{k_l[(k_l + k_w) - (1 - \varepsilon)(k_l - k_w)]}{[(k_l + k_w) + (1 - \varepsilon)(k_l - k_w)]} \tag{9.15}$$

where

k_e = Effective thermal conductivity
k_w = Thermal conductivity of wick material
k_l = Liquid thermal conductivity
ε = Screen porosity = $1 - 1.05\pi Nd/4$
N = Wire cloth number
d = Wire diameter

Referring to Chap. 2 Sect. 2.8 (Zohuri [1]), a conservative value for r_n in case of the conventional heat pipe is 10^{-5} in.

Equation 9.14 can be approximated if the value of $2\sigma/r_n$, in general, is much larger than P_c and can be shown as follows [2]:

$$\frac{Q_{\text{Boiling}_{\text{Max}}}}{L_e} = \frac{4\pi k_e T_v \sigma}{\lambda \rho_v r_n \ln\left(\frac{r_i}{r_v}\right)} \tag{9.16}$$

Figure 9.15a through 9.15d [3] demonstrate values calculated by the right-hand side of 9.16 with the pipe fluid, mesh screen, diameter ration (d_i/d_v), and operating temperature as parameters. Per Chi [3] these values calculated with the diameter of the screen wire equal to 2/3 of the wire spacing are very conservative values for the commonly designed commercial wire screens for wick, and normally for a conservative pipe design approximation, we should consider the value of $Q_{\text{Boiling}_{\text{Max}}}/L_e$ to be larger than the actual value of Q/L_e at which the pipes are to operate.

9.7 What Is the Common Heat Pipe Wick Structure?

There are four common wick structures used in commercially produced heat pipes: groove, wire mesh, powder metal, and fiber/spring. Each wick structure has its advantages and disadvantages. There is no perfect wick. Refer to Fig. 9.16 for a brief glance of actual test performance of four commercially produced wicks. Every wick structure has its own capillary limit. The groove heat pipe has the lowest capillary limit among the four but works best under gravity-assisted conditions where the condenser is located above the evaporator.

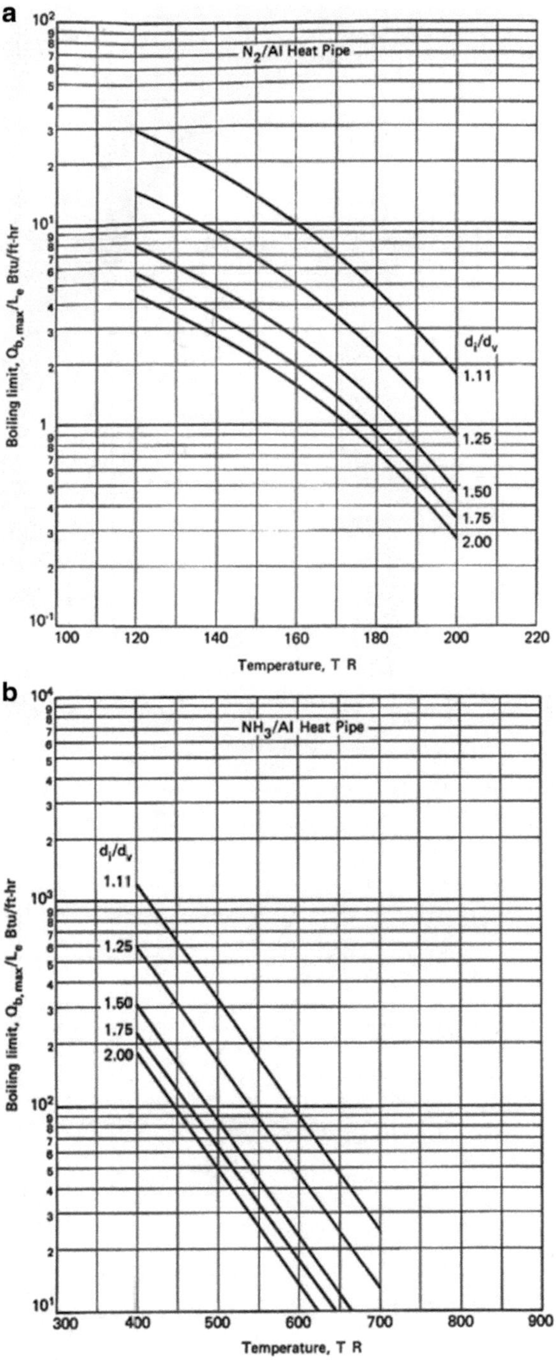

Fig. 9.16 (**a**) Boiling limit on heat flux per unit length of evaporator [3] (1 Btu/ft-h = 0.961 W/m, 1*R* = 0.5556 K); nitrogen. (**b**) Boiling limit on heat flux per unit length of evaporator [3] (1 Btu/ft-h = 0.961 W/m, 1*R* = 0.5556 K); ammonia. (**c**) Boiling limit on heat flux per unit length of evaporator [3] (1 Btu/ft-h = 0.961 W/m, 1*R* = 0.5556 K); water. (**d**) Boiling limit on heat flux per unit length of evaporator [3] (1 Btu/ft-h = 0.961 W/m, 1*R* = 0.5556 K); sodium

c

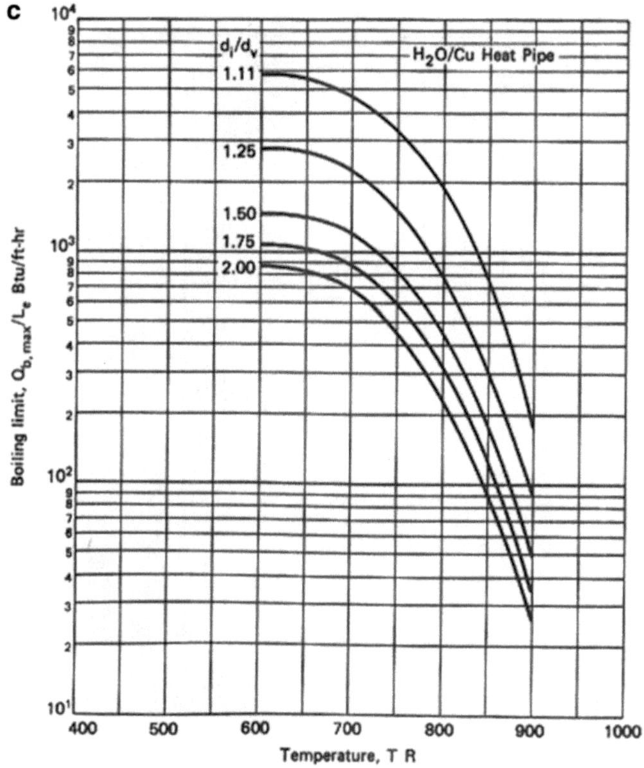

d

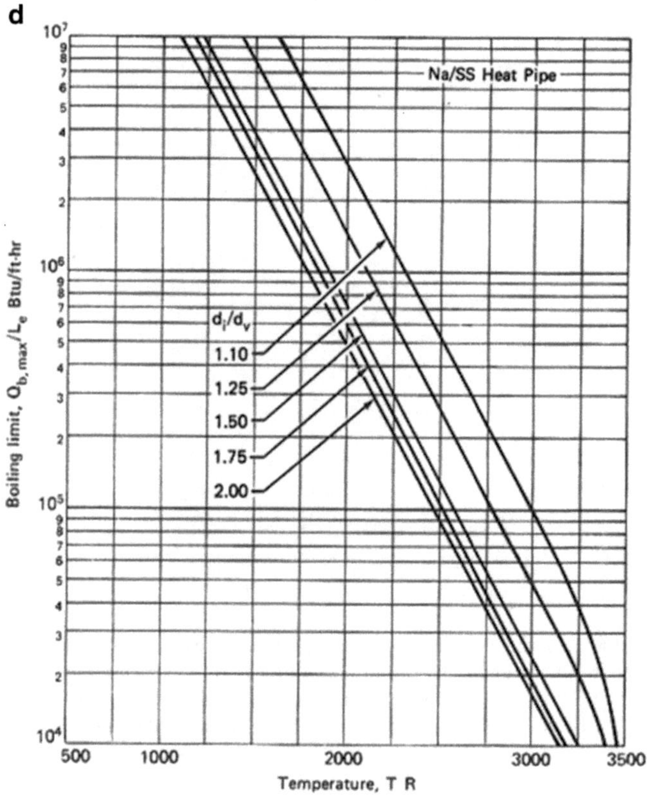

Fig. 9.16 (continued)

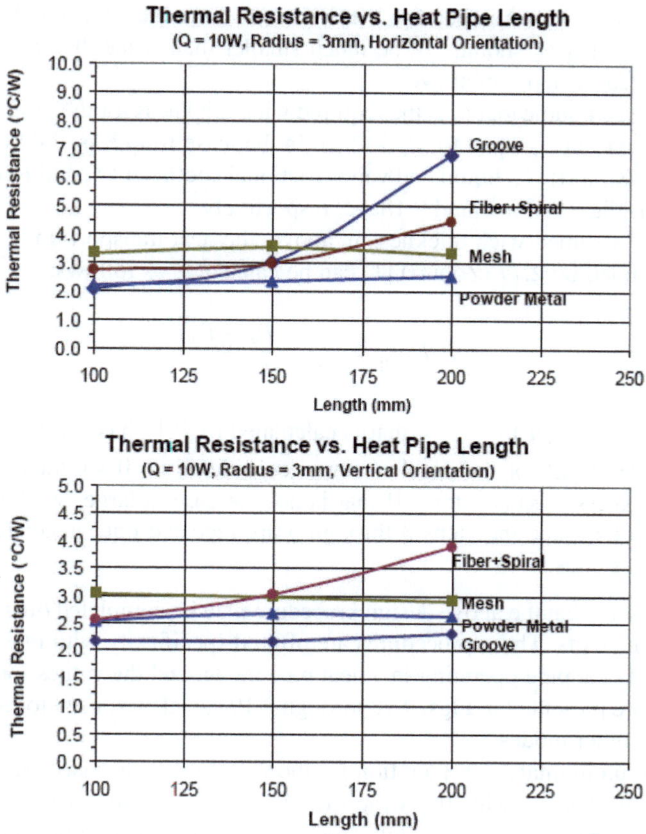

Fig. 9.17 The actual test results of heat pipe with different wick structures at horizontal and vertical (gravity assist) orientations (Enertron Corporation)

9.7.1 Wick Design

The general procedure for the design of the wrapped-screen wick for a heat pipe operating in the normal and steady-state heat pipe mode is that one can use Figs. 9.17 through 9.21 in order quickly to determine the wick dimensions using the following rule of thumbs:

1. Given the knowledge of inside pipe diameter d_i and total length L_t as well as tilting angle of ψ, the gravitational hydrostatic pressure can be calculated using the following equation, where ρ_l is the liquid density and g is the gravitation force and value of $\rho_l g$ can be read from Fig. 9.16:

$$P_g = \rho_l g(d_i \cos\psi + L_t \sin\psi) \tag{9.17}$$

2. The required mesh number for the wick can be selected using Fig. 9.15a–d considering that P_c should not be much smaller than twice the value of P_g that is calculated in step 1 in above.
3. Now we can assume that t_w is the required wick thickness for the pipe and a vapor core diameter of d_v equal to $(d_v - 2t_w)$. In that case from Eqs. 2.44 (Zohuri) [1] and 2.56 (Zohuri) [1], liquid and vapor frictional coefficient F_l and F_v can then be read from Figs. 9.17a–d and 9.18a–c, respectively.
4. Using the assumed wick thickness in above, the heat transport limit for the pipe that is defined by 2.77 (Zohuri) [1] can be calculated as follows:

$$(QL)_{\text{Capillary}_{\text{Max}}} = \frac{P_c - P_g}{F_l + F_v} \tag{9.18}$$

5. Check to see if $(QL)_{\text{Capillary}_{\text{Max}}}$ that is calculated by 9.17 has a value greater than the required (QL) for the problem under consideration. If so, then the assumed wick thickness is satisfactory. If that is not the case, a larger wick thickness is needed to be used, and steps 3 through 5 are repeated until a satisfactory wick thickness is chosen.

It can be seen that the above design procedure may be completed quickly with the aid of design charts. These procedures are offered specifically with reference to the conventional heat pipe operating in a heat pipe mode, and the values for P_c, $\rho_l g$, F_l, and F_v that are presented in Figs. 9.18 through 9.19 can also be used for the heat pipe operating in other modes.

Note that preliminary consideration for these arguments is based on wick design of wire cloth element with the wrapped screen in this section (Fig. 9.22). The principal properties of the wire cloth important for heat pipe applications are first considered from a practical point of view. The wire cloth commonly that are used for heat pipe wicks has square meshes, and the mesh size is usually specified in mesh number, which is defined as the number of meshes per linear inch measured in a direction perpendicular to the wire. Although different wire sizes can be used for the woven wire cloth, in many cases the wire size is approximately equal to the wire spacing. Approximate characteristic of wick structures, as we discussed before, indicates that the liquid flow resistance in wick structures is inversely proportional to the square of the mesh size and the capillary pumping pressure is inversely proportional to the mesh size. However, wire cloth with mesh numbers from 50 to 300 are most commonly used wicks in practice. Practical experience indicates that for successful operation of heat pipes in a gravitational field, the maximum capillary pressure must at least be about twice that of the liquid hydrostatic pressure. Figure 9.18 contains data on maximum capillary pressure as calculated by 2.7 and what is shown in Table 2.1 (Zohuri [1]) for wire screen. Figure 9.18 contains data on the hydrostatic pressure per unit vertical elevation of various liquids under a standard gravitational field of 32.2 ft/s^2 (9.81 m/s^2). As shown by Zohuri [1] in Table 2.1 in case of parallel wire wick, the effective capillary radius r_c for screen wicks is expected to be equal to half of the spacing between the wires. However, due to

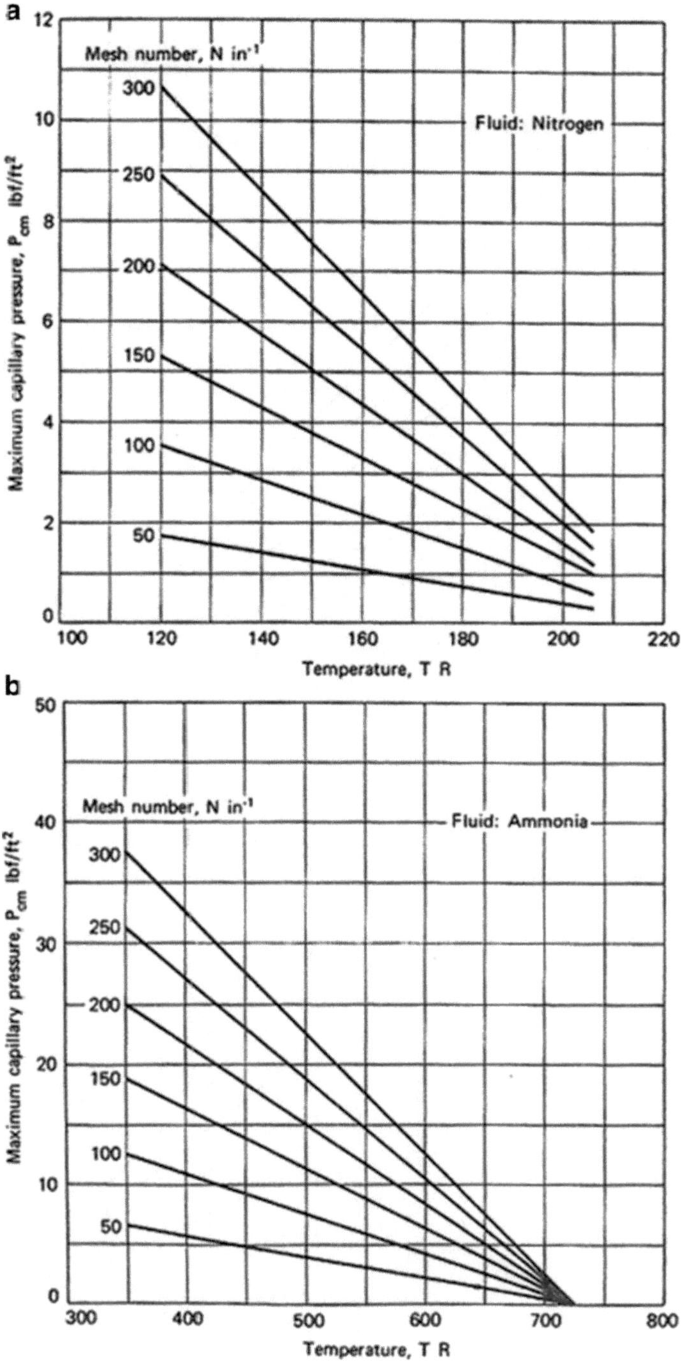

Fig. 9.18 (**a**) Maximum capillary pressure for screen wicks [3] (1 lbf/ft^2 = 47.88 N/m^2, 1R = 0.5556 K, 1 in.$^{-1}$ = 39.37 m^{-1}) nitrogen. (**b**) Maximum capillary pressure for screen wicks [3] (1 lbf/ft^2 = 47.88 N/m^2, 1R = 0.5556 K, 1 in.$^{-1}$ = 39.37 m^{-1}) ammonia. (**c**) Maximum capillary pressure for screen wicks [3] (1 lbf/ft^2 = 47.88 N/m^2, 1R = 0.5556 K, 1 in.$^{-1}$ = 39.37 m^{-1}) water

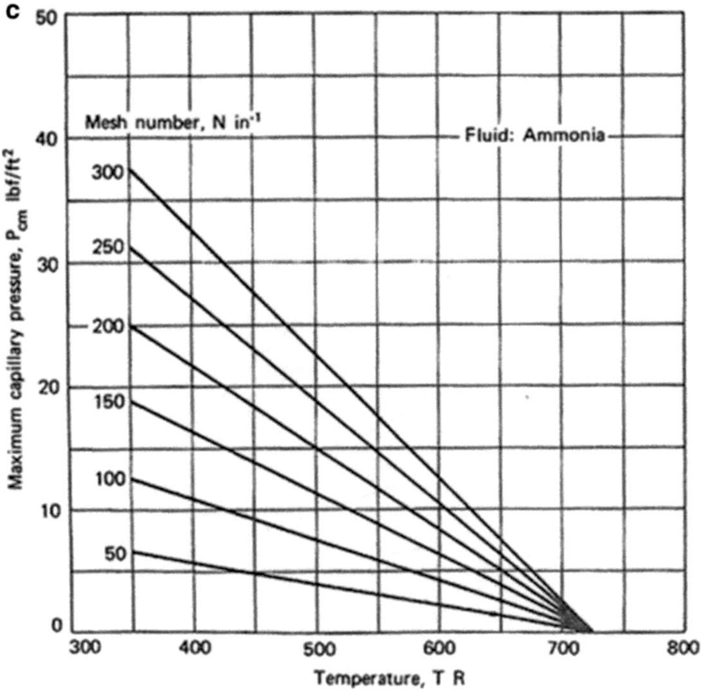

Fig. 9.18 (continued)

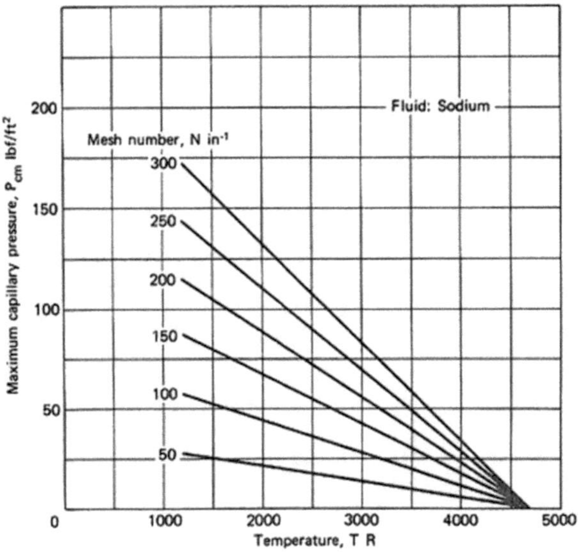

Fig. 9.19 Maximum capillary pressure for screen wicks [3] (1 lbf/ft^2 = 47.88 N/m^2, 1R = 0.5556 K, 1 in.$^{-1}$ = 39.37 m^{-1}) water

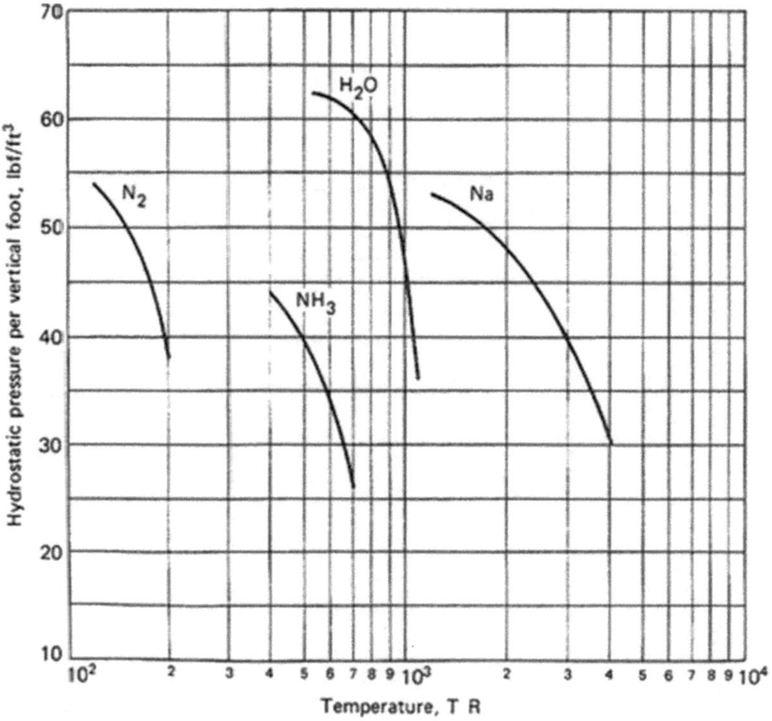

Fig. 9.20 Liquid gravitational hydrostatic pressure per unit vertical elevation versus temperature in a standard gravitational field [3] (1 lbf/ft^3 = 1.57.1 N/m^3, 1R = 0.5556 K)

staggering of the wires and the interference among adjacent layers of the screen, it has not so far been possible to determine the effective radius for screen wicks by a means of theory, but experimental data [1] from tests on single-layer screens seem to suggest that r_c is equal to half of the sum of the wire diameter d and the spacing w instead of being equal to only half of the spacing. For multilayer screens, no generalized data are yet available. It also appears that the effective radius for screen wicks can be calculated in a very conservative approach by the following equation:

$$r_c = \frac{d + w}{2} = \frac{1}{2N} \tag{9.19}$$

where N is the mesh number defined as number of wires per unit length [2].

9.8 Steady-State and Transient Regime

According to thermodynamic definition of steady state is a process which do not alter the state of a system but do change the surroundings. However, under steady-state conditions, inputs and outputs of the system remain in balance so that the

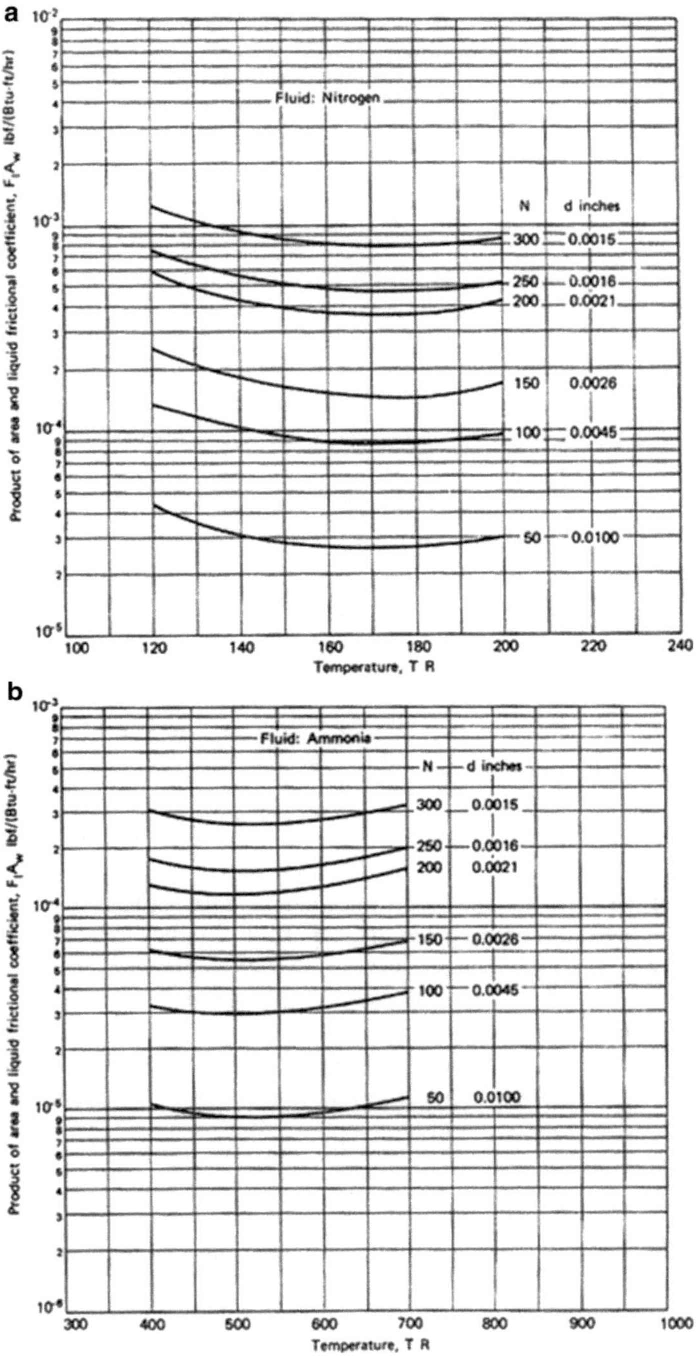

Fig. 9.21 (**a**) Liquid frictional coefficient versus temperature for liquid flow in wrapped-screen wicks [3] [1 lbf/(Btu-ft/h) = 49.82 N/W-m, 1R = 0.5556 K, 1 in. = 0.0254 m] nitrogen. (**b**) Liquid frictional coefficient versus temperature for liquid flow in wrapped-screen wicks [3] [1 lbf/(Btu-ft/h) = 49.82 N/W-m, 1R = 0.5556 K, 1 in. = 0.0254 m] ammonia. (**c**) Liquid frictional coefficient versus temperature for liquid flow in wrapped-screen wicks [3] [1 lbf/(Btu-ft/h) = 49.82 N/W-m, 1R = 0.5556 K, 1 in. = 0.0254 m] water. (**d**) Liquid frictional coefficient versus temperature for liquid flow in wrapped-screen wicks [3] [1 lbf/(Btu-ft/h) = 49.82 N/W-m, 1R = 0.5556 K, 1 in. = 0.0254 m] sodium

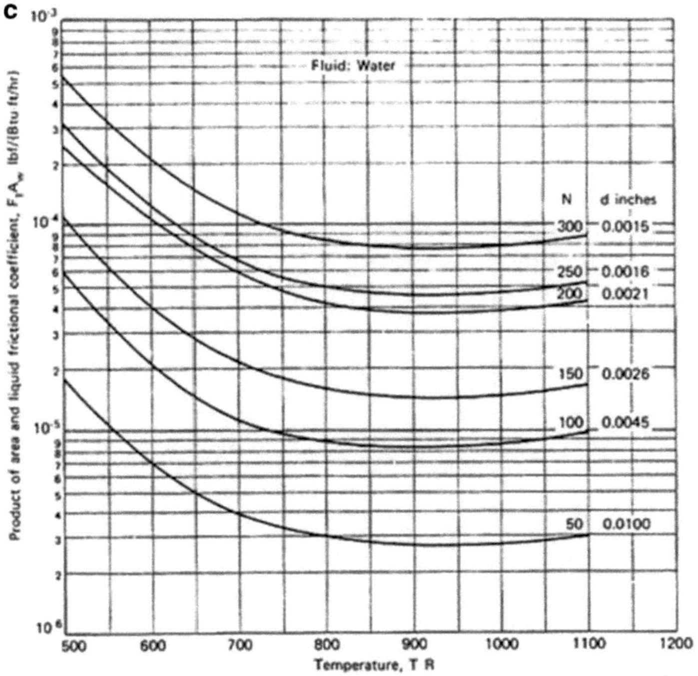

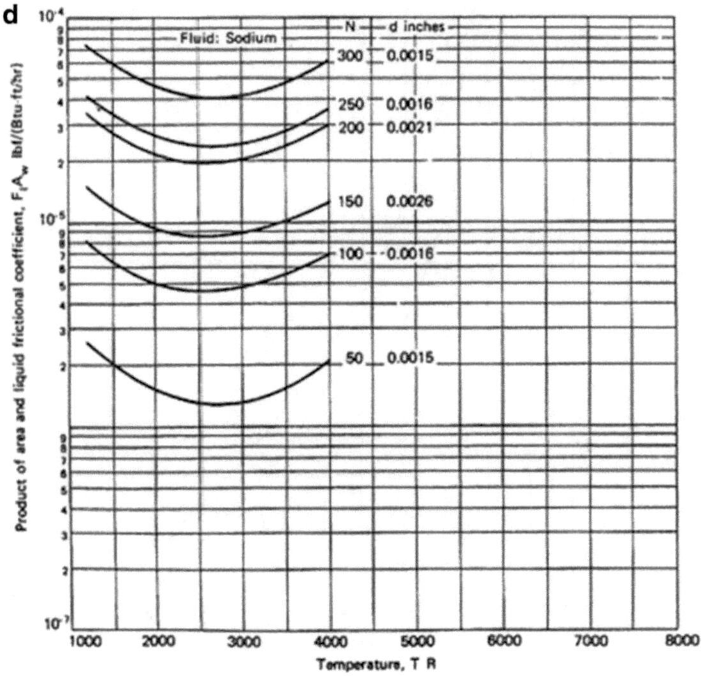

Fig. 9.21 (continued)

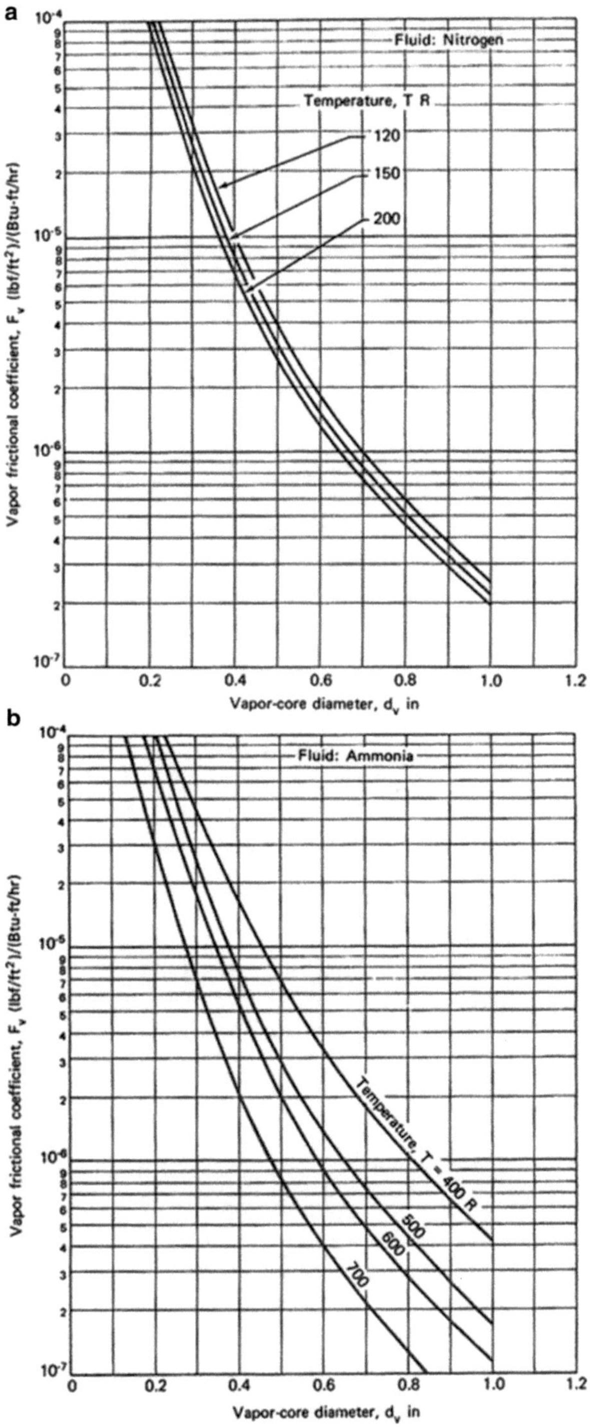

Fig. 9.22 (**a**) Vapor friction coefficient for vapor flow in round tubes [3] [1 (lb/ft^2)/(Btu-ft/h) = 536.3 (N/m^2)/W-m, 1 in. = 0.0254 m, 1R = 0.5556 K] nitrogen. (**b**) Vapor friction coefficient for vapor flow in round tubes [3] [1 (lb/ft^2)/(Btu-ft/h) = 536.3 (N/m^2)/W-m, 1 in. = 0.0254 m, 1R = 0.5556 K] ammonia. (**c**) Vapor friction coefficient for vapor flow in round tubes [3] [1 (lb/ft^2)/(Btu-ft/h) = 536.3 (N/m^2)/W-m, 1 in. = 0.0254 m, 1R = 0.5556 K] water. (**d**) Vapor friction coefficient for vapor flow in round tubes [3] [1 (lb/ft^2)/(Btu-ft/h) = 536.3 (N/m^2)/W-m, 1 in. = 0.0254 m, 1R = 0.5556 K] sodium

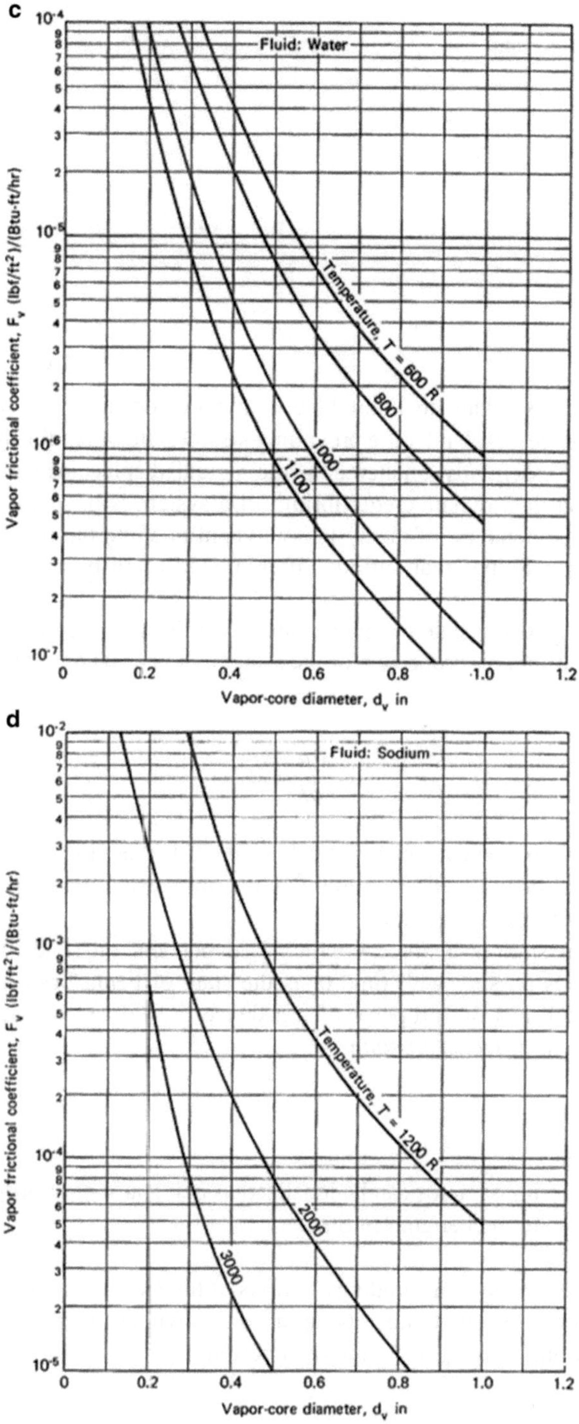

Fig. 9.22 (continued)

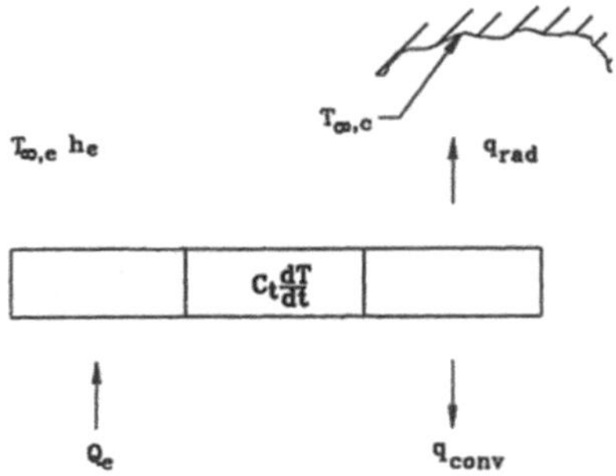

Fig. 9.23 Control volume for general lumped capacitance analysis[35]

properties of the system are not altered, but changes do occur in the surroundings as a result of such processes. Under the prevailing steady state or equilibrium conditions, system remains essentially unaltered in the virtual displacements and is a more general situation than dynamic equilibrium. If a system is in steady state, then the recently observed behavior of the system will continue into the future. In stochastic systems, the probabilities that various different states will be repeated will remain constant.

A system in a steady state has numerous properties that are unchanging in time. This implies that for any property p of the system, the partial derivative with respect to time is zero.

$$\frac{\partial p}{\partial t} = 0 \tag{9.20}$$

In many systems, steady state is not achieved until sometime has elapsed after the system is started or initiated. This initial situation is often identified as a transient state, start-up, or warm-up period.

Yet transient state is a process which is dependent on time; in other words, *transience* means passing with time or is the state of being brief and short-lived. Something which has the property of transience is said to be **transient** or often simply a *transient* or *transient state*.

Summary

1. Steady State
   Many engineering systems can be idealized as being at ***steady state***, meaning that *all* properties are unchanging in time.
2. Transient State
   Many engineering systems undergo periods of transient operation where the state changes with time, and this operation is known as ***transient state***. This is observed during start-up and shutdown periods.

9.9 Heat Pipe Steady State and Transient Analysis

Operation status of a heat pipe may either take place under steady-state or transient conditions. In steady-state mode, heat is removed from the heat pipe condenser (heat source) at same rate as heat is added to evaporator (heat sink), and heat pipe temperature remains constant.

In transient state, there is an imbalance between the added heat rate to evaporator and heat rejected from the condenser side of the heat pipe, and heat pipe temperature varies with time. This condition always takes place during heat pipe start-up from ambient (frozen state) temperature.

If the heat input rate is smaller than the heat absorption capability of the heat pipe, the start-up process will be slow and gradual. Under this condition then overheating and/or operational difficulty will not likely take place. On the other hand, start-up within heat pipe might be quite rapid if the input rate of heat is significant or much larger than the heat absorption capability. Potential start-up problems within heat pipe become a concern if the ambient temperature is below the freezing point of the heat pipe working liquid. Under this condition heat pipe does not become operational until the temperature of the evaporator section significantly is rising above this freezing point and consequently the vapor pressure inside the heat pipe is then extremely low.

When the heat pipe reaches its operational stage, its heat transport capability could initially be insufficient to prevent freeze out of condensate if there is good thermal coupling between the heat pipe condenser and the heat sink. Dry out of the evaporator wick, overheating, and possible destruction of the heat pipe could soon follow [5]. The heat pipe may fail to become in full operation even if freeze out of condensate did not occur when only a portion of condenser length is in the thermal coupling for dissipate of the incident heat load.

The heat pipe would then not necessarily overheat but could remain in a high non-isothermal state throughout start-up and into the steady-state operational phase. Under such circumstance, a *transient analysis* of the heat pipe start-up is advisable. The purpose of the transient analysis is threefold [28]:

1. To determine whether overheating or excessive thermal stress develops in the heat pipe wall
2. To determine whether heat pipe transport limits are exceeded
3. To determine whether start-up difficulty is encountered because of excessive thermal coupling between the heat pipe condenser and the heat sink

The above three objectives are achievable by calculating the time dependency of the temperature at key heat pipe locations, the actual heat transport rate, and the position of the continuum front heat that separates heated vapor from the condenser. The adequacy of the heat pipe design during start-up can then be established by comparing the temperatures and heat transport rate with temperature and heat transport limitations and by observing whether the continuum front has reached the rear of the condenser.

One should consider that the steady-state design is the starting point for the transient analysis. The steady-state design may be based on the heat load at the steady-state operating condition (if there exists one) or on the maximum expected heat load during the transient. If the maximum heat load occurs after the entire heat pipe has become functional, peak transient heat pipe temperature will probably not differ appreciably from steady-state temperatures based on the maximum heat load. If the peak heat load occurs before the entire heat pipe has become functional, predicted heat pipe temperatures based on the maximum heat load at steady-state conditions could be substantially overstated [28].

Generally speaking the heat pipe start-up process is very complex, which involves temperature variations in two- and three-dimensional space as well as over time vapor heat transport in the free molecular and continuum flow regimes, expansion of the heat pipe vapor into a hard vacuum accompanied by condensation, non-equilibrium expansion effects in a nominally saturated vapor, and solid-liquid as well as liquid-vapor phase changes. One approach to this problem is analyzed by Colwell, etc. [29].

Start-up is best accomplished by using a working fluid which is initially saturated. When this is not possible, in the case of many cryogenic or liquid metal heat pipes, the wick should be designed to give good transport during the priming operation. When a variable conductance heat pipe is required, the transient behavior will depend to a large extent on the type of variable cryogenic heat pipe (VCHP) employed and the choice of the working fluid.

Faghri 35 has provided a good mathematical modeling around the transient response to a sudden change in heat input or temperature. He presents the cases that, after a heat pipe reached a fully steady-state, it is often desirable particularly in case of variable heat pipe (VHP) to determine the period of time required to reach another steady state for a given increase in the heat increase at sink point or evaporator side of heat pipe. He describes three separate analyses and models which show the solutions for 1D and 2D heat equation in transient continuum model in cylindrical coordinate, and they are summarized here:

1. Transient lumped model
2. One-dimensional transient continuum model
3. Two-dimensional transient continuum model

Although 1D and 2D numerical models are generally more comprehensive and accurate but require considerable cup time and effort by the computer coding, the lumped analytical model, on the other hand, provides a quick and convenient approach for the heat pipe designers.

9.9.1 Transient Lumped Model

Presented by Faghri and Harley, [36] a transient lumped heat pipe model determines the average temperature as a function of time where they formulate a derivation from the general lumped capacitance analysis, which is an application of an energy balance over a control volume as shown in Fig. 9.23. The resulting energy equation for the case of both radiative and convective heat transfer from the condenser surface and an imposed heat input to the evaporator is shown below:

$$Q_{\mathrm{e}} - (q_{\mathrm{convective}} + q_{\mathrm{radiative}})S_{\mathrm{c}} = C_{\mathrm{t}}\frac{dT}{dt} \tag{9.21}$$

where

Q_{e} = The heat input
$q_{\mathrm{convective}}$ = The output heat flux by convection
$q_{\mathrm{radiative}}$ = The radiation heat flux
S_{c} = The surface area around the circumference of the cooled section
C_{t} = The total thermal capacity of the system
T = Uniformed temperature of the heat pipe
t = Time period

The total thermal capacity of the system is defined as follows:

$$C_{\mathrm{t}} = \rho V_{\mathrm{t}} c_{\mathrm{p}} \tag{9.22}$$

where

ρ = Density
V_{t} = Total volume
c_{p} = Specific heat of the system

In a heat pipe application, the total thermal capacity is defined as the sum of the heat capacities of the solid and liquid components of the heat pipe, and the heat capacity of the liquid-saturated wick is modeled accounting for both the liquid in the wick and the wick structure itself and represented by 9.23 as follows:

$$(\rho c_{\mathrm{p}})_{\mathrm{effective}} = \varphi(\rho c_{\mathrm{p}})_{\ell} + (1-\varphi)(\rho c_{\mathrm{p}})_{\mathrm{s}} \tag{9.23}$$

where φ is the porosity of the wick. For more details on this approach, reader should go through Faghri's book [6].

In Fig. 9.23, $T_{\infty,\mathrm{c}}$ is the ambient temperature surrounding the condenser, and $T_{\infty,\mathrm{e}}$ is the ambient temperature surrounding the evaporator side of the heat pipe, while h_{c} is the external convective coefficient for condenser and h_{e} is representing same value for evaporator side of the system (i.e., heat pipe).

9.9.2 One-Dimensional Transient Continuum Model

From mathematical analysis point of view, naturally any one-dimensional modeling is an easier process than a two dimensional to solve. Therefore, relative ease of any one-dimensional heat pipe modeling over the more complex two-dimensional models, it is important we take this approach first. Although either of these approaches (1D or 2D) is more involved than the lumped analysis method, it provides better understanding of axial variation of temperature along the heat pipe with time. The first presentation of a one-dimensional simulation of the transient vapor flow was performed by Jang et al. [7] to reduce the CPU time of computer required as compared to a two-dimensional vapor flow model, but it did include the thermal effects of wick and wall of heat pipe on its transient operation. Faghri [6] uses the conservation equations for vapor flow with negligible body forces that were formulated using the principles of conservation of mass and momentum in a control volume and presents the following equation applying the law of mass conservation over an elementary vapor control volume:

$$\frac{\partial \rho}{\partial t} A_{\mathrm{v}} \Delta z + \frac{\partial}{\partial z} (\rho \bar{w}) \Delta z A_{\mathrm{v}} - \rho_{\delta} v_{\delta} dS = 0 \tag{9.24}$$

where

$\bar{w}$ = Average axial vapor velocity
$A_{\mathrm{v}} = \pi R_{\mathrm{v}}^2$
$dS = 2\pi R_{\mathrm{v}} \Delta z$

Using the above definitions and dividing by $A_{\mathrm{v}} \Delta z$, the conservation of mass equation is

$$\frac{\partial \rho}{\partial t} + \frac{\partial}{\partial z} (\rho \bar{w}) = \frac{2 \rho_{\delta} v_{\delta}}{R_{\mathrm{v}}} \tag{9.25}$$

Faghri [6] analysis also continues to show that applying the law of conservation of momentum to the same control volume gives

$$\frac{\partial}{\partial t} (\rho \bar{w}) A_{\mathrm{v}} \Delta z + \frac{\partial}{\partial z} (\rho \bar{w}^2) A_{\mathrm{v}} \Delta z = \frac{\partial \sigma_{\mathrm{z}}}{\partial z} - \tau_{\mathrm{w}} dS \tag{9.26}$$

where the normal stress function is

$$\sigma_{\mathrm{z}} = -p + \frac{4}{3} \mu \frac{\partial \bar{w}}{\partial z} \tag{9.27}$$

Using the above definition for the normal stress function, A_{v}, dS and the friction coefficient and dividing by $A_{\mathrm{v}} \Delta z$ gives the conservation of axial momentum equation.

$$\frac{\partial}{\partial t}(\rho\bar{w})A_v\Delta z+\frac{\partial}{\partial z}(\rho\bar{w}^2)A_v\Delta z=-\frac{\partial}{\partial z}\left(p+\frac{4}{3}\mu\frac{\partial\bar{w}}{\partial z}\right)-\frac{f\rho\bar{w}^2}{R_v} \quad (9.28)$$

The friction coefficient f is a function of the axial Reynolds number of the vapor flow. Since the vapor flow remains laminar in evaporator, which allows the conventional pipe-flow correlations to be used.

$$f=\frac{16}{Re} \quad \text{for} \quad Re<2000 \quad (9.29)$$

If it turns out that the axial Reynolds number was greater than 2000, the flow at the condenser may be turbulent. The friction coefficient in this case can be induced from the following equation:

$$f=\frac{0.079}{Re_{0.25}} \quad \text{for} \quad Re\geq 2000 \quad (9.30)$$

One can use the ideal gas law below account for the compressibility status.

$$p=\rho R_g T \quad (9.31)$$

The conservation of energy equation in the vapor space is derived by assumption of a uniform temperature in the radial direction and negligible energy transfer from tangential and normal stresses (viscous shear and static pressure) [6]:

$$\frac{\partial T}{\partial t}=-\frac{\partial}{\partial z}(\bar{w}T)+\alpha\frac{\partial^2 T}{\partial z^2}+\frac{2v_\delta T_\delta}{R_v} \quad (9.32)$$

If the temperature is assumed to be uniform in the radial direction, then $T_\delta=T=T_v$. Therefore, the energy equation is given as follows:

$$\frac{\partial T_v}{\partial t}+\bar{w}\frac{\partial T_v}{\partial z}=+\alpha\frac{\partial^2 T_v}{\partial z^2}+\frac{2v_\delta T_v}{R_v} \quad (9.33)$$

In the evaporator, $v_\delta<0$ implies that heat is added to the vapor and conversely, in the condenser, $v_\delta>0$ then heat is removed.

It is this interfacial velocity term which couples the solutions of the energy equation in the wall-wick region and vapor space regions since the blowing/suction rate is determined from an energy balance at the liquid-vapor interface [6].

The boundary condition at the liquid-vapor interface is more complex since it couples the wick and vapor flow. The blowing or suction velocity $v_\delta\rho_\delta$ can be found from an energy balance at the interface:

$$v_\delta \rho_\delta = \frac{k_{\text{eff}}}{hf_g} \frac{\partial T_\ell}{\partial r} \begin{cases} > 0 & \text{(blowing)} \\ < 0 & \text{(suction)} \end{cases} \tag{9.34}$$

For more details of these analyses, reader should refer to Faghri's book [6].

9.9.3 Two-Dimensional Transient Continuum Model

Any operational process of heat pipe such as start-up, shutdown, and operational transients such as pulsed heat inputs is very important and requires detailed study of heat pipe transient analysis. At some time during transient operation, heat pipes have a two-dimensional behavior which cannot be accounted for by a one-dimensional model. Additionally, a two-dimensional formulation and analyses do not require any empirical relations for frictional specification and result a more accurate solution than one-dimensional models. Furthermore, a two-dimensional model is required to simulate certain important phenomena such as nonuniform input heat fluxes and multiple heat sources or sinks [6]. Extensive formulation and details including mathematical analyses are provided by Faghri [6] and reader should refer to his book.

References

1. Zohuri, B. (2016). *Heat pipe design and technology: Modern applications for practical thermal management* (2nd ed.). New York: Springer.
2. Reay, D. A., & Kew, P. A. (2006). *Heat pipes* (5th ed.). Oxford: Butterworth-Heinemann is an imprint of Elsevier.
3. Chi, S. W. (1976). *Heat pipe theory and practice*. New York: McGraw-Hill.
4. McLennan, G. A. (1983, November). *ANL/HTTP: A computer code for the simulation of heat pipe operation* (Report No. ANL-83-108). Argonne National Laboratory.
5. Colwell, G. T., Jang, J. H., & Camarda, J. C. (1987). *Modeling of startup from the frozen state*. Presented at the Sixth International Heat Pipe Conference, Grenoble, France, May 1987.
6. Faghri, A. (1995). *Heat pipe science and technology*. Washington, DC: Taylor & Francis.
7. Jang, J. H., Faghri, A., & Chang, W. S. (1991). Analysis of the one-dimensional transient compressible vapor flow in heat pipes. *International Journal of Heat and Mass Transfer, 34*, 2029–2037.

Chapter 10
Heat Pipe Manufacturing

In this chapter, heat pipe manufacturing methods are examined with the goal of establishing cost-effective procedures that will ultimately result in cheaper and more reliable heat pipes. Those methods which are commonly used by all heat pipe manufacturers have been considered, including envelope and wick cleaning, end closure and welding, mechanical verification, evacuation and charging, working fluid purity, and charge tube pinch-off. Review and evaluation of available manufacturer's techniques and procedures together with the results of specific manufacturing-oriented tests have yielded a set of recommended cost-effective specifications which can be used by all manufacturers.

10.1 Introduction

Key goal of this chapter is to give reader a basic idea of how to develop standardized manufacturing procedures which would assure the reliability of the final product. By evaluating and defining acceptable cleaning procedures, for example, the reliability of a properly fabricated heat pipe will be increased. Moreover, since each manufacturer is currently using essentially independent procedures and are built to different specifications and quality. There is a good deal of duplication of effort in the field; many individual company's procedures have been established iteratively, which is a costly and time-consuming way to develop a technology. This effort, therefore, is only the forerunner of others which will ultimately develop basic heat pipe manufacturing specifications and details for all manufacturers [1].

Essentially major infrastructure of heat pipe is based on five elements or components which are as follows:

1. Envelope
2. Wick
3. End cap

B. Zohuri, *Heat Pipe Applications in Fission Driven Nuclear Power Plants*,
https://doi.org/10.1007/978-3-030-05882-1_10

4. Fill tube
5. Working fluids

Working fluid is introduced into the pipe through the fill tube, which is subsequently sealed by a pinch-off. When heat is applied to a section of the heat pipe (evaporator), the working fluid evaporates, causing a local increase of pressure driving the vapor toward the other end of the pipe (condenser). Cooling this end causes the vapor to condense on the walls of the pipe. Finally, by capillary forces, the working fluid flows back through the wick to the evaporator section. This cycle is repeated as long as heat is supplied to the evaporator and is removed at the condenser. However, if heat is applied to the pipe at a rate higher than the wick can carry, the pipe will fail to function (dry out) [2]. A serious problem with heat pipes is the generation of non-condensable gases which will inhibit their performance.

Basically, these gases will accumulate at the cold (condenser) end of the pipe, decreasing the effective conductance of the unit, until the condenser is completely "blocked" and the pipe fails to function. A flowchart of the basic operations involved in manufacturing a heat pipe is presented in Fig. 6.22. A brief summary of the major elements of the manufacturing cycle discussed in this report follows [2].

Details of these components from manufacturing and operation point are discussed below.

10.1.1 Envelope

The envelope may be of any cross section desired by the designer, e.g., circular, square, etc., and may contain mounting flanges to simplify installation and bent in various shapes. The wick may be grooves extruded into the envelope or may be assemblies—such as Grumman's spiral artery design, made from fine wire mesh (Fig. 10.1), sintered screen, felt metal slabs, etc. The most common configuration is the round shape. The procedures for determining the tube diameter and the wall

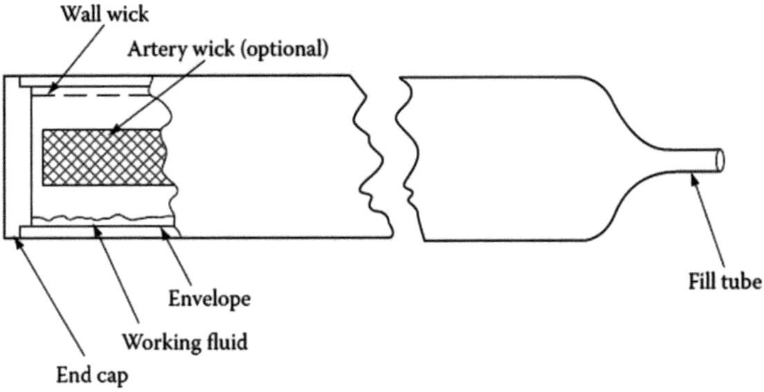

Fig. 10.1 Typical heat pipe components [2]

thickness have been discussed in Chap. 3 of book by Zohuri [1], and great deal of details can be found in Chap. 7 of Chi [2].

Heat pipe manufacturers are rarely concerned with tube manufacturing, because tubes of different materials and sizes are commercially available at reasonable cost. Both seamless and butt-welded tubes may be used as the heat pipe envelope. However, when a tube is cut to the required length, care should be taken so as not to distort the tube ends since this will make it difficult to obtain a strong joint to the end cap. Although round tubes with smooth inside surfaces are common for heat pipe applications, tube with internal fins have also become popular. Several techniques have been developed for production of tubes with internal fins. Kemme [4] was the one of the design pioneers to develop the first axial-grooved heat pipes were formed by milling grooves on a flat plate and then rolling and butt welding a tube. Today commercially integral fin tubing is now available.

10.1.2 End Cap

Although a seemingly simple operation, end closure and welding has been a problem for many heat pipe manufacturers. Common concern is porosity or cracks in the weld, which can lead to a loss of the working fluid. To minimize the probability of this failure, inspection should be performed to verify the adequacy of the seal.

Experience and research indicate that it is not uncommon for heat pipes to fail this inspection, requiring that the closure be repaired. A reliable process can lead, therefore, to a "hit-or-miss" manufacturing cycle, where the welding operation is repeated several times to repair a fault. Obviously, these added steps will adversely impact both time and cost budgets. In keeping with the overall objectives of this study, the "optimum" sealing process should be easy to perform, be repeatable, require moderately priced equipment, and be reliable and easy to inspect.

The minimum required thickness for heat pipe end cap can be determined by stress analysis as described in Chap. 3 and available computer codes mentioned give you such analysis capabilities. Careful design of the joint for welding is important and plays a great deal to weld end caps. Figure 10.2 shows four types of joint designs and they are named as follows:

(a) The butt joint
(b) The lipped butt joint
(c) The fillet joint
(d) The lap joint

Dealing with alignment of these joints is another important step, and among the four kinds, the more difficulty arises with first one (a) (the butt joints) with respect to other three. In addition, it should be noted that to obtain a strong, leakproof welded joint, the end cap must be machined to provide a welded area thickness roughly speaking equal to that of the tube wall [3]. This careful step ensures that both surfaces melt equally resulting in homogeneous fusion.

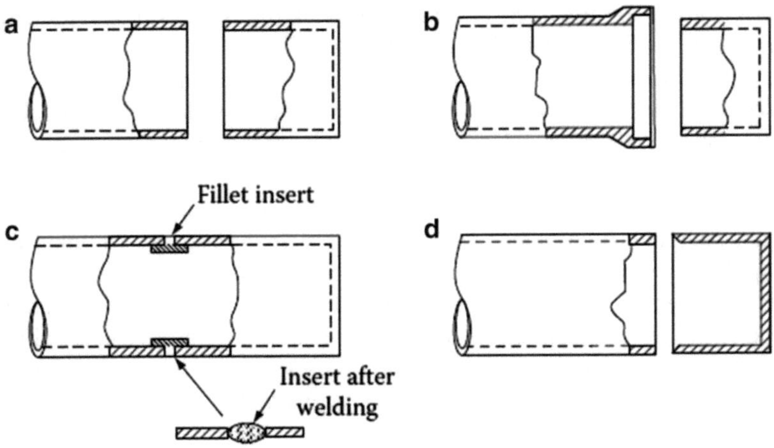

Fig. 10.2 Typical joint design for end cap welding [3]. (**a**) Butt joint. (**b**) Lipped butt joint. (**c**) Fillet joint. (**d**) Lap joint

Detail design of a heat pipe incorporating welded end caps should consider the following points:

- Fusion welding joining processes result in the best combination of joint strength and leak tightness.
- A fully mechanized fusion welding process should be given preference over a manual process. A mechanized process is significantly more consistent with respect to joint quality, e.g., strength, size of heat affected zone, and weld bead geometry.
- The use of the aluminum alloy 6061 or the stainless steel alloy 304L eliminates the need for post weld heat treatment, such as a stress relief anneals to prevent either corrosion susceptibility or embrittlement.
- Joint efficiency is a prime consideration. The 304L stainless steel alloy is not heat treatable, and a relatively high joint efficiency in the "as welded" condition can be obtained. A conservative assumption for this alloy is a weld joint strength that is 85% (automatic process) or 70% (manual process) of the minimum strengths guaranteed for the 304L product form in the annealed condition. In the case of the 6061 aluminum alloy, the condition of the material prior to welding and the use of post weld heat treatment must be taken into consideration in establishing weld joint strength.
- The geometry of the weld bead produced by the welding process selected should be known by the designer beforehand since the weld root bead cannot be machined after welding. An excessively large drop through the root bead can interfere with the operation of the heat pipe. Machining of the crown of the weld bead should be avoided. The removal of the skin material of the weld bead crown may expose interdendritic porosity and result in gas leakage from the heat pipe.
- The use of a square butt joint design is preferable.

- The end cap detail should be designed to be self-aligning during the welding operation. The end cap detail can also provide filler metal to the joint.
- The possible damage to interior details of the heat pipe resulting from welding heat should be considered and, if necessary, determined experimentally.

For more process of end cap welding along with details reader should refer to Edelstein and Haslett [2] report (Fig. 10.3).

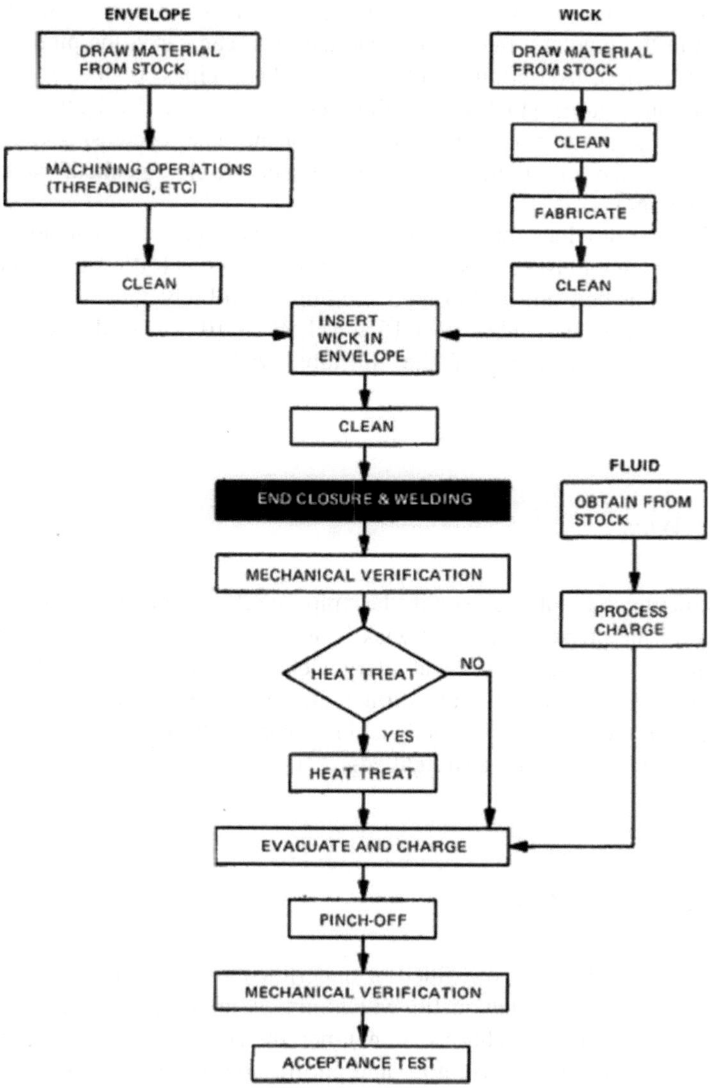

Fig. 10.3 Typical manufacturing steps for welding [3]

10.1.3 Fill Tube

An end cap and fill tube are used to complete the envelope. Working fluid is introduced into the pipe through the fill tube, which is subsequently sealed by a pinch-off. When heat is applied to a section of the heat pipe (evaporator), the working fluid evaporates, causing a local increase of pressure driving the vapor toward the other end of the pipe (condenser). Cooling this end causes the vapor to condense on the walls of the pipe. Finally, by capillary forces the working fluid flows back through the wick to the evaporator section. This cycle is repeated as long as heat is supplied to the evaporator and is removed at the condenser. However, if heat is applied to the pipe at a rate higher than the wick can carry, the pipe will fail to function (dry-out). A serious problem with heat pipes is the generation of non-condensable gases which will inhibit their performance. Basically, these gases will accumulate at the cold (condenser) end of the pipe, decreasing the effective conductance of the unit, until the condenser is completely "blocked," and the pipe fails to function [2].

The fill tube provides the only access to the heat pipe for evacuation and fluid charging, and it too must be closed and sealed. A crimp seal followed by welding has been found to be convenient. For a good crimp, the fill tube must not exceed ¼ in. (6.35×10^{-3} m) OD. Tubes of 1/16 in. (1.59×10^{-3} m) i.d. are often used. Additionally, the thickness of the fill tube wall is also critical for certain materials [3].

10.1.4 Wick

If wick structure is an integral part of a heat pipe, e.g., an axial-grooved heat pipe, the operation for the preparation of the wick is not needed. If the wick is formed from wire cloth, e.g., wrapped-screen and artery wicks, the wire cloth must first be cut to the required size. Wire cloth can be difficult to cut accurately, because it is not rigid in the plane of the material. By sandwiching a wire cloth between two flat plates, a wire cloth may be cut along a straight edge [3].

10.1.5 Working Fluid

A heat pipe should not be overfilled or underfilled with working fluid. Heat pipe working fluid must be of high quality with minimum impurities—gaseous, dissolved solid, or liquid. If these conditions are not met, the designed heat pipe performance may not be reached. Fluid inventory and purification techniques will be discussed below in a section dealing with pipe evacuation and charging [3].

10.2 Heat Pipe Manufacturing Procedures

A flowchart of the basic operations involved in manufacturing of a heat pipe is presented in Fig. 5.4. A bullet topic of the major elements are represented again, and reader can refer to Sect. 3.12.13 of Chap. 3 in book by Zohuri [1] for a brief summary of these major elements of the manufacturing cycle discussed in this report follows. These major elements are as follows:

- Envelope and wick cleaning
- End closure and welding
- Mechanical verification
- Evacuation and charging
- Fluid parity
- Charge tube pinch-off

Baseline heat pipe manufacturing procedure is summarized and illustrated in Fig. 10.4 here. Overall restrictions and assumptions unless otherwise specified in the individual manufacturing procedures, the following restrictions or assumptions shall apply [2]:

- Envelope materials considered are aluminum and stainless steel tubing.
- Working fluids are restricted to ammonia, Freon-21, and methanol.
- The pipes are intended for operation in the moderate temperature range, nominally from -100 to 200 °F (200–366 K).
- Typical pipe sizes are approximately 0.5 in. (0. 0127 m) diameter and up to about 12 ft (3.6 m) long.
- The procedures described are generally based on small quantity runs up to approximately 10–20 units. Large quantity production runs would require more automated operations, the procedures of which might be different from those described here.
- The procedures described are based on a heat pipe manufacturing sequence typical of that shown in Fig. 10.4.

10.3 Cleaning of Parts

Just as extreme care is required in selecting proper heat pipe materials to avoid compatibility problems, cleaning of the component heat pipe parts is critical to avoid similar consequences. Apparently, this is an area in which no accepted standard exists in the industry at the present time [2]. Although there are similarities in the procedures, each manufacturer uses a different process, essentially developed independently. The effectiveness of each process can only be evaluated on the basis of past experience (Fig. 10.5).

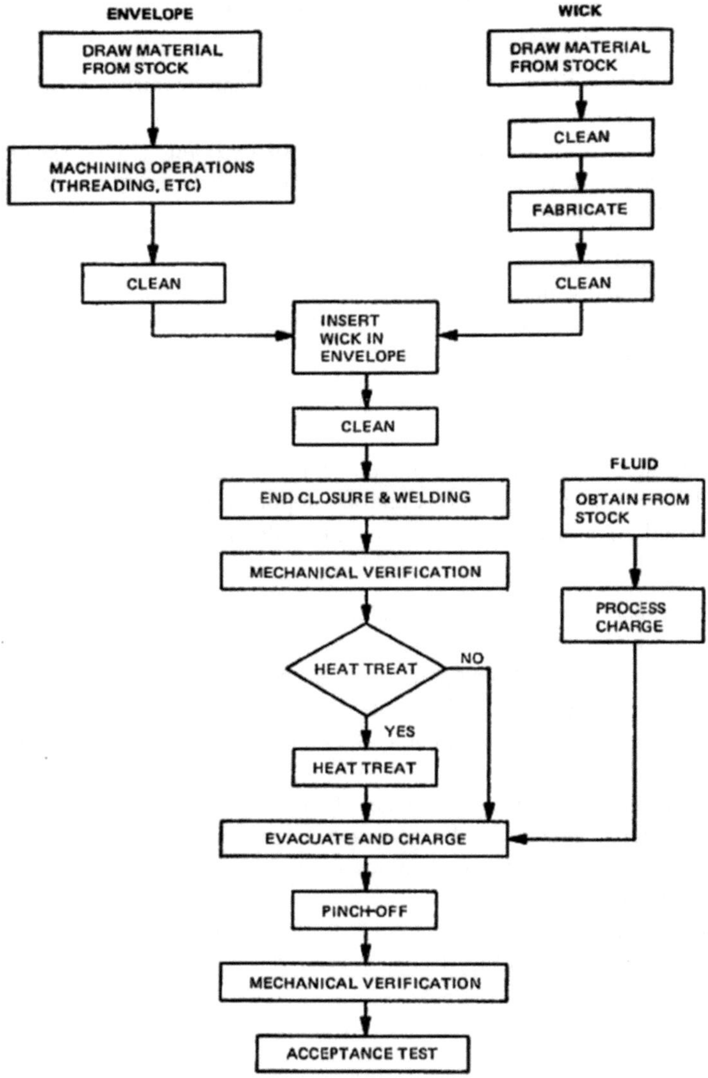

Fig. 10.4 Typical heat pipe manufacturing flowchart [3]

As seen in the heat pipe manufacturing cycle of Fig. 10.4, there are many areas where contaminants can be introduced into the heat pipe, i.e., through a dirty wick, dirty envelope, impurities in the working fluid, etc. In fact, every operation not properly performed can be a source of contamination. This section discusses envelope and wick cleaning and pretreatment for aluminum and stainless steel envelopes and stainless steel wicks. Originally, it appeared that this area should be treated as two topics, wicks and envelopes. However, during the research effort for this study, it became clear that little wick cleaning information was available. Although most

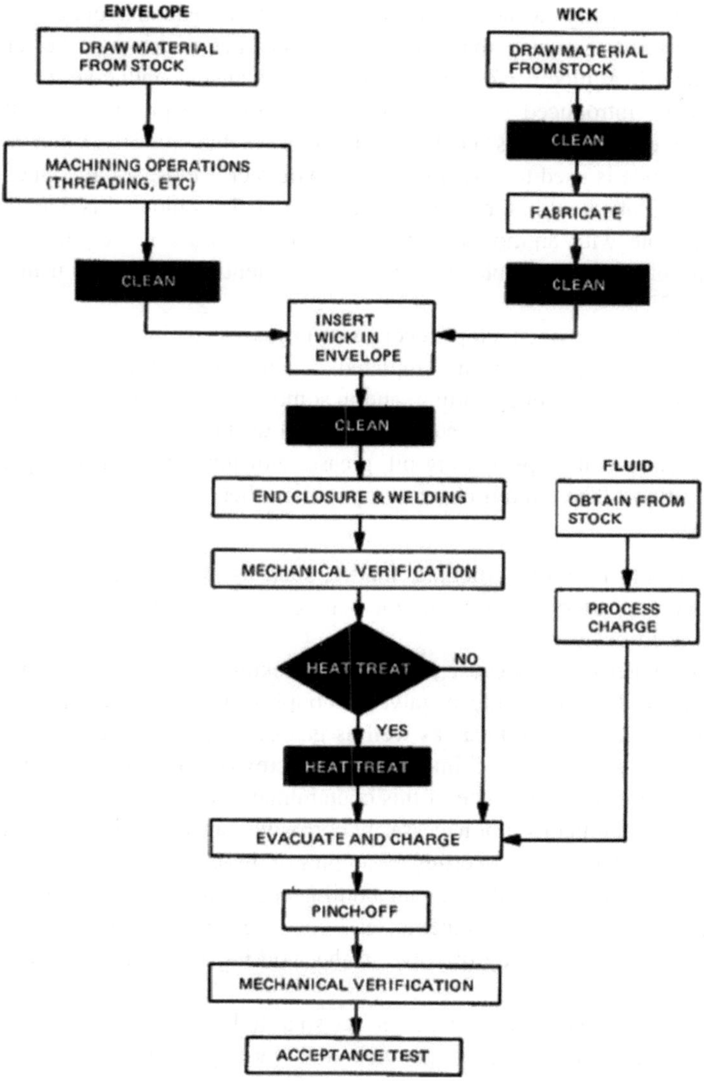

Fig. 10.5 Typical manufacturing steps for cleaning [2]

manufacturers were willing to discuss cleaning in general, few were willing to discuss wick cleaning in particular. Each manufacturer considers wick construction and fabrication details as highly proprietary, since it is this element which essentially determines heat pipe performance and separates one company from another. Of the various organizations contacted, therefore, very few were willing to supply information concerning wick cleaning.

Wick cleaning and pretreatment is at least as important as the need for envelope cleaning. Obviously, gas generation is just as likely to come from a "dirty" wick as from other improperly cleaned parts. Oil and greases imbedded in either the fine wire mesh or sintered screening material used to construct wicks must be removed to assure proper heat pipe performance. Foreign substances conducive to gas generation might be introduced in the construction process, requiring that a post wick construction cleaning process be designed to remove this impurity. For example, if a copper electrode is used to assemble a wick with tack welds, some copper particles may become imbedded in the stainless steel screen. To remove this material, which is incompatible with ammonia, a nitric acid rinse would be required. However, where possible, it is preferable to eliminate this potential problem by using tungsten electrodes [2].

The heat pipe tube or envelope receives its primary cleaning after dirty operations (such as machining) have been completed. Machining may involve preparing the tube ends for welding, tube bending, and in some cases, cutting fine circular threads on the inside surface to provide a circumferential wicking surface. An assortment of debris such as metal chips, cutting oil, grease, moisture, etc. can be expected after these operations. The overall cleaning operation, therefore, has a number of aims, namely, to:

- Mechanically remove particulate matter, such as metal chips which may clog capillary and artery surfaces and/or damage these surfaces during subsequent artery insertion.
- Remove water that can cause corrosion, attacking both aluminum and stainless steel, as well as providing a galvanic coupling between the two. Buildup of particulate reaction products, as well as gas generation, is the principal results. Loss of container structural integrity due to crevice corrosion and porosity may also result from the presence of this contaminant.
- Remove contaminants, not necessarily corrosive, but which may impair the heat pipe wicking and fluid properties. Examples of these contaminants are the variety of oils and greases used in metal cutting and removal operations, extruding, forming, etc. These contaminants may coat the internal surfaces and increase the contact angle, or may dissolve in the working fluid, changing its transport properties.
- Chemically clean and prepare the surface so as to be nonreactive with subsequent manufacturing environments, the wick, and working fluid.
- Treat the wicking surface in a manner that enhances "wettability" with the working fluid.

Failure to achieve these objectives through either improper procedures or operator error has led to heat pipe problems ranging from performance degradation to complete failure. In one reported incident, for example, water inadvertently left on the inside surface of a threaded aluminum tube was judged to be responsible for failure of a heat pipe to achieve its performance goals. Analysis led to the discovery of huge amounts of aluminum hydroxide on the inside surface which were clogging the radial flow passages.

Contaminants can also chemically react with the wall, wick, or fluid to produce non-condensable gaseous products that block the condenser and decrease heat pipe conductance. In the case of arterial wicks, gas bubbles within the wick can, and have, severely limited the heat transport capacity [4]. On the ATS program, groove failure in a single heat pipe was attributed to embrittlement and porosity caused by the presence of water in a closed pipe during heat treatment [2].

A summary of the foregoing and other problems that can be caused by improper cleaning techniques is as follows:

- Physical clogging of wall and wick capillary surfaces, thereby impairing both heat pipe transport capacity and conductance.
- Non-condensable gas generation reducing both heat pipe conductance (loss of condensation area) and transport capacity (bubbles in arterial wicks).
- Decrease in wettability of wick.
- Adverse change in fluid properties, such as surface tension, wetting angle, and viscosity.
- Loss of structural integrity of container wall due to galvanic corrosion, crevice corrosion, and porosity.

Unfortunately, many of these problems cannot be uncovered until the pipe is charged, sealed, and tested. In some cases, a long time can occur until some of these effects are noticed; by then, it is usually too late to provide corrective action. Hence, the objective must be to develop cleaning procedures which will prevent these problems from occurring and produce, therefore, a more reliable product. Moreover, in keeping with the overall aims of this study, the cleaning procedure should also be simple, inexpensive, and as free as possible from human error [2].

Table 10.1 summarizes the procedures currently employed in the heat pipe industry. A brief general description of various cleaning techniques follows, and each manufacturer's approach will be presented and analyzed. Finally, using this evaluation, recommended cleaning procedures are presented. Note that the procedures used by various manufacturers are typical. If more information is required, it is suggested that the individual company be contacted [1].

To summarize the cleaning procedure for aluminum, copper, and stainless steel, we can show the following [2]:

1. Envelope Cleaning and Pretreatment

   - Assumptions:

     - Applicable tubing: Aluminum 6061 or 6063; stainless steel 300 series
     - Tube condition: Threaded, wicked, or otherwise internally machined prior to cleaning
     - Tube size: Typically, 0.5 in. (0.0127 m) dia. up to 12 ft (3.6 m) long

   - Materials:

     - Non-etch alkaline cleaner (refer to Table 10.2).
     - Chromated deoxidizer (refer to Table 10.3).
     - Filtered air.
     - Anhydrous isopropyl alcohol.

Table 10.1 Summary of heat pipe cleaning procedures is currently in use [1]

Manufacturer	Envelope cleaning		Wick cleaning (stainless steel)
	Aluminum	Stainless steel	
Dynatherm	• Solvent		
	• Acid		
NASA/GSFC	• Solvent		
	• Acid		
Grumman	• Solvent	• Solvent	• Solvent
	• Alkaline/acid	• Passivation	• Passivation
TRW	• Solvent	• Ultrasonic	• Ultrasonic
	• Alkaline/acid	• Vacuum fire	• Vacuum fire
	• Ultrasonic		
DWDL/MDAC	• Solvent		
ESRO/MBB	• Ultrasonic		• Ultrasonic
GE	• Alkaline/acid	• Alkaline/acid	
University of Stuttgart		• Ultrasonic	
		• Passivation	
NASA/MSFC	• Solvent	• Solvent	
	• Alkaline/acid	• Alkaline	
		• Passivation	

Table 10.2 Examples of non-etch alkaline cleaners [1]

Material	Concentration	Temperature
Ridoline No. 53 (Amchem Products Co.)	2–10 oz/gal	140–180 °F
Oakete No. 164 (Oakite Products Co.)	2–10 oz/gal	140–180 °F
Kelite Spray White (Kelite Corp)	40–60% by volume	Ambient
A-38 (Pennwatt Corp)	4–8 oz/gal	160–180 °F

Table 10.3 Examples of chromated deoxidizer solutions (immersion type) [1]

Material	Concentration	Temperature	Immersion Time
Mixture of: Chromated deoxidizer replenisher No. 17[a] (Amchem Products Co.)	2–6 oz/gal	Ambient to 120 °F	5–30 min
Nitric acid 42° Be	10–20% by volume		
Mixture of: Chromated deoxidizer replenisher No. 17[a]	2–6 oz/gal	Ambient	5–30 min
Sulfuric acid 66′ Be	4–7% by volume		

[a]Deoxidizer makeup No. 7 to be used for initial makeup

Table 10.4 Examples of passivating solutions [1]

Material	Concentration	Temperature	Immersion time
Nitric acid	35–65% by volume	Ambient	30 min to 2 h
Mixture of: Sodium dichromate or potassium dichromate	1–4 oz/gal	Ambient	30 min to 2 h
Nitric acid	15–30% by volume		

- Dry nitrogen.
- Passivating solution (refer to Table 10.4).

- Procedure—aluminum tubes:
  - Clean in cold 1,1,1-trichloroethane with bristle brush on wire extension. Periodically clean brush between strokes.
  - Flush internal surface with cold trichloroethane; dry with filtered air and cap pipe ends.
  - Immerse in non-etch alkaline cleaner for 5 min minimum. Refer to Table 5.2 for materials and temperatures.
  - Follow with a two (2) minute tap water rinse, raising and lowering tube during rinsing.
  - Immerse in chromated deoxidizer. Refer to Table 5.3 for material, time, and temperature.
  - Follow with a two (2) minute tap water rinse, raising and lowering tube during rinsing.
  - Thoroughly dry inside surface with forced filtered air.
  - Rinse with anhydrous isopropyl alcohol.
  - Force dry with clean, filtered, dry nitrogen heated to 160 °F.
  - Cap pipe ends.
  - If applicable, insert artery, rinse with isopropyl alcohol, and dry as in step 9.
  - If heat treat is required after welding, then:

    Evacuate pipe for 4 h at 600 °F and leak check.
    Seal evacuated heat pipe.
    Perform heat treat operations on sealed pipe.

- Procedure—stainless steel:
  - Clean in cold 1,1,1-trichloroethane with bristle brush on wire extension. Periodically clean brush between strokes.
  - Flush internal surface with cold trichloroethane; dry with filtered air and cap pipe ends.
  - Immerse in passivating solution. Refer to Table 10.3 for materials, temperatures, and time.

  - Follow with a two (2) minute tap water rinse, raising and lowering tube during rinsing.
  - Thoroughly dry inside surface with forced filtered air.
  - Rinse with anhydrous isopropyl alcohol.
  - Force dry with clean, filtered, dry nitrogen heated to 160 °F.
  - Cap pipe ends. If applicable, insert artery, rinse with isopropyl alcohol, and dry as in step 7.

- General notes:
  - The cleaning procedure must be as free from operator error as possible, since improperly executed procedures may also lead to unwanted contamination. Training of personnel regarding the cleanliness requirements for heat pipes is mandatory. So are adequate safeguards and inspection points during the steps of the procedure; these must be consistent with good quality assurance practices.
  - Do cleaning and charging operations in fairly rapid sequence and in proximity to one another. Avoid storing the pipe for long duration, which increases the likelihood of contamination. Performing these operations near each other lessens the danger of contamination during transportation.

For more details of this operation reader should refer to Edelstein and Haslett [1] report.

10.4 Assembly of Heat Pipes

Assembly of heat pipe parts includes welding of end caps and fill tube and forming and inserting of wick if applicable. Since parts have been thoroughly cleaned, they should be assembled immediately following the cleaning if practicable. Otherwise, cleaned heat pipe parts should be stored in a clean dry atmosphere to prevent contamination by vapor, smoke, and dust suspended in air. Plastic gloves should be worn while handling parts to prevent contamination by skin oils and acids.

- Wick Forming and Insertion [4]
  Manual forming and insertion of wrapped-screen wicks can be accomplished as follows. The assembled wick must not contain wrinkles, and to prevent this, the screen may be wrapped on a cleaned mandrel. The total diameter of mandrel and wrapped screen should be only slightly less than the heat pipe inside diameter, so that the residual stress in the coiled screen will force it against the pipe wall when it is released from the mandrel. Obviously, the screen ends must be even, and the screen must be properly positioned so that the installation of end caps will not interfere with or crush the screen. To ensure physical contact between the screen layers and the pipe wall, a tapered plug or ball may be forced through the wick. A helical spring with an unstretched diameter slightly larger than the wick inside diameter is sometimes used to hold the screen layers in contact. It too can be

installed with the aid of a mandrel. In this case, the spring is stretched, while it is held by the mandrel. The length of the stretched spring must not greatly exceed that of the installed length or when it is released from the mandrel, axial forces may displace the screen. With the wick properly positioned in the pipe, the end caps are next welded on [2].

- End Cap Installation [2]

  End cap installation: If the fill tube is not an integral part of the end cap, it should first be welded to its end cap. End caps with or without a fill tube are usually welded to the pipe ends. High-quality welded joints are required at all seams since porosity or cracks in the weld can lead to a loss of the working fluid. To minimize the probability of this failure, inspections should be performed to verify the adequacy of the seal. A number of welding techniques are available. However, gas torch welding, e.g., using oxyacetylene gases, is generally not recommended, because of the presence of flux. Oxygen and filler metals tend to recontaminate the cleaned parts. Tungsten inert gas (TIG) welding, either manual or automatic, and electron beam welding (EBW) have been found to be satisfactory for heat pipe welding.

  The TIC welding is an electric arc welding process using a sharp-tipped tungsten electrode surrounded by an annular shield of inert gas flowing from a torch tip. Filler metals are not generally used for heat pipe welding, but they may be integral parts of the end caps, e.g., the lip of the lipped butt joint shown in Fig. 10.6 can serve as filler metal. In addition, this process does not employ a flux. Hence, TIC welding does not contaminate the cleaned heat pipe part. The EB

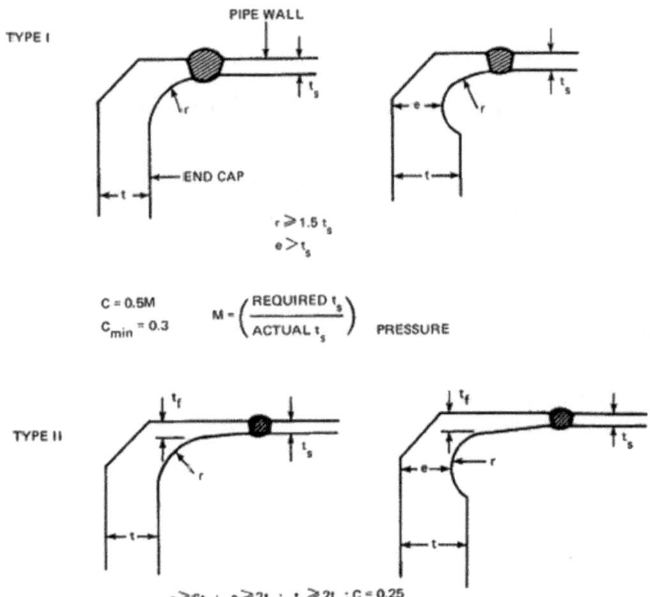

Fig. 10.6 End cap design details [1]

weld is made in a vacuum chamber, and it eliminates the formation of the surface compounds from the metal and air. In addition, the EBW offers minimal heat input with maximum thermal density. It enables one to produce a welded joint with a minimum heat affected zone, and consequently, the joint properties may approach those of the parent metal. It is, therefore, ideal for heat pipe welding. However, the investment cost in equipment for EBW may exceed that for automatic TIC by over 100% and that for manual TIC by over 2000%. Hence, the choice of welding process is dependent upon the availability of the equipment; and the investment in equipment depends greatly upon the quantity of production and the required quality of the products. Nevertheless, TIC and EB welding have been found satisfactory for heat pipe welding [2].

- End Closure and Welding [1]
  - End closure design guidelines. Observe the following general guidelines:
    - Acceptable end cap designs for flat circular heads may be Type I or II as shown in Fig. 10.6.
    - End cap thickness can be determined from Figs. 10.7 and 10.8.
    - The welded end should have a square butt joint design (see Fig. 10.9).
    - Self-alignment during welding should be provided, as per the lipped butt joint for stainless steel (Fig. 10.10), the consumable filler insert for aluminum (Fig. 10.11), or an equivalent design.
    - Full weld penetration should be achieved.
    - Machining of the crown of the weld bead should be avoided.

10.4.1 Summary of Assembly of Heat Pipes Procedures

The following is a list of the procedures described above, which should be followed during heat pipe assembly [5]:

1. Select container material.
2. Select wick material and form.
3. Fabricate wick and end caps, etc.
4. Clean wick, container, and end caps.
5. Outgas metal components.
6. Insert wick and locate.
7. Weld end caps.
8. Leak check welds.
9. Select working fluid.
10. Purify working fluid (if necessary).
11. Degas working fluid.
12. Evacuate and fill heat pipe.
13. Seal heat pipe.

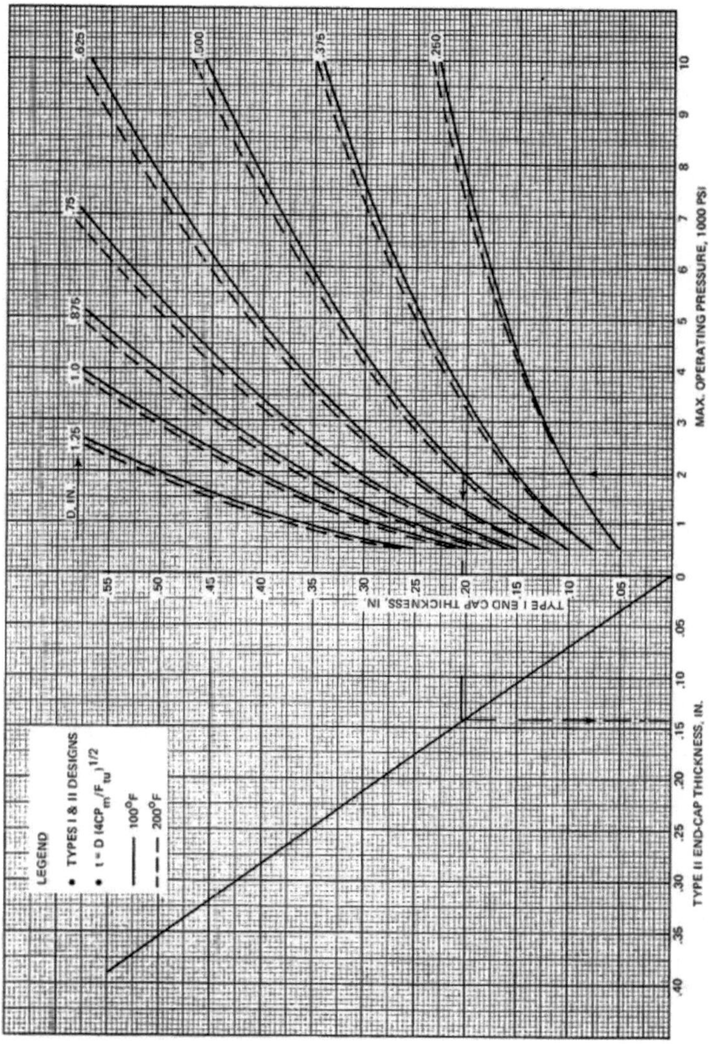

Fig. 10.7 End cap design curve 6061-T6 aluminum (as welded) [2]

It may be convenient to weld the blank end cap before wick insertion, and in cases of sintered and diffusion bonded wicks, the outgassing may be done with the wick in place in the container.

For the manufacturer considering the production of a considerable number of identical heat pipes, for example, 50 or more units following prototype trials, a number of the manufacturing stages may be omitted. Outgassing of metal components may be unnecessary, and it may be found that, depending upon the filling and evacuation procedure used, the fluid degassing may be eliminated as a separate activity [5].

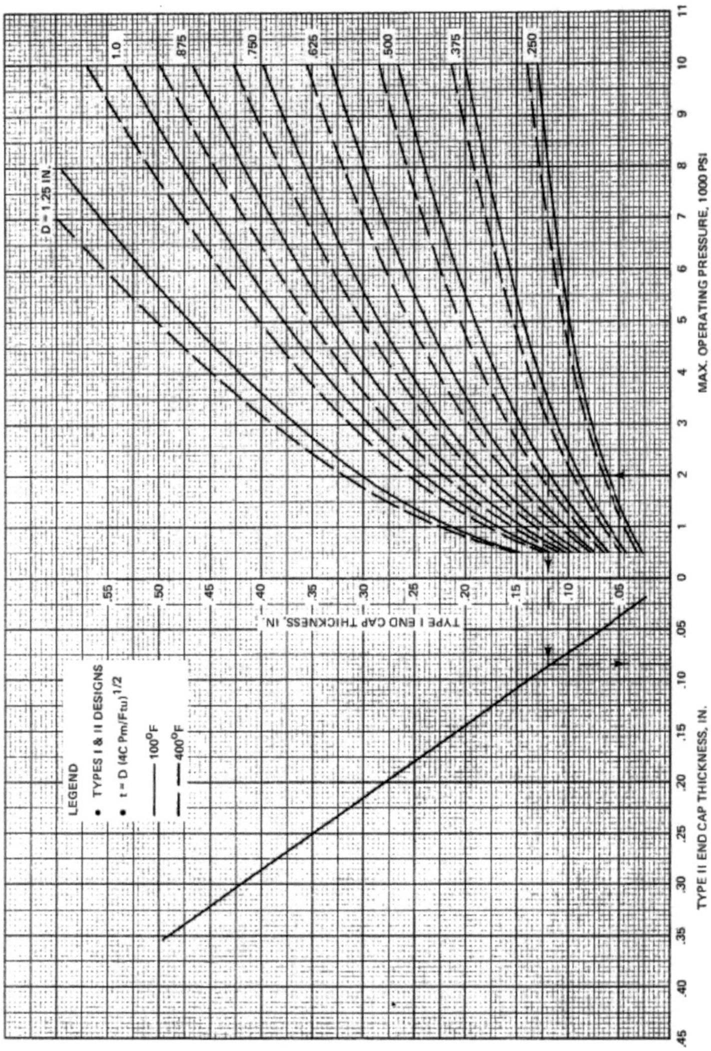

Fig. 10.8 End cap design curves, 304 stainless steel (as welded) [2]

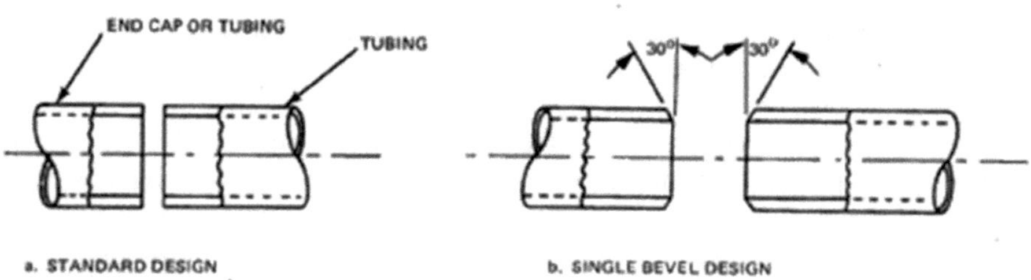

Fig. 10.9 Square butt joint designs [2]. (**a**) Standard design, (**b**) single level design

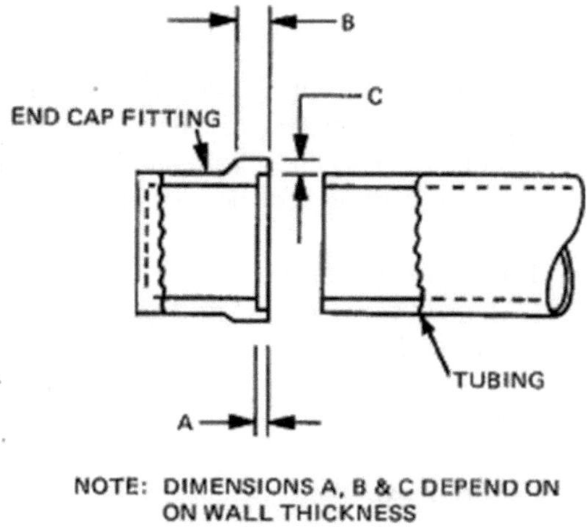

Fig. 10.10 Lipped butt joint design [2]

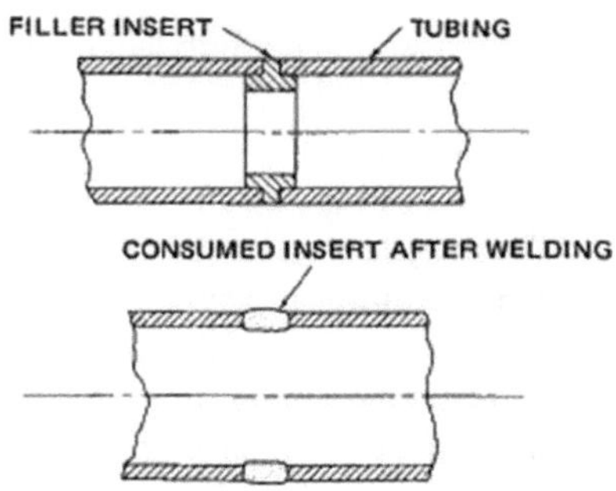

Fig. 10.11 Square butt joint with consumable filler insert [2]

Prior to charging, a heat pipe must be evacuated to remove materials that may subsequently appear as unwanted non-condensable or that chemically react with the working fluid forming undesirable corrosion product.

The non-condensable are due not only to the free gas in the pipe but also to the molecules absorbed on the metal surface. Removal of free gases in the pipe can be done simply by pumping down with a vacuum pump. Removal of absorbed gas requires the evacuation of the pipe at elevated temperatures. The time required to absorb surface contaminants is usually reduced with the increased temperature. However, metals may lose their strength at high temperature [3].

Heat Pipe Assemblies Design Guidelines

Orientation with Respect to Gravity

For the best performance, the application should have gravity working with the system; that is, the evaporator section (heated) should be lower, with respect to gravity, than the condenser (cooling) section. In other orientations where

(continued)

gravity is not aiding the condensed liquid return, the overall performance will be degraded. Performance degradation depends on a number of factors including wick structure, length, and working fluid of the heat pipe along with heat flux of the application. Careful design can minimize the performance loss and allow an accurate prediction performance.

Temperature Limits

Most pipes use water and methanol/alcohol as the working fluids. Depending on the wick structure, pipes will operate in environments with as low as −40 °C. Upper temperature limits depend on the fluid, but 60–80 °C is the average limit.

Heat Removal

Heat can be removed from the condenser using air cooling in combination with either extrusion, bonded-fin heat sinks, or flat-fin stock. Enclosing the condenser in a cooling jacket allows liquid cooling.

Reliability

Heat pipes have no moving parts and have demonstrated life of over 20 years. The largest contributor to heat pipe reliability comes from control of the manufacturing process. The seal of the pipe, purity of the materials used in the wick structure, and cleanliness of the internal chamber have measurable effect on the long-term performance of a heat pipe. Any leakage will eventually render the pipe inoperable. Contamination of the internal chamber and wick structure will contribute to the formation of non-condensable gas (NCG) that will degrade performance over time. Well-developed processes and rigorous testing are required to ensure reliable heat pipes.

Forming or Shaping

Heat pipes are easily bent or flattened to accommodate the needs of the heat sink design. Forming heat pipes may affect the power handling capability as the bends and flattening will cause a change in fluid movement inside the pipe. Therefore, design rules that take into consideration heat pipe configurations and the effect on thermal performance ensure the desired solution performance.

Effects of Length and Pipe Diameter

The vapor pressure differential between the condenser end and the evaporator end controls the rate at which the vapor travels from one end to the other. Diameter and length of the heat pipe also affect the speed at which the vapor moves and must be considered when designing with heat pipes. The larger the diameter, the more cross-sectional area available to allow vapor to move from

(continued)

the evaporator to the condenser. This allows for greater power carrying capacity. Conversely, length when in opposition to gravity has a negative effect on heat transport as the rate at which the working fluid returns from the condenser end to the evaporator end is controlled by the capillary limit of the wick which is an inverse function of the length of the pipe. Therefore, shorter heat pipes carry more power than longer pipes when used in application not assisted by gravity.

Wick Structures
Heat pipe inner walls can be lined with a variety of wick structures. The four most common wicks are:

(a) Groove
(b) Wire mesh
(c) Sintered powder metal
(d) Fiber/spring

The wick structure provides a path for liquid to travel from condenser to the evaporator using capillary action. Wick structures have performance advantages and disadvantages depending on the desired characteristics of the heat sink design. Some structures have low capillary limits making them unsuitable for applications where they must work without gravity assist.

10.5 Evacuation and Charging

Prior to charging, a heat pipe must be evacuated to remove materials that may subsequently appear as unwanted non-condensable. Evacuation and charging are two processes that are closely related.

Figure 5.12 shows a flowchart for the evacuation and charging procedure used by heat pipe manufacturers, e.g., TRW, Grumman, and McDonnell Douglas, as presented by Edelstein and Haslett [2].

Initially, the pipe is evacuated, typically on a 4-ino Veeco pumping station at a station pressure of about 10^{-6} mmHg. Heaters wrapped around the pipe provide the elevated temperatures at which evacuation occurs. A schematic of the setup is shown in Fig. 10.13. The temperatures shown in Fig. 10.12 are representative for aluminum tubes. TRW bakes out at about 325 °F for 16 h, while Grumman bakes out at lower temperatures of 170–250 °F, but for a longer time of up to 48 h. Although higher evacuation temperatures may be beneficial in removing additional adsorbed molecules, loss of mechanical properties of aluminum can occur at temperatures in the 300 °F range. Figures 10.12 and 10.13 show the effects of elevated temperature exposure on the room-temperature ultimate and yield strength of aluminum 6061-T6 [1]. It is seen that loss of strength begins to occur at a temperature of 380 °F for 1/2 h exposure. The close relation between evacuation and charging can be seen in this

Fig. 10.12 Overall evacuation and procedure used by TRW, Grumman, and DWDL/McDonnell Douglas [2]

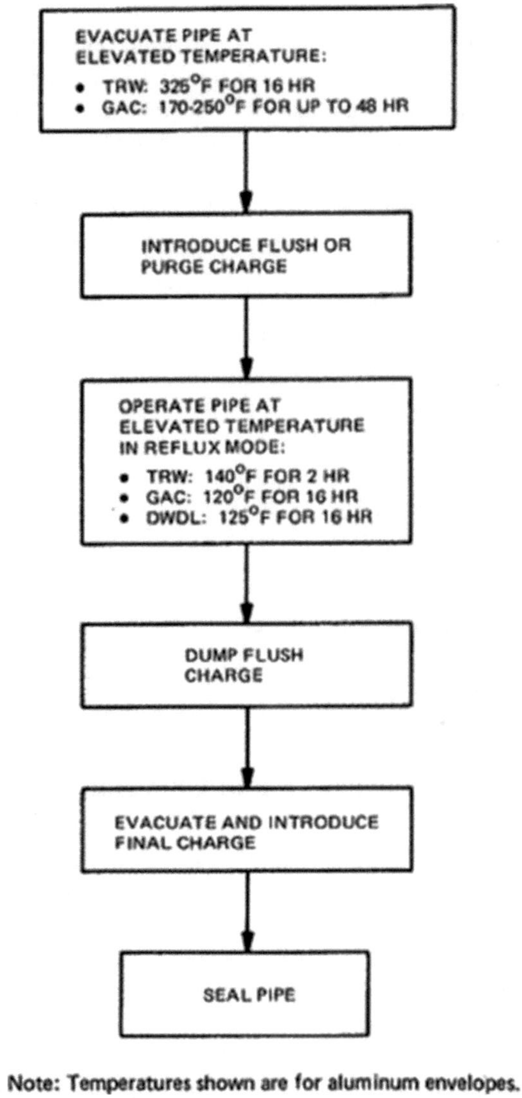

Fig. 10.12; hence, evacuation and charging are often performed in the same equipment setup.

Typically, a 1/16 in. (0.159 cm) i.d. tube, about 4 in. (10:160 cm) long, serves as the charge (pinch-off) tube. Based on the results presented in the previous section, this size tube is mildly restrictive during pump down for 1/2 in. (1.270 cm) d pipes. A technique which avoids the use of a small diameter charge tube will be presented later.

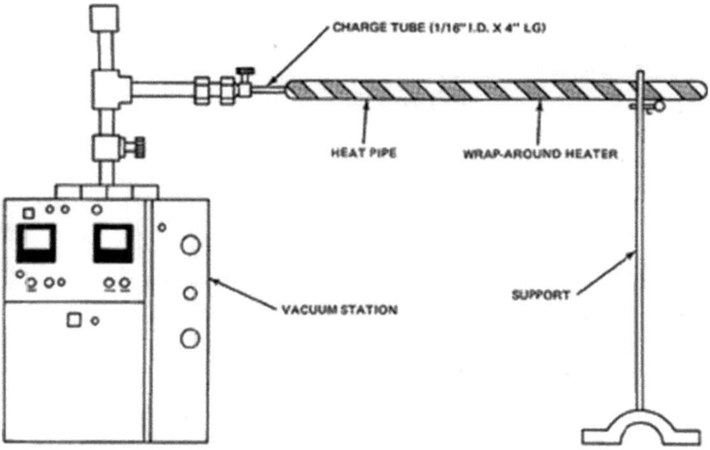

Fig. 10.13 Heat evacuation setup [2]

Figure 10.14 shows a possible setup for combined evacuation and charging. Apossible procedure is as follows: with valve B closed and valves A and C open, the pipe is first pumped down at the ambient temperature; and then, the pumping is continued while the pipe is heated. The temperature of the heat pipe and the pumping time depend, of course, on the pipe material and its eventual operating temperature as described in the previous paragraph. This process is sometimes called vacuum bake out. After completion of this vacuum bake out process, the pipe is flushed with a small amount of fluid. For this purpose, the fluid in the charge bottle is first heated above the fluid boiling temperature at the pressure of the vacuum system. Momentary opening of valve B then allows a small amount of flush charge to be dumped into the pipe. After the pipe is flushed in this manner once or twice, the pipe is ready for charging [3].

10.5.1 Fluid Charging

Details of the charging process depend on the state of the working fluid at the ambient temperature. If the fluid is in the gaseous state at room temperature, such as the case for the cryogenic heat pipe fluid, the charge can be introduced from a cylinder containing gas of high quality. The amount of charge can be measured by the gas pressure in the pipe at room temperature. Hence, the charging process consists of closing valve A and opening valves B and C. When the required amount of fluid has been charged into the pipe, valves B and C are closed. The pipe is then ready for pinching-off and final sealing.

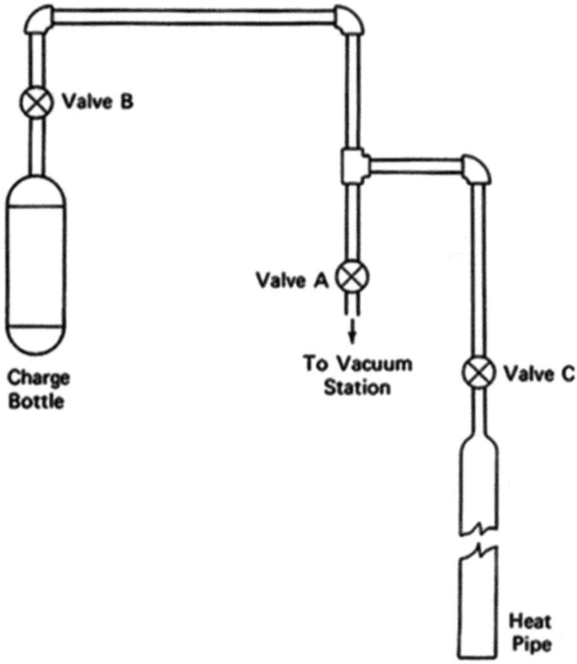

Fig. 10.14 Schematic of equipment for heat pipe evacuation and charging [3]

10.5.2 Fluid Purity and Inventory

Should a manufacturer use almost 99.999% pure ammonia costing $40 per pound, or will 99.99% pure ammonia costing $2.50 per pound be satisfactory? What impurities are present in the ammonia? How will they affect heat pipe performance? These are some typical questions asked by both manufacturers and users of heat pipes concerning the requirements for the purity of heat pipe working fluids. These considerations can strongly affect both product reliability and cost [2].

The most prominent manifestation of impurities in a heat pipe is the accumulation of non-condensable gas in the condenser zone, with consequent loss of heat pipe conductance. Depending on the design and operating conditions, the presence of gas may not be serious and may go completely unnoticed. In other cases, significant blockage can occur; for some artery pipes, loss of pumping capacity can result [2].

Impurities may be in the pipe in the form of adsorbed gas molecules even before the working fluid is introduced. It may be brought in during the fluid transfer operation, or it may be present in the fluid itself. Often, the fluid purchased from the supplier may have significantly higher impurity levels than what was nominally specified [2].

Certain techniques are available to purify the fluid to a higher state than it is "as received" from the supplier. Additional cleaning may be necessary to satisfy functional requirements, or as a safeguard against uncertain impurity levels in the suppliers fluid [2].

We can now discuss effects of impurities on heat pipe performance. In general, the impurities found in working fluids, such as ammonia, may consist of:

- Gases, such as nitrogen, oxygen, argon, carbon dioxide, carbon monoxide, and methane
- Water
- Miscellaneous materials such as oils, hydrocarbons, and nonvolatile solids

Of the miscellaneous materials, the most detrimental effect on heat pipe performance can be loss of wettability of the wick due to oily residues. As previously mentioned a significant amount (130 ppm) of an oily contaminant was found in the liquid phase of ammonia and identified to be dioctyl phthalate. These materials, some of which may be soluble in the working fluid, may also adversely affect fluid properties, such as surface tension, wetting angle, and viscosity [2].

The presence of water in an aluminum or stainless steel pipe can cause corrosion, resulting in the loss of structural integrity. However, the quantities of water generally found in working fluids can be minimized by various purification techniques, some of which will be discussed later. The resultant water quantities, in terms of parts per million, are usually small enough so as not to present a serious corrosion loss-of-strength problem, since the corrosive reaction will generally cease when the water is consumed. The reaction products, however, may be far more serious in terms of non-condensable gas generated [2].

Fluid used for the heat pipe must obviously be of high purity. However, non-condensable may be dissolved in the supposedly pure liquids and solids. For example, non-condensable that are transferred to the charge bottle during vapor transfer may be removed by subjecting the fluid to repeated freeze-thaw cycles [2]. A schematic of this processing technique is shown in Fig. 10.15. The fluid is frozen in the charge bottle using, for example, liquid nitrogen for ammonia and ambient air for sodium.

It is also necessary that the heat pipe is neither underfilled nor overfilled with its working fluid. An underfill may result in degradation of its performance. On the other hand, an overfill can result in condenser blockage. Its slight overfill can be tolerated; the fluid inventory can be calculated by Eq. 10.1. The fluid inventory calculated by this equation is larger than what is required, because it discounts the

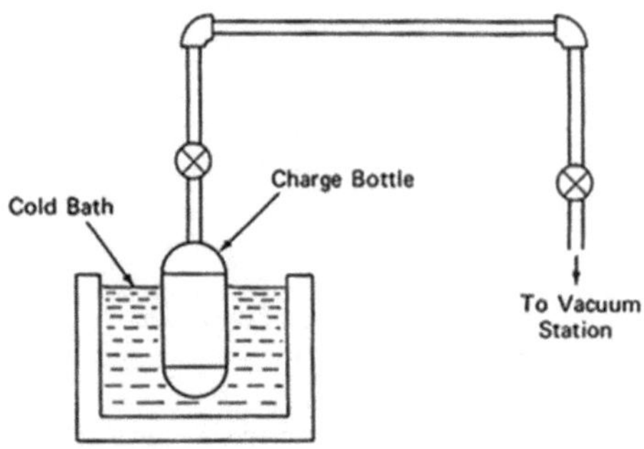

Fig. 10.15 Schematic of fluid purification process [3]

meniscus recession. Nevertheless, the depletion of liquid due to this recession is usually negligibly small:

$$m = A_v L_t \rho_v + A_w L_t \varepsilon \rho_l \tag{10.1}$$

where

m = Fluid inventory
A_v = Vapor cross-sectional area
A_w = Wick cross-sectional area
L_t = Total heat pipe length
ρ_v = Vapor density at pipe operating temperature
ρ_l = Liquid density at pipe operating temperature
ε = Wick porosity

10.5.3 Analysis of Gas Blockage

In Fig. 10.16, the volume occupied by the non-condensable is, from the ideal gas law:

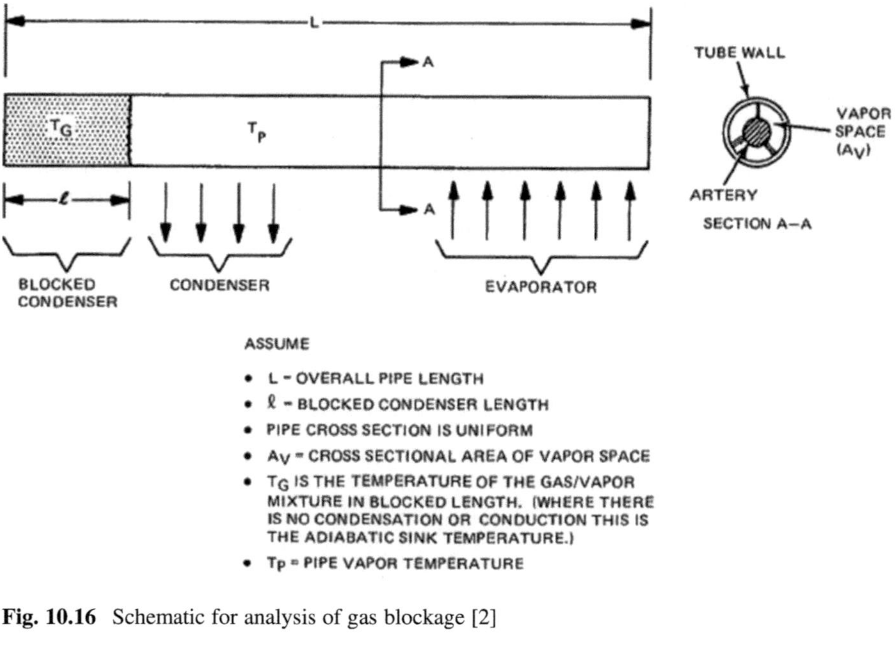

Fig. 10.16 Schematic for analysis of gas blockage [2]

$$N = n_G R T_G / P_G \tag{10.2}$$

where

n_G = Moles of non-condensable gas in pipe
R = Gas constant
T_G = Temperature of the gas/vapor mixture in blocked length
P_G = Partial pressure of inert gas in length, ℓ

P_G, furthermore, is the difference between the working fluid pressure in the active and inactive portions of the pipe, i.e.,

$$P_G = P(T_p) - P(T_G) \tag{10.3}$$

where

$P(T_p)$ = Working fluid pressure at temperature T
$P(T_G)$ = Gas pressure at temperature T
ℓ = Blocked condenser length
T_p = Pipe vapor temperature

From Fig. 10.16, the gas volume can be related to the blocked length by

$$V = \mathrm{A_v}.\ell \tag{10.4}$$

Substituting Eqs. 10.3 and 10.4 in Eq. 10.2 and solving for ℓ yields

$$\ell = \frac{n_G R T_G}{\mathrm{A_v}\{P(T_p) - P(T_G)\}} \tag{10.5}$$

The moles of non-condensable can now be defined in terms of a working fluid impurity level, f,

$$f = \frac{n_G}{n_p} \tag{10.6}$$

where

n_p = Moles of working fluid in pipe
n_G = Moles of non-condensable gas in pipe
f = Mole fraction of non-condensable in working fluid

Substituting Eq. 10.6 into Eq. 10.5 and dividing by the overall pipe length, L, yields the blockage as a fraction of the total pipe length,

$$\frac{\ell}{L} = \frac{fn_{\mathrm{p}}RT_{\mathrm{G}}}{LA_{\mathrm{v}}\{P(T_{\mathrm{p}}) - P(T_{\mathrm{G}})\}} \tag{10.7}$$

If n'_{p} is defined as the heat pipe charge per unit length, i.e., $n'_{\mathrm{p}} = n_{\mathrm{p}}/L$, then Eq. 10.7 becomes

$$\frac{\ell}{L} = \frac{fn'_{\mathrm{p}}RT_{\mathrm{G}}}{A_{\mathrm{v}}\{P(T_{\mathrm{p}}) - P(T_{\mathrm{G}})\}} \tag{10.8}$$

This expression now relates the blockage to the operating conditions, T_{p} and T_{G}; to pipe design parameters n_{p}, A_{v} and working fluid; and to an impurity level factor, f.

10.5.4 Effect of Heat Pipe Design and Operating Conditions on Gas Blockage

In Fig. 10.17, the pipe blockage is presented as a function of the temperature difference between the operating and nonoperating sections for 0.5 in. (1.270 cm) axial groove ammonia heat pipe. An impurity content, f, of 0.0001 was assumed which is equivalent to 100 ppm (molar). The gases can be considered as being present in the charge fluid or as gas remaining in the pipe after evacuation.

The curve shows that for a fixed temperature difference ($T_{\mathrm{p}} - T_{\mathrm{G}}$), the blockage decreases at higher temperatures. Also, the blocked length increases for small differences in temperature between the pipe and gas temperature (thermal sink), such as in the case of isothermalizing heat pipes. As an example, consider a 10-ft (3 m) long axial groove/ammonia heat pipe whose temperature (T_{p}) is 420 °R, attached to a sink (T_{G}) at 400 °R, i.e., $T_{\mathrm{p}} - T_{\mathrm{G}} = 20$ °R. The blockage produced by an impurity level of 100 ppm would be 4.2% or 0.42 ft (5.0 in.) (12.7 cm). For an impurity level of 10 ppm (not shown) the blockage, which is proportional to f, would be 0.50 in. (1.27 cm).

The effect of a different pipe configuration is shown in Fig. 10.18, where a 0.5-in., (1.270 cm) spiral artery design is compared to the 0.5-in. groove design. Because of its smaller vapor space, the spiral design produces a greater blockage then the groove under the same conditions.

The effect of different working fluids is shown in Fig. 10.19. Ammonia, acetone, and Freon-21 are plotted for the 0.5 in. axial groove design. It is seen that ammonia results in a smaller blocked length than acetone, or Freon-21. This is because the pressure of ammonia changes more rapidly with temperature than the other fluids. The value of these curves is that it allows the designer, for his particular application, to make a judgment regarding the allowable quantity of gaseous impurities or, conversely, to design the condenser length so as to accommodate some estimated amount of gas that may be present (Fig. 10.20).

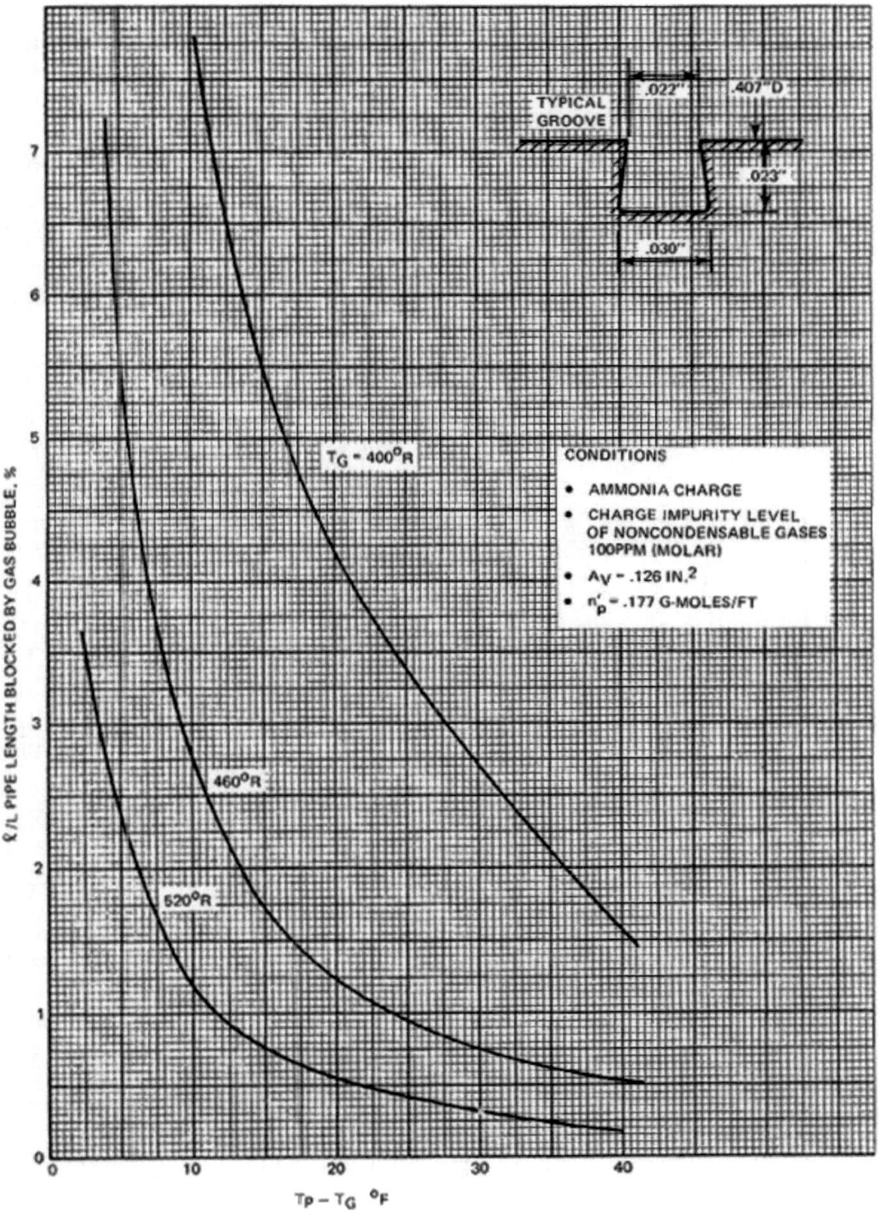

Fig. 10.17 Gas blockage in typical grooved heat pipe [2]

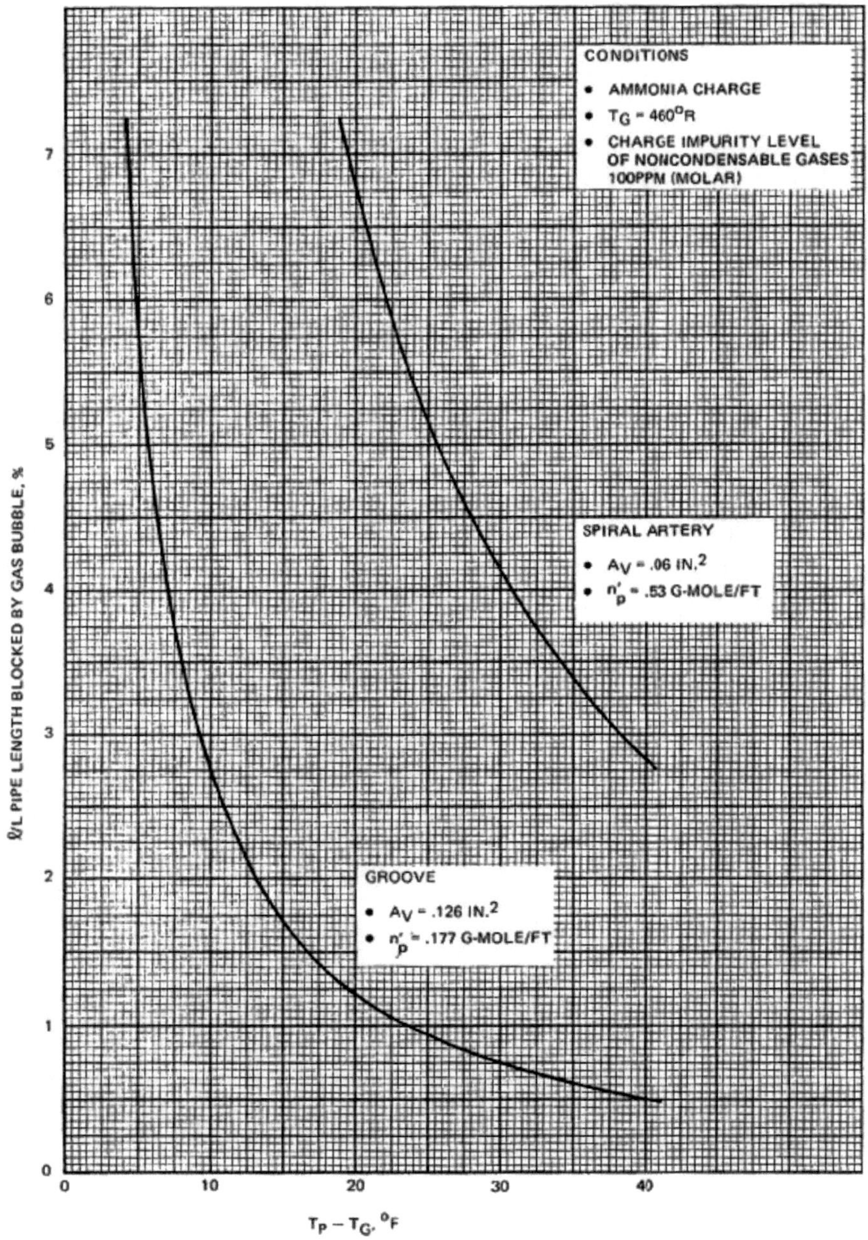

Fig. 10.18 Effect of heat pipe configuration on gas blockage [2]

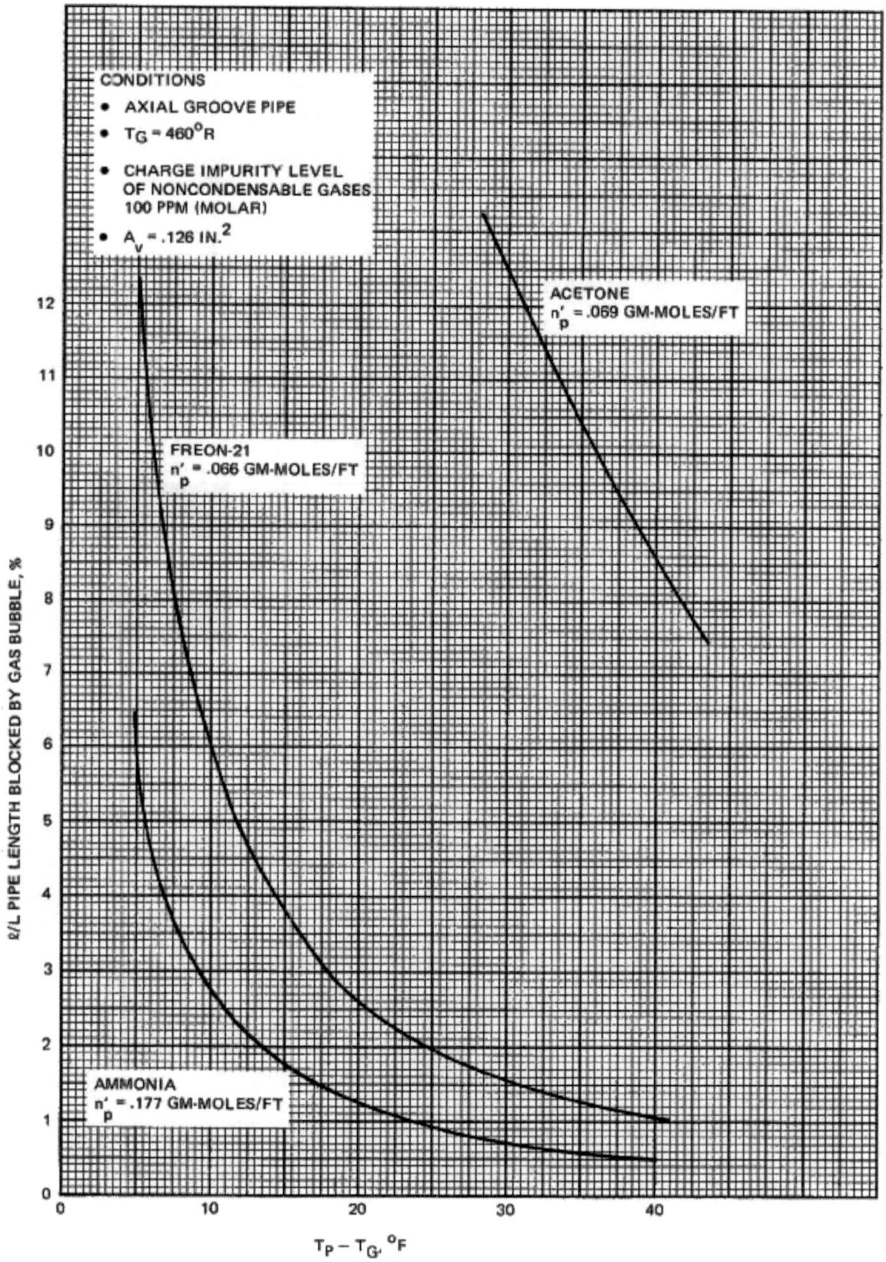

Fig. 10.19 Effect of working fluid on gas blockage [2]

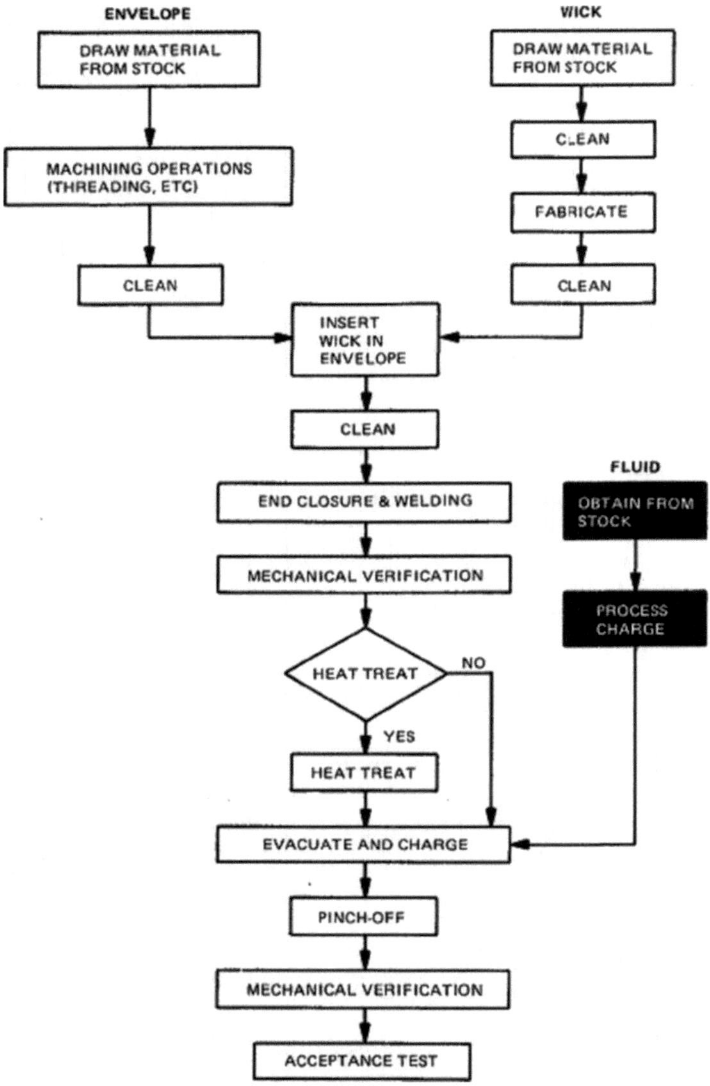

Fig. 10.20 Typical manufacturing steps for charging process [2]

10.6 Full Tube Closure

The manufacture of the heat pipe is near its completion with the exception that the pipe is still attached to a closed valve at the fill tube. It is necessary to sever the pipe from the valve and to form a permanent seal at the fill tube. In this process, no gas must enter the pipe and no fluid should be lost from the pipe. A closure technique, which has proved both economical and reliable, consists of:

1. Crimp seal (flatten and pinch) the fill tube to form a temporary leakproof closure.
2. Sever the valve from the pipe by making a cut in the flattened area of the fill tube on the valve side of the crimp seal.
3. Weld the cut end of the fill tube by TIG or EB welding and remove the crimping tool.

10.7 Heat Pipe Testing Techniques

Per Chi [2], testing of the heat pipe may answer many questions. A simple test on the wetting of the wick and on the absence of leakage will insure that the pipe will operate as a heat pipe. A heat conduction test with an appropriately designed heat source and sink may verify the heat transfer capability and characteristics of the heat pipe. A test on the life of a heat pipe may include recording the performance of sample heat pipes over a long period of time, or an accelerated life test may be performed with metallurgical examination of materials over predetermined intervals of time. In addition, tests may be required to answer questions on transient characteristics of the pipe. We will concentrate our discussion of the pipe testing on verifying the mechanical soundness of the pipe, proper wetting of the wick, and operating characteristics of the pipe.

10.7.1 Mechanical Sounders

A sound structural design that has been properly verified by nondestructive tests is paramount to reliable, long-term heat pipe operation. The ASME pressure vessel code is recommended when specifying allowable design stresses, proof pressure, and burst pressure. Simplified methods are also presented by Edelstein and Haslett [1] for including the stress effects due to internal pressure, end caps, thermal expansion, saddle attachments, pipe bends, and dynamic loading. Cost-effective methods of leak detection for ammonia, Freon, and methanol used during pre-charging and post-charging operations include X-ray examination, pressurization underwater, helium detection, and copper sulfate/ethylene glycol (for ammonia) [2].

The pipe must be able to withstand the maximum vapor pressure. For this reason, the pipe envelope and end caps have to be properly designed, and joints must be made with high-quality welding before a design and production procedure can be finalized for large quantity production. Experimental heat pipes may be made and subjected to high vapor pressure by heating the pipe so as to assure that the pipe envelope, the end caps, and the joints will be able to withstand the designed vapor pressure with appropriate margins of safety [3].

10.7.2 Wick Wetting

Unfortunately, many of these problems cannot be uncovered until the pipe is charged, sealed and tested. In some cases, a long time can occur until some of these effects are noticed; by then, it is usually too late to provide corrective action. Hence, the objective must be to develop cleaning procedures which will prevent these problems from occurring and produce, therefore, a more reliable product. Moreover, in keeping with the overall aims of this study, the cleaning procedure should also be simple, inexpensive, and as free as possible from human error [2].

The simplest way to determine if the working fluid has wet the wick is to oscillate the heat pipe along its axis by hand. If the fluid slugs the end caps, it is fairly obvious that the liquid is riding on the wick in the vapor space. This preliminary test works fairly well for pipes having working fluid in the liquid state at room temperature. However, one can shake the pipe with vigor and break the force of surface tension. Liquid within the vapor space will result. This method is limited to qualitative testing of wick wetting. X-ray radiography may sometimes be used to observe uniformity of the liquid inside the wick or the presence of excessive liquid at the bottom of the pipe [3].

10.7.3 Performance Versification

If a pipe passes the mechanical and wetting tests, it can be assured that the pipe will operate as a heat pipe under favorable conditions. In order to establish pipe performance such as the maximum heat transfer capability and effective thermal conductance of the pipe, tests using the setup illustrated schematically in Fig. 10.21 may be used for the moderate-temperature heat pipe. The pipe may be placed at any desired orientation. The heat is supplied by an electrical tape wound around the evaporator section of the pipe [3].

Either the evaporator or the condenser temperature may be maintained constant regardless of the variation of the rate of heat supply by controlling the water inlet

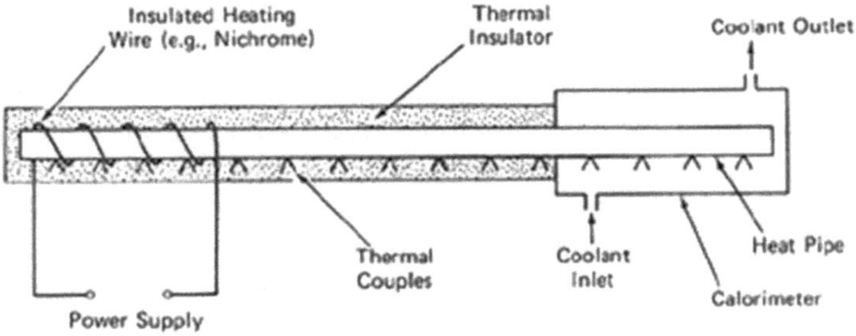

Fig. 10.21 Testing setup for moderate-temperature heat pipe [2]

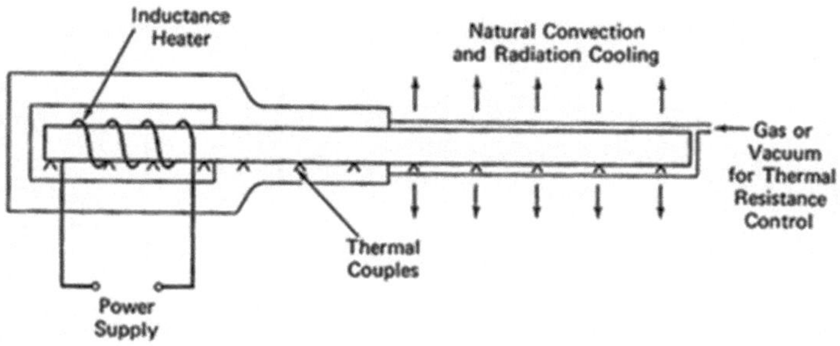

Fig. 10.22 Testing setup for liquid metal heat pipe [3]

temperature and flow rate. Thermocouples installed along the heat pipe wall measure the axial wall temperature along the pipe at different heat transfer rates, which are indicated by the electrical power input. As the electrical power input is gradually raised to a certain limit, it may be observed that the temperature indicated by the thermocouple at the end of the evaporator rises suddenly above that indicated by other thermocouples at the evaporator. This sudden rise in temperature at the end of an evaporator indicates partial dry-out of the evaporator. Hence, both heat transport limit and temperature characteristics of heat pipes may be measured using apparatus shown in Fig. 10.21.

The operating temperature and the heat transfer capability for the liquid-metal heat pipes are usually much higher than those for the moderate-temperature heat pipes. For increasing the power supply capability, the electrical heating tape shown in Fig. 10.21 can be replaced by an electrical induction coil. When this setup is used for testing liquid-metal heat pipes, radiation and natural convection of heat to the atmosphere provides a convenient heat sink. In order to control the pipe operating temperature, the thermal resistance of the condenser section may be controlled by providing a concentric annular gap as shown in Fig. 10.22.

References

1. Zohuri, B. (2016). *Heat pipe design and technology: Modern applications for practical thermal management* (2nd ed.). New York: Springer.
2. Edelstein, F., & Haslett, R. (1974, August). *Heat pipe manufacturing study*. Final Report prepared by Grumman Aerospace Corp. for NASA, Contract No. NASS5-23156.
3. Chi, S. W. (1976). *Heat pipe theory and practice*. Washington, DC: Hemisphere Publishing Corporation.
4. Kemme, J. E. (1966). *Heat pipe capability experiment* (Report No. LA-3585-MS). Los Alamos Scientific Laboratory.
5. Rhodes, R. A., Jr. (1973). *Procedures for the construction of a screen wick heat pipe*. A Lecture Note on Heat Pipe, U. S. Army Mobility Equipment R and D Center.

Appendix A: Dimensional Equivalents and Physical Constants

In this appendix we have provided some basic dimension and physical unit constants.

Dimensional Equivalents

Length	1 ft = 12 in. = 30.48 cm = 0.3048
	1 m = 100 cm = 39.37 in. = 3.28 ft
Mass	1 lbm = 0.03108 slug = 453.59 g = 0.45359 kg
	1 kg = 100 g = 0.06852 slug = 2.205 lbm
Time	1 h = 3600 s
	1 s = 2.778×10^{-4} h
Force	1 lbf = 4.448×10^{5} dyne = 4.448 N
	1 N = 10^{5} dyne = 0.2249 lbf
Angle	1 degree = 1.745×10^{-2} rad
	1 rad = 57.30 degree
Temperature	1 deg F = 1 deg R = 0.5556 deg C = 0.5556 deg K
	1 deg K = deg C = 1.8 deg R = 1.8 deg F
	deg F = 1.8 deg C + 32
	deg C = 0.5556 (deg F − 32)
	deg R = deg F + 459.69
	deg K = deg C + 273.16
	deg R = 1.8 deg K
	deg K = 0.5556 deg R
Energy	1 Btu = 777.66 ft-lbf = 252 cal = 1.054×10^{10} erg = 1054 J
	1 J = 10^{7} erg = 0.239 cal = 0.7375 ft-lbf = 9.485×10^{-4} Btu
Power	1 Btu/h = 2.778×10^{-4} Btu/s = 2.929×10^{6} erg/s = 0.2929 W
	1 W = 10^{7} erg/s = 9.481×10^{-4} Btu/s = 3.414 Btu/h

(continued)

B. Zohuri, *Heat Pipe Applications in Fission Driven Nuclear Power Plants*,
https://doi.org/10.1007/978-3-030-05882-1

Pressure	1 lbf/ft^2 = 6.944 × 10^{-3} lbf/in.^{2} = 4.78.8 dyne/cm^2 = 47.88 N/m^2
	1 lbf/in.^{2} = 144 lbf/ft^2 = 68,948 dyne/cm^2 = 6894.8 N/m^2
	1 N/m^2 = 10 dyne/cm^2 = 1.450 × 10^{-4} lbf/in.^{2} = 2.089 × 10^{-2} lbf/ft^2
Area	1 ft^2 = 1.44 in.^{2} = 929 cm^2 = 0.0929 m^2
	1 m^2 = 104 cm^2 = 1550 in^2 = 10.76 ft^2
Volume	1 ft^3 = 1728 in.^{3} = 2.832 × 104 cm^3 = 0.02832 m^3
	1 m^3 = 106 cm^3 = 6.102 × 104 in.^{3} = 35.31 ft^3
	1 gal (US liquid) = 0.13368 ft^3 = 0.003785 m^3
Density	1 gal (US liquid) = 0.13368 ft^3 = 0.003785 m^3
	1 lbm/ft^3 = 0.03108 slug/ft^3 = 1.602 × 10^{-2} g/cm^3 = 16.02 kg/m^3
	1 kg/m^3 = 10^{-3} g/cm^3 = 0.00194 slug/ft^3 = 0.06242 lbm/ft^3
Viscosity (dynamic)	1 lbm/ft-h = 8.634 × 10^{-6} slug/ft-s = 4.134 × 10^{-3} g/cm-s = 4.134 × 10^{-4} kg/m-s
	1 kg/m-s = 10 g/cm-s = 2.089 × 10^{-2} slug/ft-s = 2.419 × 10^3 lbm/ft-h
Thermal conductivity	1 Btu/ft-h-F = 2.778 × 10^{-4} Btu/ft-s-F = 1.730 × 105 erg/cm-s-K = 1.730 W/m-K
	1 W/m-K = 10^5 erg/cm-s-K = 1.606 × 10^{-4} Btu/ft-s-F = 0.578 Btu/ft-h-F
Surface tension	1 lbf/ft = 1.459 × 10^4 dyne/cm = 14.59 N/m
	1 N/m = 10^3 dyne/cm = 0.06854 lbf/ft
Latent heat of vaporization	1 Btu/lbm = 32.174 Btu/slug = 2.32 × 10^7 erg/g = 2.324 × 10^3 J/kg
	1 J/kg = 10^4 erg/g = 1.384 × 10^{-2} Btu/slug = 4.303 × 10^{-4} Btu/lbm
Heat transfer coefficient	1 Btu/ft^2-h-F = 5.674 × 10^3 erg/cm^2-s-K = 5.674 W/m^2-K
	1 W/m^2-K = 10^3 erg/cm^2-s-K = 0.1762 Btu/ft^2-h-F

Physical Constants

Gravitational acceleration (standard), g = 32.174 ft/s^2 = 980.7 cm/s^2 = 9.807 m/s^2

Universal gas constant, $\bar{R}$ = 1545.2 ft-lb/mol-R = 1.987 Btu/lbm-mol-R = 8.314 × 10^7 erg/g-mol-K = 8.314 × 10^3 J/kg-mol-K

Mechanical equivalent of heat, J = 777.66 ft-lbf/Btu = 4.184 × 10^7 erg/cal 1 N-m/J

Stefan-Boltzmann constant, $\overline{\sigma}$ = 0.1713 × 10^{-8} Btu/ft^2-h-R4 = 5.670 × 10^{-5} erg/cm^2-s-K^4 = 5.657 × 10^{-8} W/m^2-K^4

Appendix B: Electromagnetic Pump

An electromagnetic pump is a pump that moves liquid metal (or any electrically conductive liquid) using electromagnetism. A magnetic field is set at right angles to the direction the liquid moves in, and a current is passed through it. This causes an electromagnetic force that moves the liquid. Applications include pumping liquid metal through a cooling system.

Introduction

Electromagnetic pumps have been used for pumping liquid sodium in auxiliary circuits such as fill and drain and purification circuits of sodium-cooled fast breeder reactors. Despite their low efficiency, these pumps are used in fast reactors because of their high reliability and low maintenance due to absence of moving parts. Besides, EM pumps can be used for pumping impure sodium. For example, Indira Gandhi Centre for Atomic Research (IGCAR) has developed electromagnetic pumps of various capacities and successfully used them in experimental facilities.

Sodium is used as a coolant in fast breeder reactors because of its excellent neutronic and heat transfer characteristics. Sodium is a fairly good conductor of electricity also, and this has led to development of many electromagnetic sensors and devices for use in liquid sodium. One such device is the electromagnetic pump which is used to pump liquid sodium in auxiliary circuits of a fast reactor and in various test facilities.

Though centrifugal pumps are used for pumping sodium in the primary and the secondary circuits of the reactor, electromagnetic pumps have been preferred in the auxiliary circuits. Low efficiency of these electromagnetic pumps prohibits their use in the main sodium circuits of a fast reactor. But their ability to operate even with impurities in sodium and their high reliability and almost maintenance-free operation makes them an ideal choice. These electromagnetic pumps work on the principle that

B. Zohuri, *Heat Pipe Applications in Fission Driven Nuclear Power Plants*,
https://doi.org/10.1007/978-3-030-05882-1

whenever a current-carrying conductor is placed in a perpendicular magnetic field, a force acts upon it.

There are various types of electromagnetic pumps which can be mainly classified as conduction electromagnetic pumps and induction electromagnetic pumps [1]. In conduction electromagnetic pumps [1, 2] the electric current in sodium flows via conduction from an external circuit which requires physical connection of the external circuit to the duct where as in induction pumps, the current is induced in sodium without connecting an external circuit to the stainless steel (SS) duct. Since no physical contact is there with SS duct in case of induction pumps, they are considered to be more reliable compared to conduction pumps. Both the conduction and induction types of pumps have been developed and are in operation in Indira Gandhi Centre for Atomic Research (IGCAR). Induction pumps are mainly of two types—the Flat Linear Induction Pump (FLIP) and the Annular Linear Induction Pump (ALIP). FLIP has a flat duct with stator normally on upper and lower part of the duct. It also has rectangular copper bars welded on its sides for providing a low resistance short circuit path and thereby increasing its efficiency. Besides, the flat duct is less suitable for high-pressure applications; therefore it was decided to use ALIP for Secondary Sodium Fill and Drain Circuit (SSFDC) of Prototype Fast Breeder Reactor (PFBR) presently under construction at Kalpakkam [3]. This pump has a flow rate of 170 m^3/h and can deliver a head of 4 kg/cm^2. This paper describes the design data of the pump and details of the testing of pump carried out at the Steam Generator Test Facility (SGTF) in IGCAR.

Electromagnetic Pump Working Principle

Liquid metal loops are used for heat removal and for the study of certain magneto-fluidic phenomenon like MHD (magnetohydrodynamic) effects. These loops operate at high temperatures and carry fluids that are invariably toxic in nature. Ensuring the purity of fluid in a closed loop application needs nonintrusive pumps and electromagnetic pumps. We have designed and analyzed a prototype electromagnetic pump to be used in mercury loop for carrying out various studies. This electromagnetic pump is designed using permanent magnets which are mounted on periphery of rotor, which is rotated using DC motor. The liquid metal flows in a semicircular duct surrounding the rotor. See Fig. B.1, where surface plot of magnetic field density superimposed with contours of magnetic potential of an electromagnetic pump simulated using multiphysics software COMSOL®.

A magnetic field ($\underset{\text{rc}}{\vec{b}}$) always exists around the current ($\vec{I}$)-carrying conductor. When this currents-carrying conductor is subjected to an external magnetic field ($\underset{\text{rap}}{\vec{B}}$), the conductor experiences a force perpendicular to the direction of $\vec{I}$ and $\underset{\text{rap}}{\vec{B}}$. This is because the magnetic field produced by the conductor and the applied magnetic field

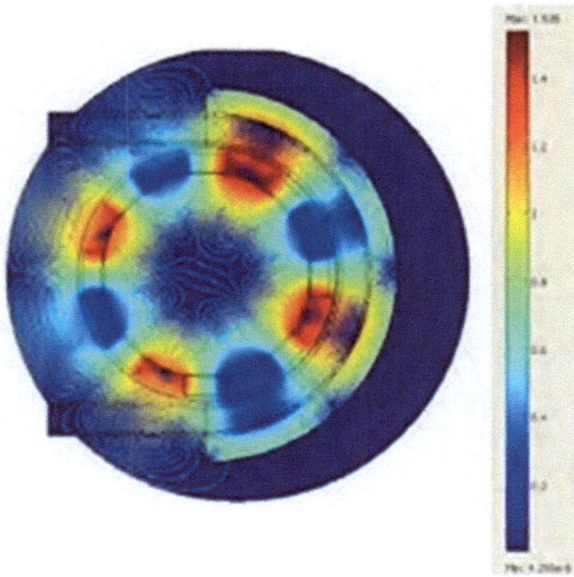

Fig. B.1 Schematic of heat removal liquid metal loops

attempt to align with each other. A similar effect can be seen between two ordinary magnets.

This principle is used in an electromagnetic pump. The current is fed through a conducting liquid. Two permanent magnets are arranged to produce a magnetic field $\overrightarrow{B}_{rap}$ as shown in the Fig. B.2. The supplied current has a current density ($\overrightarrow{J}$) and the magnetic field associated with this current can be called as "reaction magnetic field ($\overrightarrow{b}_{rc}$)." The two magnetic fields $\overrightarrow{B}_{rap}$ and $\overrightarrow{b}_{rc}$ attempt to align with each other; thus this causes mechanical motion of the fluid as illustrated in Fig. B.2.

As we stated at the introduction of this appendix, a special type of liquid metal thermomagnetic device is the Annular Linear Induction Pump (ALIP). It is known that electromagnetic pumps have a number of advantages over mechanical pumps: absence of moving parts, low noise and vibration level, simplicity of flow rate regulation, easy maintenance, and so on. However, while developing induction pumps, in particular ALIPs, we are faced with a significant problem of magnetohydrodynamics (MHD) instability arising in the device. The manifestation of the instability does not allow linear induction pump development in a certain range of flow rate or the development of high efficiencies under certain flow rates and dropping pressure conditions.

Linear induction pumps use a traveling magnetic field wave created by three-phase currents and the induced currents and their associated magnetic fields that generate a Lorentz force (see Fig. B.3). The three-phase winding arrangement for the solenoids usually follows the sequence *AA ZZ BB XX CC YY* where *A*, *B*, and *C* denote the balanced three-phase winding and *X*, *Y*, and *Z* the opposite phase; for

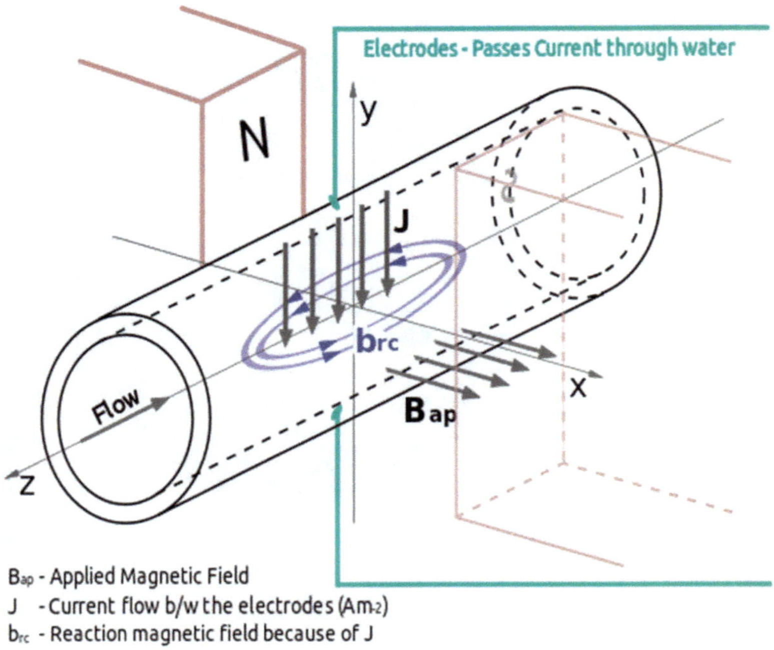

Fig. B.2 Schematic of electromagnetic pump

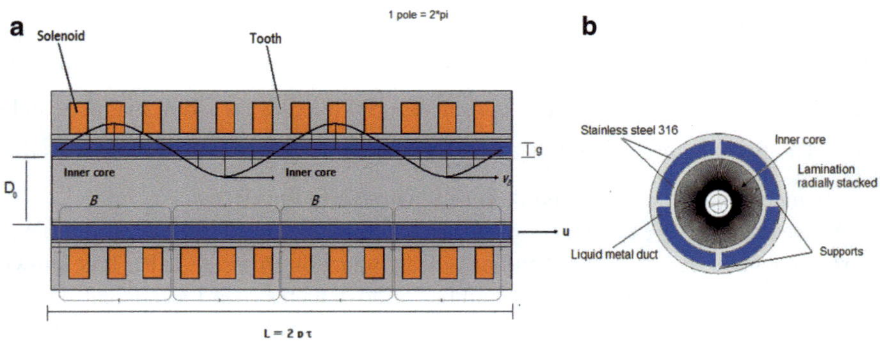

Fig. B.3 Cross section of an ALIP in a conceptual representation

a direct balanced system, if $A = 0^{\circ}$, $\vec{B} = 120^{\circ}$, and $C = 240^{\circ}$, then $X = 180^{\circ}$, $Y = 300^{\circ}$, and $Z = 60^{\circ}$. The correct winding sequence for the solenoids is obtained by arranging the sequence by the rising phase: *AA ZZ BB XX CC YY*. The complex flow behavior in this type of device includes a time-varying Lorentz force and pressure pulsation due to the time-varying electromagnetic fields and the induced convective currents that originate from the liquid metal flow, leading to instability problems along the device geometry. The determination of the geometry and of the electrical configuration of a thermomagnetic device gives rise to an inverse

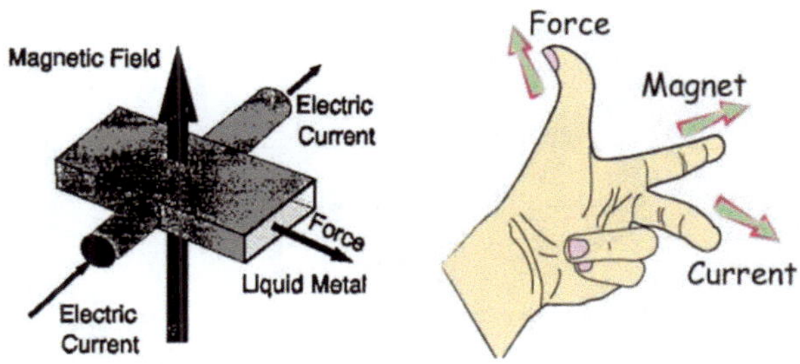

Fig. B.4 Illustration of basic principles of electromagnetic pump

magnetohydrodynamic field problem. When the requirements of the design are defined, this problem can be solved by an optimization technique.

Figure B.3a shows ALIP, Annular Linear Induction Pump; Fig. B.3b shows the cross section of the ALIP, and it is an adaptation of a diagram originally drawn by Dong Won Lee, KAERI, Korea Atomic Energy Research Institute. This is adapted with permission [4].

The objective function which must be maximized in the optimization problem is derived from the main design requirement. Usually for an MHD device, this is the efficiency. Other design requirements can be taken into account as constraints. For a nonlinear system, such as for linear induction pumps, the main objective functions are low weight and high efficiency, and so more than one maximum can exist. In this case a technique for global optimization must be used.

In a typical electromagnetic pump, the following basic principles do work as follows and is illustrated in Fig. B.4 as:

- Liquid metal are conductors.
- Based on the Fleming's left-hand rule.
- Current flowing vertically through the liquid metal experiences a force ($\vec{F} = I\vec{L} \times \vec{B}$).

As it is illustrated in Fig. B.5, more details and types of EM pump are revealed including different components involved with infrastructure of these types of pumps that is used as part of cooling system in liquid metal breeder fissionable nuclear reactors.

Types of electromagnetic pumps include:

- Conduction pump: In this case current is directly conducted into fluid through electrodes. It has two variants, alternative current (AC) and direct current (DC).
- Induction pump: Current in the conducting fluid is induced by a traveling magnetic field.
- Thermoelectric pump: Current flowing through the liquid metal is derived directly from the thermal power contained in the hot liquid metal flow, such as the one that we can encounter in liquid metal fast breeder reactor (LMFBR) of Generation III like French-built Phoenix-II or liquid metal breeder of Generation IV, such as Molten Salt Reactor (MSR) sodium-cooled fast reactor (SFR). See Chap. 2 of this book.

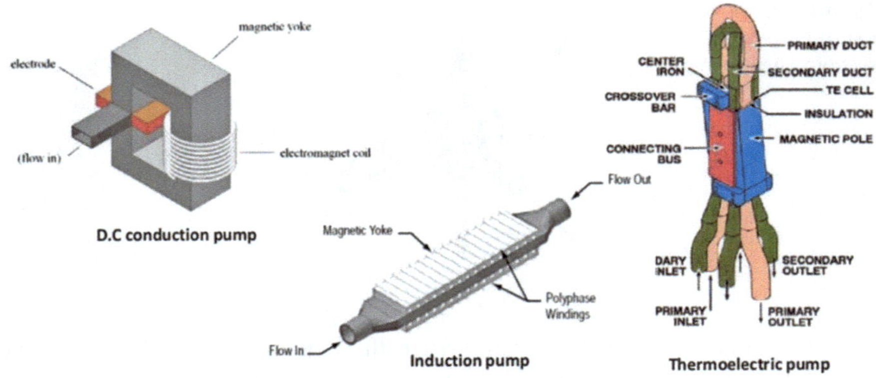

Fig. B.5 Detailed components of electromagnetic pump

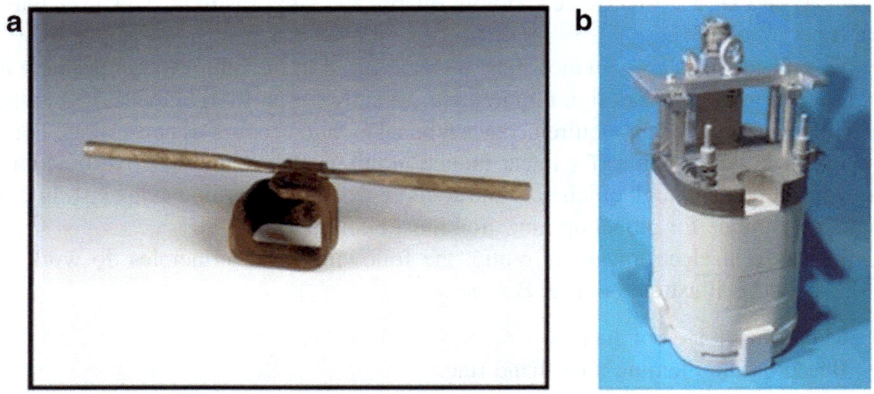

Fig. B.6 Simple applications of EM pump illustration. (**a**) Image of EM Pump Used in Nuclear Reactor (Courtesy Handbook of Nuclear Engineering by Daniel Gabriel). (**b**) Assembled EM Pump for Foundry (Courtesy CMI Novacast Inc.)

As far as holistic observation of EM pump application is concerned, Fig. B.6 presents such general application of device as two main driven scenario and they are:

- Cooling of nuclear reactor
- Pouring and transportation of high-temperature metals in foundry

If we take a look at the electromagnetic pump form inside point of view as it is depicted in Fig. B.7, we notice that the electrical current is induced by transformer action.

The transformer's primary coil T in Fig. B.7 is connected to an AC single-phase power source. The transformer pole pieces are arranged in the shape of a picture frame and server as the carrier of magnetic flux. The transformer secondary winding S around the bottom leg of the picture frame as it is shown in Fig. B.7 is molten metal

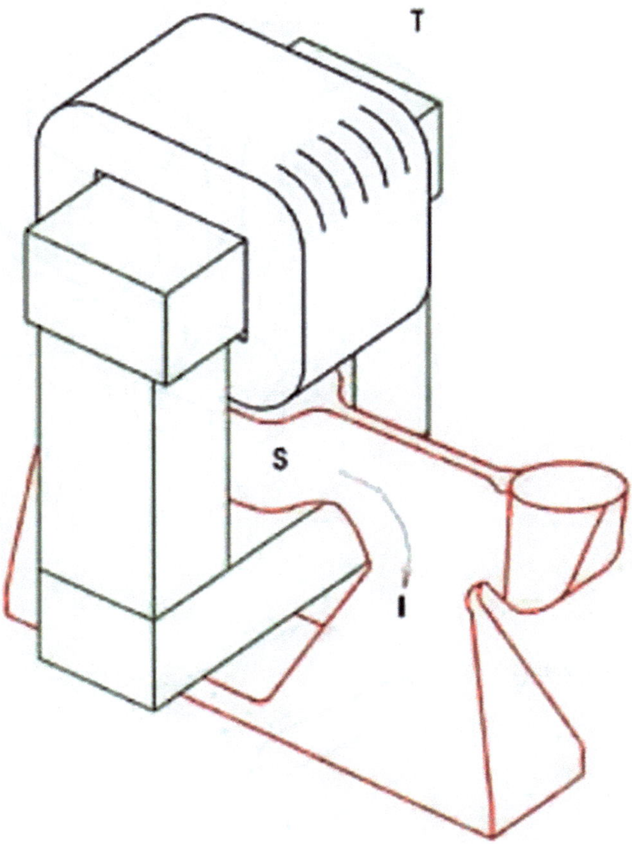

Fig. B.7 Schematic of electromagnetic pump from inside

and is formed by channels in ceramic parts. The turns ratio amplifies the electrical input current to produce very high amperage I in the molten metal.

Figure B 8 is a demonstration of a sectional view of electromagnet construction that is added, which consists of a C-shaped pole piece and two excitation coils.

The opening in the C straddles the necked-down section of the molten secondary turn, so the magnetic field H crossing the pole gap is perpendicular to the secondary current I, resulting in force Q to move metal through the pump.

In Fig. B.9, we see the complete electromagnetic pump from inside perspective of infrastructure.

It is surrounded by or encapsulated in ceramic parts to protect it from molten metal contact to prevent any corrosions or any other side effects from liquid metal of internal nuclear core. Pump output Q is varied by controlling input power and can be regulated from almost dropwise flow up to full bore delivery.

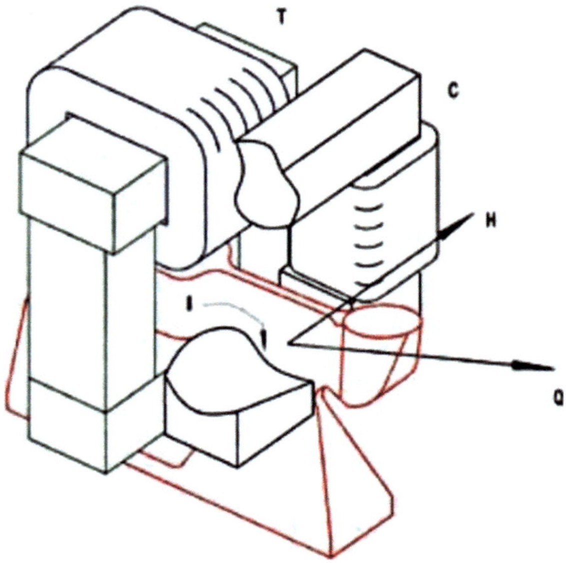

Fig. B.8 Further schematic of electromagnetic pump from inside

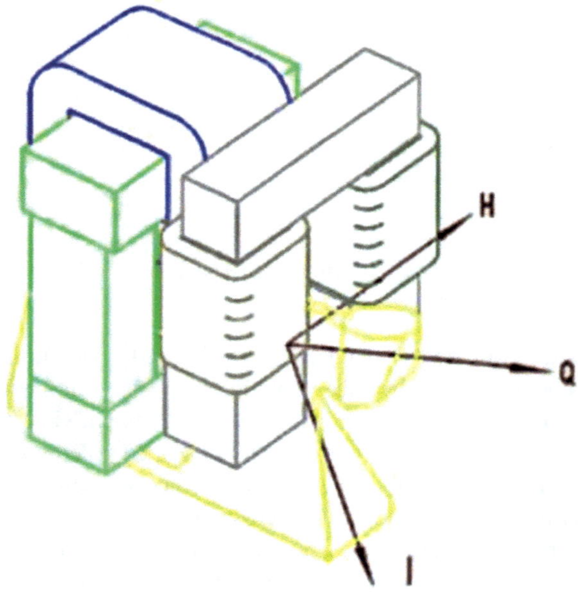

Fig. B.9 Complete schematic of electromagnetic pump from inside

Working Principle of Annular Linear Induction Pump (ALIP)

Like all other electromagnetic pumps, Annular Linear Induction Pump (ALIP) works on the principle that whenever a current-carrying conductor is placed in a perpendicular magnetic field, a force acts on the conductor. The magnitude of this force is given by $F = BIL$, where "I" is the current through the conductor, L is the length of the conductor, and B is the magnetic flux density in which the conductor is placed. In the case of ALIP, there is an annular region in which sodium flows, outside this annular region there are copper windings which are excited by three-phase AC supply and produces a linearly moving magnetic field. This linearly moving magnetic field, according to Faraday's law of induction, induces a current in the liquid metal (Fig. B.10). The interaction of this current and moving magnetic field produces a force on the sodium resulting in the pumping action as illustrated in Fig. B.10, where it shows the working principle of Annular Linear Induction Pump [5].

Since the working principle of ALIP is similar to an induction motor, the equivalent circuit of ALIP (Fig. B.11) is also similar. At the same time, there are some differences as compared to an inductor motor. Typical slip in induction motors

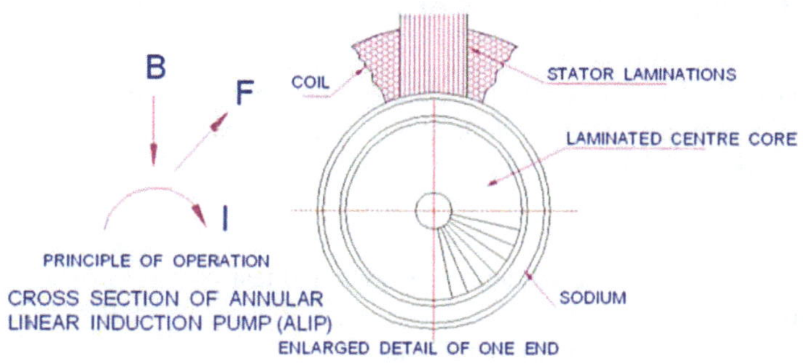

Fig. B.10 Working principle of Annular Linear Induction Pump [5]

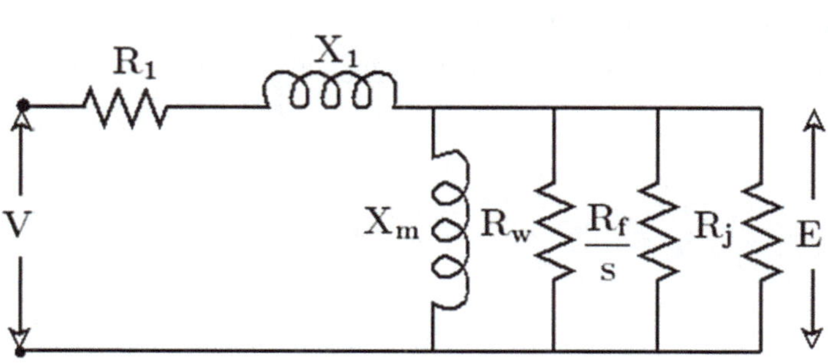

Fig. B.11 Equivalent Circuit of ALIP [5]

is 0.05 or less whereas for ALIP typical slip is in the range of 0.4–0.9 which leads to higher slip losses. The presence of ducts for containing sodium not only introduces additional resistive elements in the equivalent circuit but also leads to higher air gap compared to that in an induction motor. These features lead to reduction in power factor and in efficiency when compared to that of an induction motor. Besides, end effects and hydraulic losses also lead to reduction in efficiency [5].

In Fig. B.11, the following parameters do apply as:

R_1—Resistance of stator winding
X_1—Leakage reactance of stator winding
X_m—Magnetic reactance
R_f—Fluid resistance
R_j—Resistance of inner duct
R_w—Resistance of outer duct
E—Air gap electromagnetic force (EMF)
s—Slip

Note that: The winding in the ALIP is circular pancake type and is different from conventional winding used in rotating induction motors.

Advantages and Limitations of Electromagnetic Pumps

There exist some advantages to utilization of this pump in any application, where these EM pumps are installed, while along with such pros, there exist some limitation for these pumps and they are all listed below.

1. Advantages

   - No moving parts, no vibrations, or wear and tear
   - No seals, no splits
   - Less maintenance and more reliable
   - Safe to be used

2. Limitations

   - Power losses due to back EMF and Ohmic heating
   - Limited uses since very few liquids are good conductors of electricity

References

1. Publishing Co. Inc, New York. 1987.
2. Nashine, B. K., et al. (2007). Performance testing of indigenously developed DC conduction pump for sodium cooled fast reactor. *Indian Journal of Engineering and Material Sciences, 14*, 209–214.
3. Chetal, S. C., et al. (2006). The design of the prototype fast breeder reactor. *Nuclear Engineering and Design, 236*, 852–860.
4. Maidana, C. O., & Nieminen, J. E. (2017). First studies for the development of computational tools for the design of liquid metal electromagnetic pumps. *Nuclear Engineering and Technology, 49*(1), 82–91.
5. Sharma, P, Sivakumar, L. S., Rajendra Prasad, R., Saxena, D. K., Suresh Kumar, V. A., Nashine, B. K., et al (2011). Design, development and testing of a large capacity annular linear induction pump. *Asian Nuclear Prospects 2010, Energy Procedia, 7*, 622–629, Elsevier.

B. Zohuri, *Heat Pipe Applications in Fission Driven Nuclear Power Plants*,
https://doi.org/10.1007/978-3-030-05882-1

Index

B. Zohuri, *Heat Pipe Applications in Fission Driven Nuclear Power Plants*,
https://doi.org/10.1007/978-3-030-05882-1

MIX
Papier aus verantwortungsvollen Quellen
Paper from responsible sources
FSC® C105338

If you have any concerns about our products,
you can contact us on
ProductSafety@springernature.com

In case Publisher is established outside the EU,
the EU authorized representative is:
Springer Nature Customer Service Center GmbH
Europaplatz 3, 69115 Heidelberg, Germany

Printed by Libri Plureos GmbH
in Hamburg, Germany